# New Frontiers in Chemical Biology
## Enabling Drug Discovery

**RSC Drug Discovery Series**

*Editor-in-Chief:*
Professor David Thurston, *London School of Pharmacy, UK*

*Series Editors:*
Dr David Fox, *Pfizer Global Research and Development, Sandwich, UK*
Professor Salvatore Guccione, *University of Catania, Italy*
Professor Ana Martinez, *Instituto de Quimica Medica-CSIC, Spain*
Dr David Rotella, *Wyeth Research, USA*

*Advisor to the Board:*
Professor Robin Ganellin, *University College London, UK*

*Titles in the Series:*
1: Metabolism, Pharmacokinetics and Toxicity of Functional Groups: Impact of Chemical Building Blocks on ADMET
2: Emerging Drugs and Targets for Alzheimer's Disease; Volume 1: Beta-Amyloid, Tau Protein and Glucose Metabolism
3: Emerging Drugs and Targets for Alzheimer's Disease; Volume 2: Neuronal Plasticity, Neuronal Protection and Other Miscellaneous Strategies
4: Accounts in Drug Discovery: Case Studies in Medicinal Chemistry
5: New Frontiers in Chemical Biology: Enabling Drug Discovery

*How to obtain future titles on publication:*
A standing order plan is available for this series. A standing order will bring delivery of each new volume immediately on publication.

*For further information please contact:*
Book Sales Department, Royal Society of Chemistry, Thomas Graham House, Science Park, Milton Road, Cambridge, CB4 0WF, UK
Telephone: +44 (0)1223 420066, Fax: +44 (0)1223 420247, Email: books@rsc.org
Visit our website at http://www.rsc.org/Shop/Books/

# New Frontiers in Chemical Biology
## Enabling Drug Discovery

Edited by

**Mark E. Bunnage**
*Pfizer Global Research and Development, Sandwich, Kent, UK*

RSCPublishing

RSC Drug Discovery Series No. 5

ISBN: 978-1-84973-125-6
ISSN: 2041-3203

A catalogue record for this book is available from the British Library



Published by The Royal Society of Chemistry,
Thomas Graham House, Science Park, Milton Road,
Cambridge CB4 0WF, UK

Registered Charity Number 207890

For further information see our website at www.rsc.org

# *Preface*

*"Few scientists acquainted with the chemistry of biological systems at the molecular level can avoid being inspired."*

Donald J. Cram

Despite ever-increasing investment in biomedical research, there has been a significant decline in the number of new drug approvals in recent years. In an effort to improve productivity, and expand accessible drug space, the drug discovery community has increasingly looked to advance alternative therapeutic paradigms, such as biotherapeutics (antibodies, vaccines, nucleic acids, peptides, *etc.*), in addition to more classical small molecule medicinal chemistry approaches. However, independent of chosen therapeutic modality, improving future drug discovery productivity also critically depends on reducing the high levels of attrition currently seen in "proof-of-concept" Phase II clinical trials. This will require the development of a much deeper knowledge of biological systems in order to identify and validate those biomolecular targets for which there is the highest possible confidence of disease-relevance in humans.

Chemical biology is an emerging field at the interface between chemistry and biology that utilises the tools and techniques of chemical synthesis to study and influence biological systems. Recent developments in chemical biology have great potential to help address the drug discovery productivity challenges expressed above. For example, chemical biology studies have already led to the identification of novel targets with exciting therapeutic potential and it is clear that the field will prove a key enabler of improved target discovery and validation in the future. Moreover, the precise synthetic manipulation of biological molecules germane to many chemical biology approaches is also now fuelling a new wave of chemically modified biologics with unique properties ("chemologics") that emerge from the union of medicinal chemistry design with biotherapeutics research. In these, and many other ways, chemical biology is

RSC Drug Discovery Series No. 5
New Frontiers in Chemical Biology: Enabling Drug Discovery
Edited by Mark E. Bunnage
© Royal Society of Chemistry 2011
Published by the Royal Society of Chemistry, www.rsc.org

beginning to emerge as a key discipline within 21st century drug discovery and the purpose of this book is to highlight a number of the important developments in this regard. Since the field of chemical biology is both vast and growing, it should be stressed that this book is not intended to be exhaustive. Instead, the combination of general overviews in this book, coupled with the more in depth treatments of selected topics, is intended as an introduction to the field for the reader.

I would like to thank all of the contributors to the chapters in this book for their outstanding efforts and support of this project. I would also like to thank the staff at RSC, especially Gwen Jones and Katrina Harding, for their patience and support in bringing the book to completion. It is very much hoped that this book will provide a useful resource for scientists, both in industry and academia, who are looking to build their awareness of chemical biology and how this exciting field of science may enable future drug discovery.

*Mark E. Bunnage*
*Executive Director*
*Worldwide Medicinal Chemistry*
*Pfizer Inc., Sandwich, Kent, UK*

# Contents




RSC Drug Discovery Series No. 5
New Frontiers in Chemical Biology: Enabling Drug Discovery
Edited by Mark E. Bunnage
© Royal Society of Chemistry 2011
Published by the Royal Society of Chemistry, www.rsc.org

# *The Chemical Genetic Approach: The Interrogation of Biological Mechanisms with Small Molecule Probes*

MARTIN FISHER[a,b] AND ADAM NELSON[a,b,*]

[a] School of Chemistry, University of Leeds, Leeds, UK, LS2 9JT; [b] Astbury Centre for Structural Molecular Biology, University of Leeds, Leeds, UK, LS2 9JT

## 1.1 Introduction to Chemical Genetics

The chemical genetic approach[1] exploits small molecule probes to modulate the function of biological macromolecules, and hence to reveal insights into biological mechanisms. The term "chemical genetics" stems from the analogy of the approach to classical genetics. Chemical genetics exploits small molecule probes to modulate protein function; in contrast, classical genetics modulates protein function indirectly through mutation of the corresponding gene. The features of the chemical genetics and classical genetics are rather different, making the two approaches broadly complementary.

The use of small molecule probes can confer a number of advantages over conventional genetic manipulations. First, the effects of small molecules on the function of a target protein are conditional and, in addition, are usually rapid (usually diffusion-controlled) and reversible.[2] Second, by varying the

---

RSC Drug Discovery Series No. 5
New Frontiers in Chemical Biology: Enabling Drug Discovery
Edited by Mark E. Bunnage

Published by the Royal Society of Chemistry, www.rsc.org

concentration of a small molecule probe, it may be possible to tune the activity of a target protein in a refined way.[3] Third, the chemical genetic approach can sometimes allow multiple functions of an individual protein to be independently modulated.[4,5] Fourth, by using more than one small molecular probe in combination, the function of multiple proteins may be independently modulated.[6] Finally, the approach is useful when classical genetics is difficult to employ: for example, in organisms (especially mammals) with diploid genomes and a slow reproductive rate.[7] The chemical genetic approach is particularly useful for investigating tightly coordinated, dynamic biological mechanisms, especially when classical genetics may render an organism inviable.[8]

One of the obvious features of classical genetics is its incredible specificity: mutation of a specific gene can allow a single protein (from a family of closely related proteins) to be targeted. In addition, classical genetics is highly general, and can be used to study the function of many proteins. However, classical genetics does not generally allow the *conditional* modulation of a specific protein (unless, for example, a temperature-sensitive[9] mutant is used). RNA interference (RNAi) is a popular alternative to classical genetics which works by inhibiting the synthesis of a specific protein by targeting a specific RNA transcript.[10] However, RNAi is also poorly suited to time-sensitive studies because the modulation is made at the mRNA level: degradation of the protein target needs to occur before the effects of modulation are observed.

The chapter is not an exhaustive review of chemical genetics: the field, and its impact on our understanding of fundamental biological mechanisms, has already been comprehensively reviewed elsewhere.[1] Rather, the chapter will emphasise the approaches that may be used to discover small molecule probes, the solutions that address the challenges raised by chemical genetics, and the insights into fundamental biological mechanisms that may be revealed.

### 1.1.1   Forward and Reverse Chemical Genetics

The chemical genetic approach comes in two broad guises that are contrasted in Figure 1.1: forward chemical genetics and reverse chemical genetics.[1] In the forward chemical genetic approach, the target of the small molecule modulator is not known at the outset of the investigation. The active ligand is, therefore, identified on the basis of the observation of a specific phenotypic change that it induces. A major challenge of the forward chemical genetic approach is the subsequent identification of the protein that the active ligand targets.[11] The interrogation of biological mechanisms using the forward chemical genetic approach (and the associated challenges associated with target identification[11]) are described in Section 1.3.2. The forward chemical genetic and the forward genetic approaches are broadly analogous in that a phenotypic change to a biological system is induced in both cases: through modulation using either small molecule probes (sometimes known as "perturbogens"[12]) (forward chemical genetics) or random mutagenesis (forward genetics).

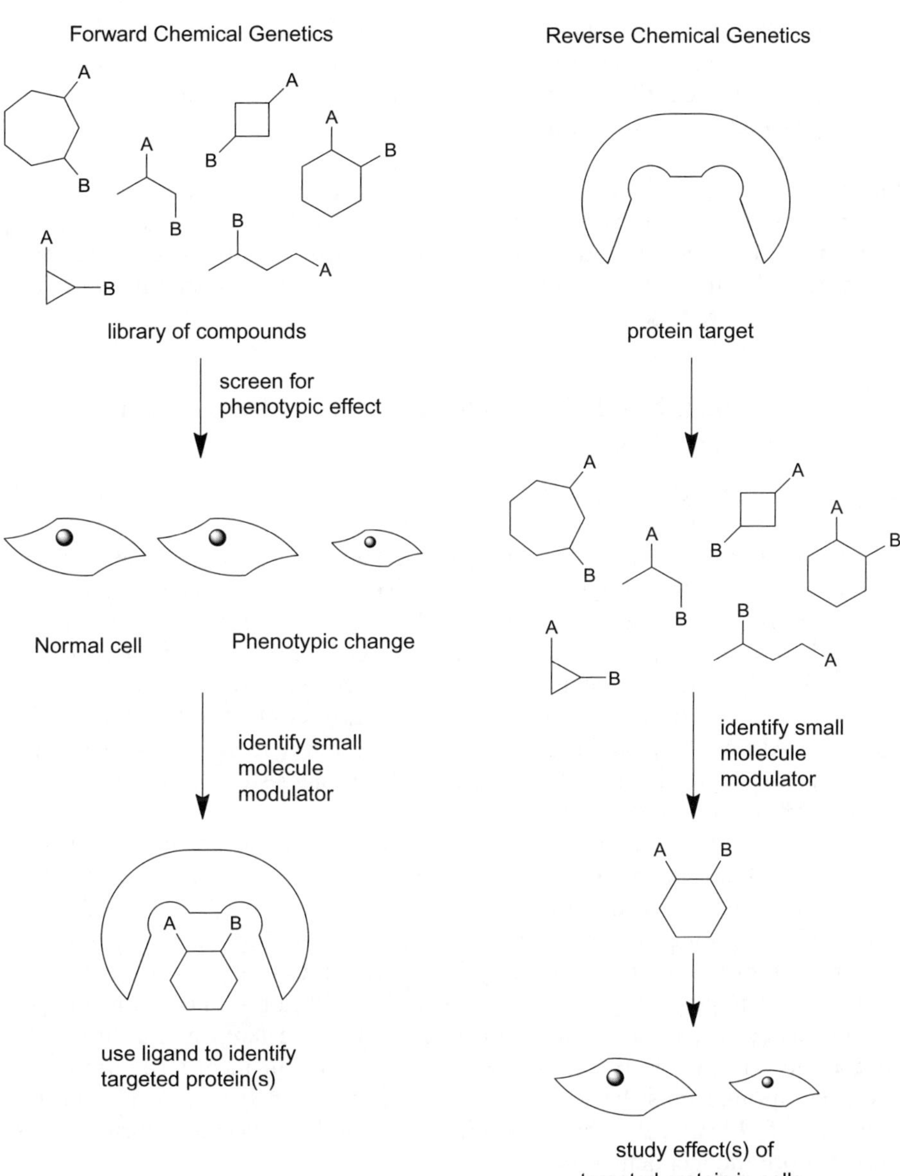



**Figure 1.1** Schematic overview of forward and reverse chemical genetics. In forward chemical genetics (left panel), a small molecule probe is discovered on the basis of its phenotypic effect; the small molecule probe may then be subsequently used to identify the protein responsible for the phenotypic effect. In contrast, in reverse chemical genetics (right panel), a small molecule probe is discovered on the basis of its ability to modulate a specific protein; the small molecule probe may then be used to modulate the cellular function of that specific protein, and the phenotypic effects that result.

In contrast, the reverse chemical genetic approach begins with the discovery of a small molecule modulator of a specific target protein. Most usually, the small molecule is identified on the basis of the modulation of the activity of the purified protein; however, assays of the cellular activity of a specific protein are also possible. Both reverse chemical genetics and reverse genetics modulate the function of a specific protein: either directly by the binding of a ligand (reverse chemical genetics), or indirectly through targeted mutagenesis of the corresponding gene. The discovery of ligands for use in reverse chemical genetics, and some of the insights into biological mechanism that have thereby been revealed, are described in Section 1.3.1.


### 1.1.2   Screening Small Molecule Libraries: Some General Considerations

The forward and reverse chemical genetic approaches both require the identification of an appropriate small molecule modulator. Most usually, active ligands are initially discovered using a high-throughput primary screen. In such cases, it is, of course, imperative that active molecules can be reliably distinguished from inactive molecules, and the performance of an assay must be assessed using appropriate statistical parameters (such as plate-based $Z$ factors).[13] In addition, it is necessary to correct screening results for systematic error by normalisation to controls (for example, to controls on each screening plate).[14] Although guidelines for reporting data from high-throughput screening of small molecule libraries have been suggested, appropriate standards have not yet been formalised.[15]

Having identified a "positive" in a primary screen, further investigation is necessary to confirm its validity.[15,16] Greater confidence is gained if an active compound is related structurally to other active compounds identified in the assay (allowing preliminary structure-activity relationships to be formulated). It is important to confirm that the active compound did not interfere with the primary screen: possible problems can include inhibition of a coupling enzyme in an assay (rather than the target protein), or the interference of a fluorescent compound with a fluorescence-based readout. The chemical structure of the active compound must be verified, and then, ideally, resynthesised, repurified and retested. Ideally, the activity of the compound should be confirmed using one or more independent assays.


## 1.2   Synthetic Strategies for Exploring Biologically Relevant Chemical Space

A challenge in chemical biology is to design small molecule libraries that span large tracts of biologically relevant chemical space.[17] The chemical space defined by small molecules is vast: it has been estimated that there are $>10^{60}$

compounds with molecular weight less than 500 Da (only a tiny fraction of which have been prepared by chemical synthesis).[18] How then, can the biologically relevant sub-fractions of chemical space be identified and targeted?

One approach is to design libraries around known scaffolds that have been biologically validated such as those found in drug molecules[19] or natural products.[20] In biology-oriented synthesis (BIOS),[20] for example, natural product scaffolds are regarded as pre-validated starting points for ligand design since they have been selected through evolution to interact with protein binding sites. In a related approach, libraries may be designed around scaffolds that are closely related to those that are already known to be biologically relevant.[21]

Alternatively, a fragment-based approach may be used in ligand design.[22] Essentially, a fragment-based approach dramatically reduces the chemical space that needs to be searched. Emphasis is placed on ligands with high "ligand efficiency" (binding energy per heavy atom) rather than high absolute affinity. The approach allows a much higher proportion of the relevant chemical space to be explored, whilst leaving ample scope (and providing a better starting point) for ligand optimisation.

The diversity of chemical libraries may derive from the high substitutional, stereochemical and scaffold diversity of its members. Combinatorial chemistry is focused on the synthesis of chemical libraries with only high substitutional diversity. In general, combinatorial syntheses exploit a rather limited palette of chemical transformations to vary the substitution of library members. The synthesis of compound libraries with high substitutional diversity is described in Section 1.2.1.

Varying the configuration of library members is more challenging, but has been hugely facilitated by the many reliable asymmetric transformations that have been developed in the past 30 years. In many cases, the high enantioselectivity of these transformations can dominate over substrate-based diastereoselectivity, and allow the configuration of compounds to be varied systematically. The synthesis of a compound library with both high substitutional and stereochemical diversity is described in Section 1.2.2.

Charting chemical space systematically, however, does require the preparation of libraries with high scaffold diversity. Historically, chemists' exploration of chemical space has been highly uneven and unsystematic, and much emphasis has been placed on a small number of molecular scaffolds.[23] (Note that the ~ 30 million cyclic organic compounds in the CAS registry are overwhelmingly dominated by a remarkably small number of molecular scaffolds:[1] thirty (of the 2.5 million!) molecular scaffolds are found in 17% of the compounds, and 0.25% of the scaffolds are found in 50% of the compounds.) A key goal of diversity-oriented synthesis[24] (DOS) is to address this problem through the preparation of libraries of small molecules that populate broad tracts of chemical space. In Section 1.2.3, the progress that has been made towards developing synthetic strategies that allow the systematic variation of scaffolds of ligands will be described.

## 1.2.1 Synthesis of Compound Libraries with High Substitutional Diversity

The combinatorial chemistry approach has been extended to the diversification of complex scaffolds, including natural product[25] and natural product-like[26] scaffolds. For example, a library of conformationally restricted 1,3-dioxanes **4** with high substitutional diversity has been prepared (Scheme 1.1).[26] A split-pool approach was used to diversify three substituents in the final compounds **4**. The supported epoxides **1** were opened with either a thiol or an amine nucleophile to yield 1,3-diols **2**, which were converted into Fmoc-substituted 1,3-dioxanes. Some of the nucleophiles used in the first step were hydroxy-substituted, leading to the formation of mixed acetals in the second step: these acyclic acetals were, therefore, hydrolysed by treatment of the resins with PPTS in THF-MeOH. Following removal of the Fmoc group ( → **3**), the resulting free amines were converted into amides, ureas, thioureas or sulfonamides ( → **4**), and the final compounds cleaved from the beads. Using material cleaved from a single bead, and with knowledge of the substituent added into the final step, it was possible to use mass spectrometry to identify possible combinations of substituents added in the first two diversification steps. The library has been exploited in the discovery of small molecule tools: uretupamine

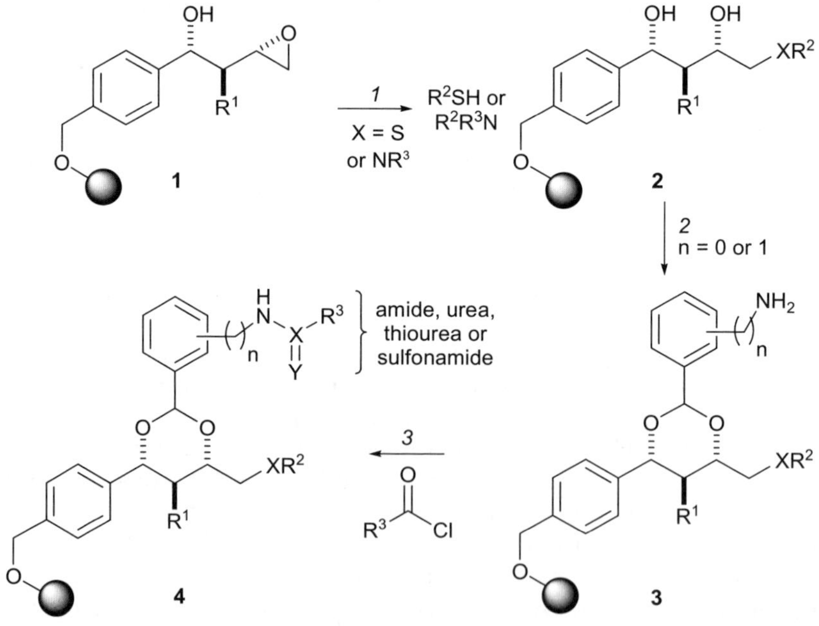

**Scheme 1.1** Schreiber's split–pool synthesis of 1,3-dioxanes with high substitutional diversity. *Reagents and conditions*: (1) $R^2SH$ or $R^2R^3N$; (2) (i) $ArCH(OMe)_2$, $Me_3SiCl$, HCl, dioxane; (ii) PPTS, THF–MeOH; (iii) piperidine; (iv) $Me_3SiCl$; (3) amide, urea, thiourea or sulfonamide formation.

(see Section 1.3.1.1),[27a,b] which targets Ure2p, a repressor of metabolic genes, and tubacin,[27c,d] a class II histone deacetylase inhibitor. In addition, derivatisation of the scaffold of the alkaloid, galanthamine, yielded secramine, a potent inhibitor of protein trafficking from the Golgi apparatus to the plasma membrane;[25] crucially, although secramine's structure was inspired by galanthamine, it had an entirely distinct biological function.

## 1.2.2 Synthesis of Compound Libraries with High Stereochemical Diversity

The synthesis of stereochemically diverse compound libraries has been facilitated by the development of a wide range of reliable asymmetric transformations. For example, an asymmetric hetero-Diels-Alder reaction was used to prepare a library of dihydropyrancarboxamides in which both substitution and configuration was varied (structures **7** and their enantiomers) (Scheme 1.2).[28] The library of 4320 structures was encoded using a binary encoding protocol which involved the attachment of tags to individual macrobeads. The asymmetric step (**5** → **6**) served a number of purposes: it generated the molecular scaffolds; it introduced some stereochemical diversity; and it incorporated the variable substituent, $R^1$. Deprotection of the ester, amide formation, and release from solid support gave the final compounds. The library enabled the discovery of haptamide A, a ligand which targeted a subunit of the transcription factor, Hap3p.[28]

## 1.2.3 Synthesis of Compound Libraries with High Scaffold Diversity

The development of robust and reliable synthetic methods that yield a range of diverse small molecule scaffolds has proved extremely demanding. In this section, we describe the progress that has been made towards developing synthetic strategies that allow the scaffolds of small molecules to be systematically varied: in particular, the development of "folding" pathways (Section 1.2.3.1), "branching" pathways (Section 1.2.3.2) and oligomer-based approaches (Section 1.2.3.3).

### 1.2.3.1 Use of "Folding Pathways" to Introduce Scaffold Diversity

The "folding pathway" approach uses common reaction conditions to transform a range of substrates into products with alternative molecular scaffolds. The substrates are encoded to "fold" into the alternative scaffolds through strategically placed appending groups (sometimes known as σ-elements[29]). This strategy leads to scaffold diversification under substrate control.

Schreiber has developed a folding pathway which exploits the Achmatowicz reaction (Scheme 1.3).[29] The fate of oxidation of the furan substrates **10**

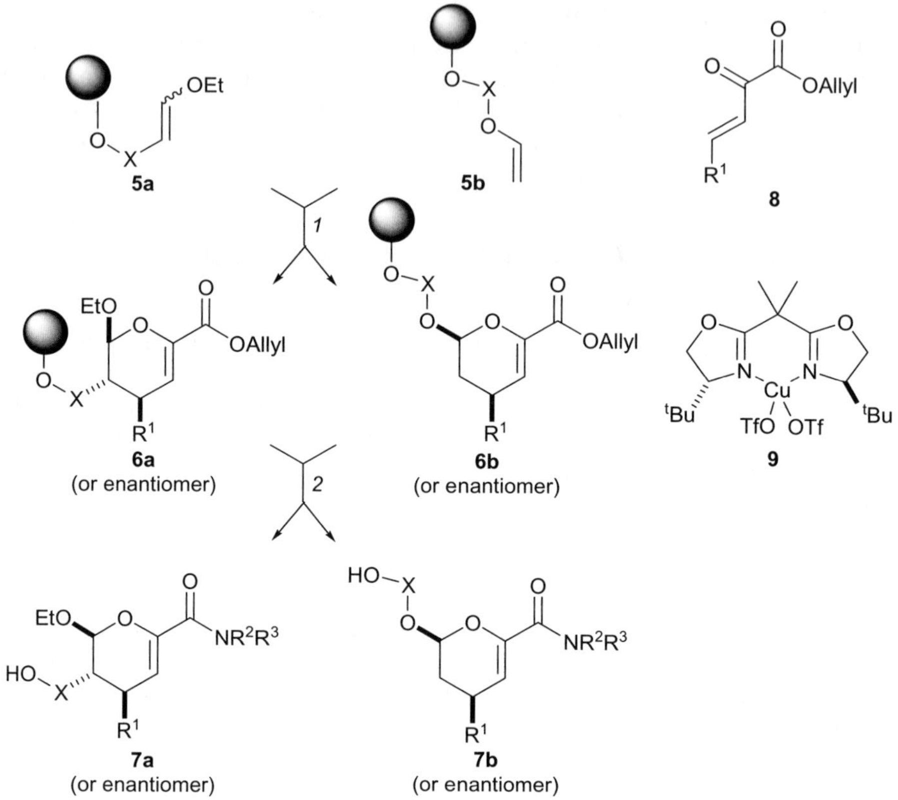

**Scheme 1.2** Schreiber's synthesis of dihydropyrans with high substitutional and stereochemical diversity. *Reagents and conditions*: (1) **8**, 20 mol% **9** (or enantiomer), THF, rt; (2) (i) Pd(PPh$_3$)$_4$, thiosalicylic acid, THF, rt; (ii) R$^2$R$^3$NH, PyBOP, iPr$_2$NEt, CH$_2$Cl$_2$–DMF, rt; (iii) HF·pyridine, THF, rt then Me$_3$SiOMe.

depends on the functionalisation of groups elsewhere in the molecule (the σ-elements). Hence, oxidation of the furan **10c**, which does not bear any free hydroxy groups, simply generated the ene-dione **11c**. With suitably positioned nucleophilic groups in the starting material, however, a *cis*-enedione intermediate could be intercepted. Hence, upon oxidation, the furyl alcohol **10b** folded, and eliminated water, to yield the alkylidene-pyran-3-one **11b**. In contrast, with two free hydroxyl groups, **10a** folded to yield the bicyclic ketal **11a**. The combination of solid phase chemistry and common reaction conditions made the strategy amenable to split-pool synthesis, which significantly increased the efficiency of library generation.

Schreiber used substrate-based control to create alternative indole alkaloid-like scaffolds (Scheme 1.4).[30] An α-diazo-β-keto-carbonyl group and an indole ring were strategically placed at different positions on the scaffolds of the starting materials (**12**). The densely functionalised starting materials were

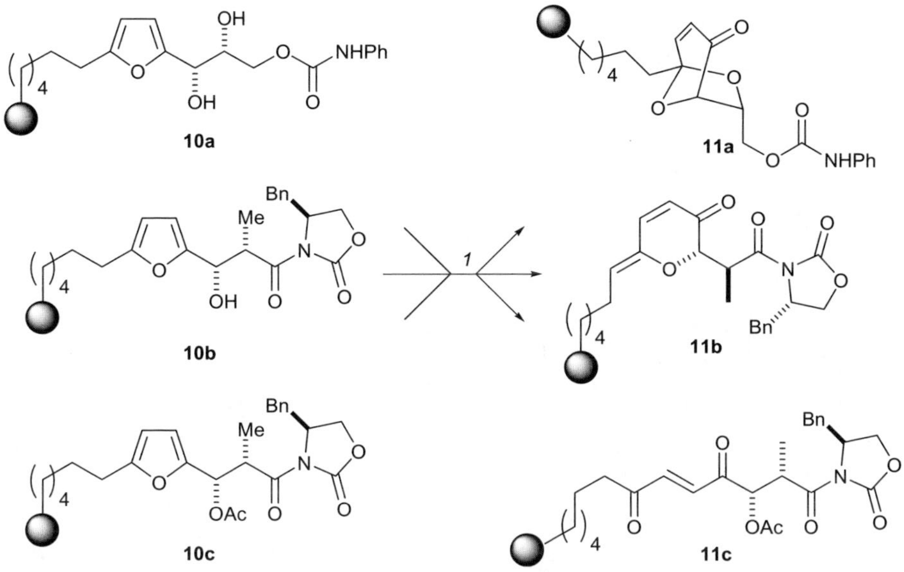

**Scheme 1.3** Schreiber's approach to pre-encoding skeletal information using the Achmatowicz reaction. *Reagents and conditions*: (1) *N*-bromosuccinimide, NaHCO$_3$, NaOAc, THF–H$_2$O (4:1), rt, 1 h; PPTS, CH$_2$Cl$_2$, 40–45 °C, 20 h, **11a**: 33% (64%); **11b**: 35% (86%); **11c**: 81% (>90%). Numbers in parentheses are purities determined by LC-MS.

assembled either using an Ugi four-component coupling reaction or by alkylation of a common scaffold; subsequent conversion into a diazo compound yielded the cyclisation precursors. The substrates, **12**, were treated with a catalytic amount of rhodium(II) octanoate dimer in benzene (80 °C). Presumably, generation of a carbonyl ylid was followed by intramolecular 1,3-dipolar cycloaddition with the indole ring to give the alternative polycyclic scaffolds, **13**.

The "folding" pathway strategy has been shown to be useful in the preparation of small molecule libraries based on a few diverse scaffolds. In addition to the examples highlighted here, radical chemistry has been harnessed to "fold" precursors into a range of fused and bridged amine scaffolds.[31] The scope of the "folding" pathway strategy has, however, been enormously expanded by using an oligomer-based approach to prepare precursors for a "folding" step (see Section 1.2.3.3).

## 1.2.3.2 Use of "Branching Pathways" to Introduce Scaffold Diversity

The "branching pathway" strategy involves the conversion of common precursors into a range of distinct molecular scaffolds through careful choice of the

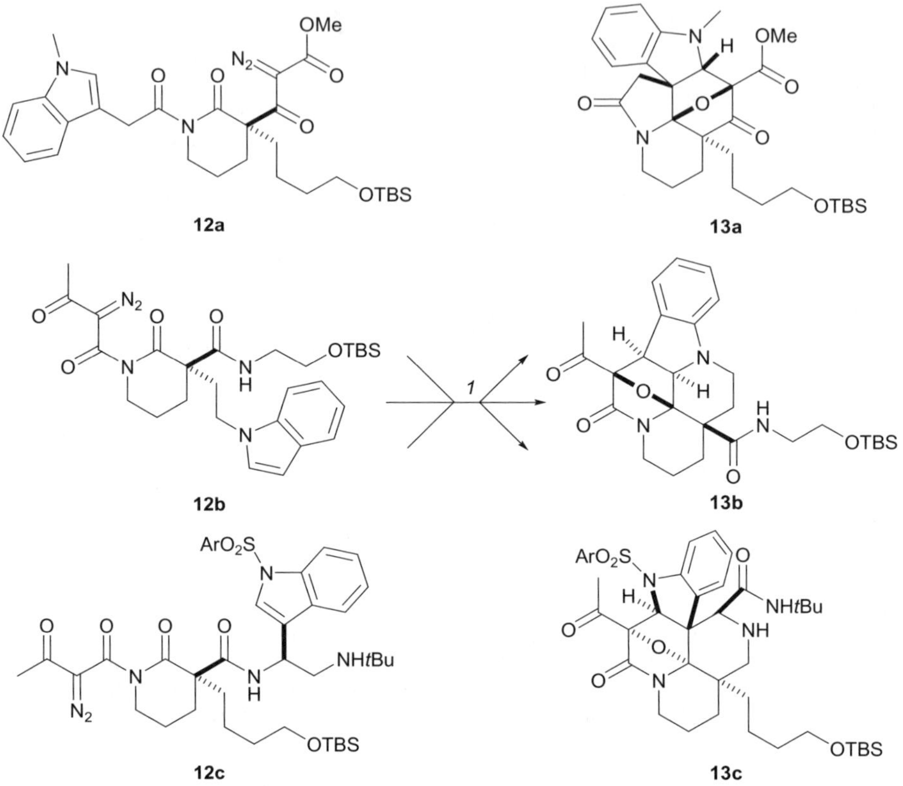

**Scheme 1.4**   Schreiber's approach to pre-encoding skeletal information via a rhodium-catalysed tandem cyclisation–cycloaddition strategy. *Reagents and conditions*: (1) $Rh_2(O_2CC_7H_{15})_4$, benzene, 50 °C; **13a**: 74%; **13b**: 73%; **13c**: 57%.

reaction conditions. Ideally, a range of flexible precursors would be designed which could participate in complementary reactions leading to alternative scaffolds.

An outstanding example of a "branching pathway" exploited complementary cyclisation reactions to yield alternative molecular scaffolds (Scheme 1.5).[32] A four-component Petasis condensation reaction was used to assemble flexible cyclisation precursors (*e.g.* **14**). Alternative cyclisation reactions were then used to yield products with distinct molecular scaffolds: Pd-catalysed cyclisation (→ **15a**); enyne metathesis (→ **15b**); Ru-catalysed cycloheptatriene formation (→ **15c**); Au-catalysed cyclisation of the alcohol onto the alkyne (→ **15d**); base-induced cyclisation (→ **15e**); Pauson–Khand reaction (→ **15f**); and Miesenheimer [ 2,3]-sigmatropic rearrangement (not shown). Four of these cyclisation reactions could be used again to convert the enyne **15e** into molecules with four further scaffolds (**17a–d**). In addition, Diels–Alder reactions with 4-methyl-1,2,4-triazoline-3,5-dione converted the dienes **15b**, **17a** and **17d**

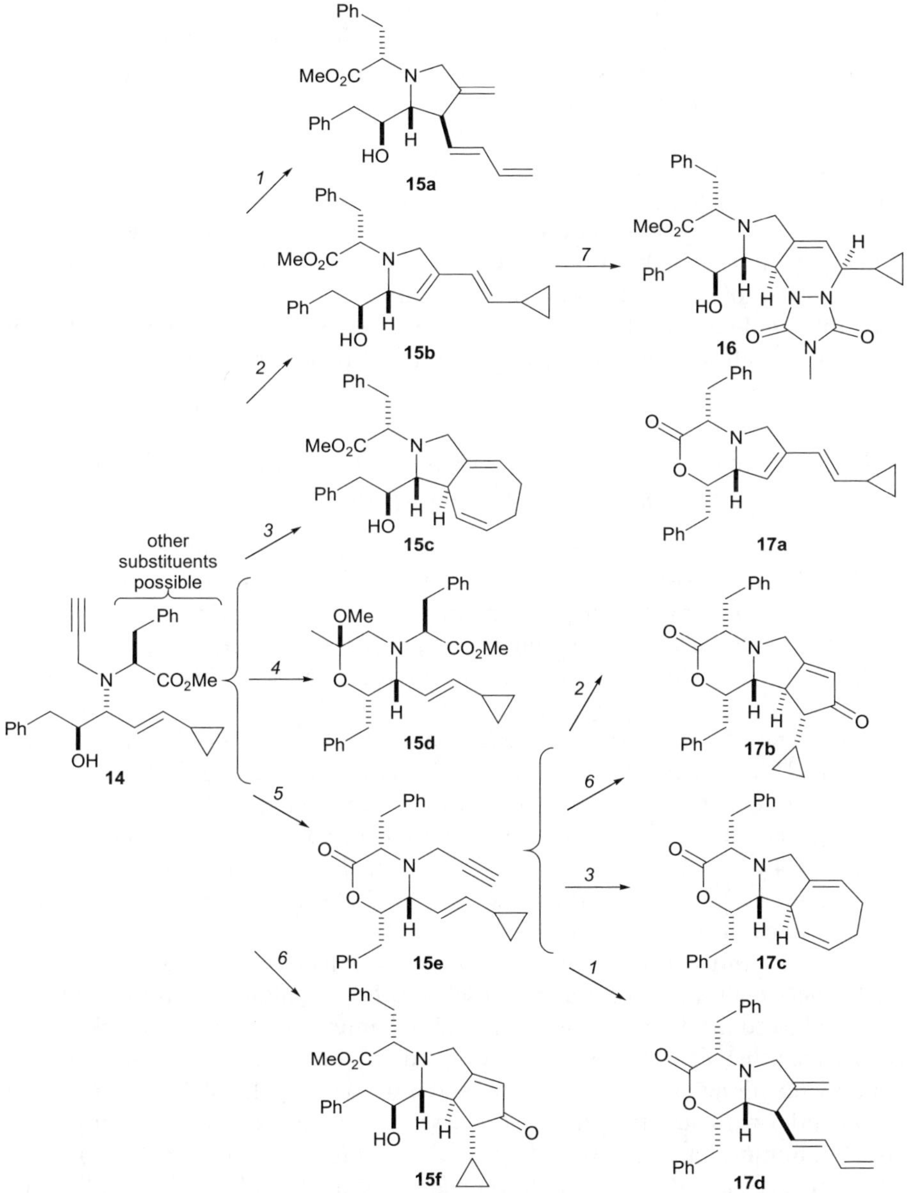

**Scheme 1.5** Schreiber's branching pathway exploiting complementary cyclisation reactions. *Reagents and conditions*: (1) 10 mol% Pd(PPh$_3$)$_2$(OAc)$_2$, benzene, 80 °C, [**15a**: single diastereomer, 81%; **17d**: single diastereomer, 70%]; (2) 10 mol% Hoveyda–Grubbs 2nd gen. cat., CH$_2$Cl$_2$, reflux, [**15b**: *trans:cis* 85:15, 89%; **17a**: *trans:cis* 85:15, 87%]; (3) 10 mol% CpRu(MeCN)$_3$PF$_6$, acetone, rt, [**15c**: single diasteromer, 85%; **17c**: single diasteromer, 91%]; (4) 10 mol% NaAuCl$_4$, MeOH, rt, single diasteromer, 80%; (5) NaH, toluene, rt, 88%; (6) Co$_2$(CO)$_8$, Et$_3$NO, NH$_4$Cl, benzene, rt, [**15f**: dias. >90:10, 85%; **17b**: single diasteromer, 85%]; (7) 4-methyl-1,2,4-triazoline-3,5-dione, CH$_2$Cl$_2$, rt, single diasteromer from *trans* diene, 72%.

into the corresponding polycyclic products (*e.g.* **16**; other products not shown). The key to this powerful synthetic approach lay in the design of precursors (*e.g.* **14**) which were effective substrates in a wide range of efficient and diastereoselective cyclisation reactions.

The "branching pathway" strategy has been shown to allow the preparation of small molecule libraries based on a range of diverse scaffolds. The strategy requires the careful design of flexible precursors that may be converted selectively into different scaffolds under alternative reaction conditions. The development of effective branching pathways may also require considerable optimisation of the individual steps leading to alternative scaffolds. It is probably unlikely that the approach can be extended such than many scores of alternative scaffolds may be generated from individual precursors.


### 1.2.3.3   Introducing Scaffold Diversity by Combining Building Blocks: the "Build–Couple–Pair" Strategy

An extremely powerful strategy for preparing small molecule libraries with high scaffold diversity exploits simple building blocks in combination. The approach requires the building blocks to be prepared ("built") and then connected ("coupled"). Finally, pairs of functional groups are reacted ("paired") intramolecularly to yield new ring systems in the final scaffolds. The so-called "build–couple–pair" strategy has been reviewed.[24b]

The scope of the approach is extremely broad and, indeed, some folding pathways[30] (*e.g.* Scheme 1.4; Section 1.2.3.1) and branching pathways[32] (*e.g.* Scheme 1.5; Section 1.2.3.2) can be considered to exemplify the "build–couple–pair" strategy. For example, the four component Petasis reaction illustrated in Scheme 1.4 allowed simple building blocks to be combined; complementary cyclisation reactions were then use to "pair" functional groups to yield a diverse range of product scaffolds.[32]

The cyclisation reactions of *N*-acyl iminium ions have yielded a wide range of distinct small molecule scaffolds (Scheme 1.6).[33] Initially, peptide synthesis was used to connect a range of functionalised amino acid building blocks. Each of the resulting peptides, **18**, was designed to contain a masked aldehyde, a suitably positioned secondary amide, and a pendant nucleophile. Treatment of the peptides with acid triggered the release of an aldehyde, the formation of an *N*-acyl iminium ion, and cyclisation to yield a final scaffold (*e.g.* **19a–f**).

Metathesis cascades have underpinned the synthesis of diverse small molecule libraries.[34,35] Metathesis is a superb "pairing" reaction for the "build–couple–pair" approach: first, it can yield many different ring systems and, second, alkenes (and alkynes) are compatible with the many reactions that may be used to connect building blocks. Metathesis has been used to prepare a library of natural product-like molecules (Scheme 1.7).[35] Initially, unsaturated building blocks were attached iteratively to fluorous-tagged linker to yield metathesis precursors **20**. Crucially, alternative attachment reactions were used such that the building blocks were connected through bonds that either did, or

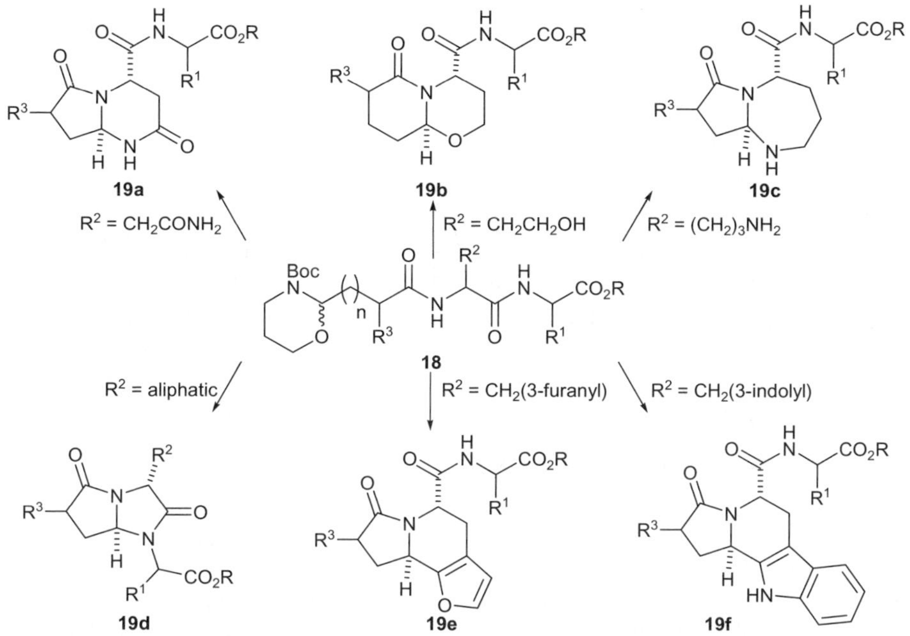

**19a**   $R^2 = CH_2CONH_2$

**19b**   $R^2 = CH_2CH_2OH$

**19c**   $R^2 = (CH_2)_3NH_2$

**18**

$R^2$ = aliphatic    $R^2 = CH_2$(3-furanyl)    $R^2 = CH_2$(3-indolyl)

**19d**   **19e**   **19f**

**Scheme 1.6** Cyclisation of *N*-acyl iminium ions in the synthesis of a range of small molecule scaffolds. *Reagents and conditions*: $H^+$.

did not, remain as a vestige in the final products. Finally, metathesis cascades were used to "reprogramme" the scaffolds and to release the final molecules, **21**, from the fluorous tag. The resulting library had unprecedented scaffold diversity (over eighty distinct scaffolds).[36]

The "build–couple–pair" is an extremely powerful and flexible approach for preparing small molecule libraries based on a diverse range of scaffolds. The approach is highly general, and many different reactions have been used to connect ("couple") building blocks. Variants of the Ugi multi-component reaction are, however, still over-represented amongst the reactions that have been used to connect building blocks.[32,37]

A wide range of other reactions have been exploited in the final cyclisation ("pairing") step: in addition to the examples illustrated here, lactamisations,[38,39] metal-catalysed cyclisations[37,39,40] and cycloadditions[37] have, for example, been exploited to yield final product scaffolds. At its most powerful, the "build–couple–pair" strategy can allow the combinatorial variation of the scaffolds of small molecules. However, a significant challenge will be to identify reactions other than olefin metathesis that have the broad scope and chemoselectivity needed to yield scores of different ring systems. It is certainly possible that the overall approach may, in the future, be used to prepare small molecule libraries based on hundreds, or even thousands, of distinct molecular scaffolds.

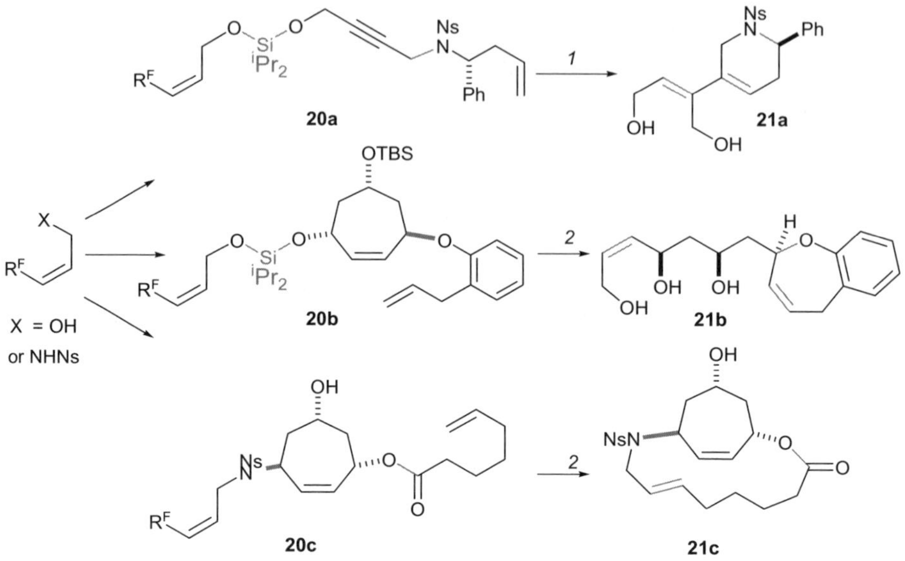

**Scheme 1.7** Synthesis of natural product-like molecules with unprecedented scaffold diversity. Initially, building blocks were added iteratively to a fluorous-tagged linker, with intermediates purified by fluorous–solid phase extraction. Metathesis cascades were used to reprogramme the scaffolds and to release final products from the fluorous-tagged linker. *Reagents and conditions*: (1) Grubbs' first-generation catalyst, **21a**: 23%; **21b**: 56%; (2) fluorous-tagged Hoveyda–Grubbs second-generation catalyst, **21c**: 33%.

## 1.2.4 Emerging Approaches for Ligand Discovery Using Large Small Molecule Libraries

Combinatorial biochemical techniques,[41] such as RNA display, phage display and aptamer SELEX, can allow large (*e.g.* $10^8$–$10^{13}$) libraries of macromolecules to be screened on the basis of binding to a target protein. These techniques enable the discovery of novel macromolecules that bind to, and hence modulate the function of, the target protein. In contrast, most screens of small molecule function are restricted to much smaller libraries, essentially because individual molecules usually need to be separately purified before they are (usually individually) assayed: even in large pharmaceutical companies, high-throughput screens are restricted to smaller (generally $<10^6$) compound libraries.

Extremely powerful methods are now emerging to increase the efficiency of screens for small molecule binders.[42–44] These methods enable the synthesis of extremely large small molecule libraries in a manner that is broadly analogous to combinatorial biochemical approaches. In some cases, an evolutionary approach is possible,[42,44] with increasing active compounds being selectively fed into subsequent rounds of chemical synthesis. Large libraries of constrained peptides have been encoded by phages, and have been screened to yield an inhibitor that was specific to human plasma kallikrein.[42] A broader range of chemistry, which nonetheless still needs to be compatible with aqueous conditions (and DNA), is possible using DNA-encoded[43] or DNA-programmed[44] organic synthesis.

A split-pool approach has been used to prepare a library of 800 000 000 DNA-encoded small molecules (see Scheme 1.8 for an example for a synthesis of a DNA-encoded library).[43] After each diversification step, the small molecules were derivatised with a DNA tag to encode the substituent that had been added. At the end of the synthesis, the small molecules – each of which was

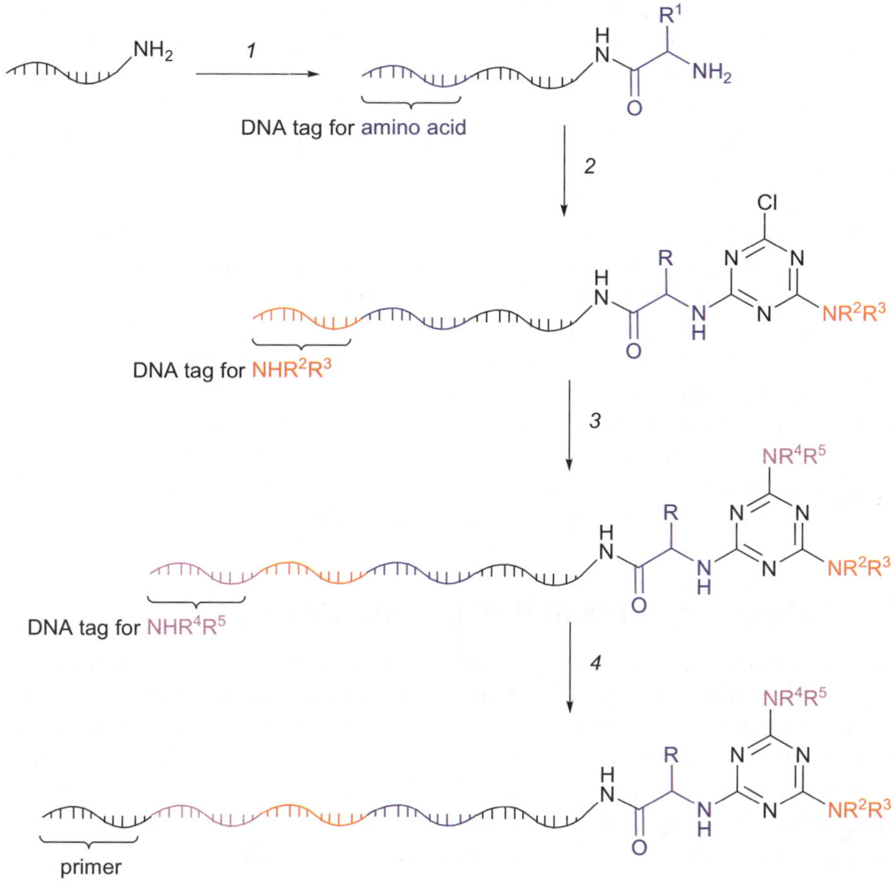

**Scheme 1.8**  Synthesis of a DNA-encoded small-molecule library. The DNA tag is not drawn to scale. *Reagents and conditions*: (1) (a) attach DNA tag; (b) acylation with Fmoc-protected amino acid; (c) purify; (d) deprotect; (2) (a) attach DNA tag; (b) cyanuric chloride; (c) substitute with R$^2$R$^3$NH; (3) (a) attach DNA tag; (b) substitute with R$^4$R$^5$NH; (c) purify; (4) ligate primer.

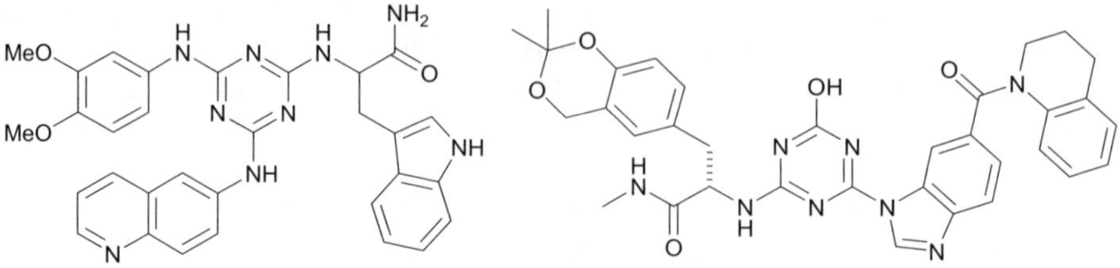

**Figure 1.2**    Structures of some inhibitors of Aurora A kinase (**22**) and p38 MAP kinase (**23**) discovered using a DNA-encoded small-molecule library.

derivatised with a unique DNA tag – were pooled. The library was then interrogated by affinity selection (*i.e.* ability to bind the target protein). The identity of the binding molecules was deconvoluted by high-throughput sequencing of their DNA tags, to yield extremely rich structure-activity relationships for the active compounds. The general approach has been used to identify inhibitors of two enzymes: Aurora A kinase (*e.g.* **22**) and p38 MAP kinase (*e.g.* **23**) (Figure 1.2).

DNA can also be used to *programme* a series of organic transformations (rather than simply to record the history of a synthesis).[44] In this format, all of the transformations may be performed in a single pot, with the synthesis of each individual molecule being determined by base-pairing between the DNA-tagged substrate and its complementary DNA-tagged reagent.

DNA-programmed synthesis lends itself to an evolutionary approach: the tags for small molecule binders may be amplified after each round of affinity selection, and fed into a subsequent generation of synthesis. The approach has been used to discover, from a library of 100 000 000 compounds, ligands for the *N*-terminal SH3 domain of the proto-oncogene Crk.[44a]

## 1.3   Discovery of Small Molecule Probes

In this section, the discovery of small molecules for the interrogation of biological mechanisms will be described. An appropriate small molecule tool may be discovered either on the basis of modulation of the activity of a specific protein (reverse chemical genetics, Section 1.3.1) or its phenotypic effect (forward chemical genetics, Section 1.3.2). Emphasis will be placed on the approaches that may be used to discover useful small molecule tools, and the insights into biological mechanisms that may be accrued.

### 1.3.1   Discovery of Small Molecule Probes Using a Reverse Chemical Genetic Approach

The reverse chemical genetic approach begins the discovery of a small molecule modulator of a specific target protein. The small molecule is usually identified

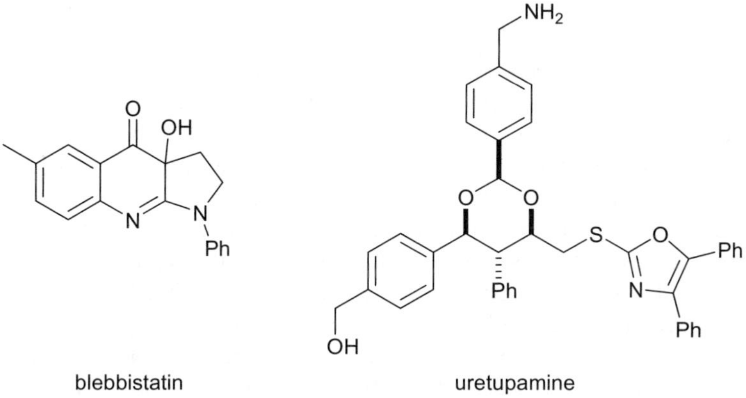

**Figure 1.3** Structures of blebbistatin and uretupamine.

using an *in vitro* assay of the activity of a purified protein. However, it is also possible to identify ligands that modulate the cellular function of a specific protein.

Blebbistatin (Figure 1.3) was discovered using a high-throughput screen for small molecules that inhibit non-muscle myosin II.[8b] Blebbistatin was then used to study myosin II-dependent cell processes. The compound blocked cell blebbing rapidly (within ∼2 min) and reversibly, and disrupted directed cell migration and cytokinesis in vertebrate cells. Blebbistatin was also used in combination with other ligands to demonstrate that exit from the cytokinetic phase of the cell cycle is dependent on ubiquitin-mediated protein degradation.

Small molecule microarrays[45,46] can allow the discovery of small molecule binders to a specific protein: the methodology is accessible to many academic groups, and can allow high-throughput screening of large compound libraries. A small molecule microarray was used to discover uretupamine (Figure 1.3) which binds to Ure2p, the central repressor of genes involved in nitrogen metabolism.[27a,b] A 3780-member library of small molecules was used to create a small molecule microarray which was then challenged with fluorescently tagged Ure2p. Transcription profiling was used to determine the effects of uretupamine on the regulation of genes in yeast. Uretupamine was found to modulate only a subset of the functions of Ure2p: the regulation of the glucose-sensitive subset of the downstream genes. The example illustrates that reverse chemical genetics can be a more refined tool for interrogating cellular protein function than reverse genetics since the effects of uretupamine were more specific than those resulting from deletion of the *URE2* gene.

Small molecule microarrays were also used to discover robotnikinin (Section 1.4), a ligand that binds to the extracellular Sonic Hedgehog (Shh) protein.[47] Approximately 10 000 diversity-oriented synthesis compounds and natural products were immobilised on a single microscope slide. The small molecule microarray was challenged with the 20 kDa N-terminal fragment (ShhN) of Shh: some structurally related macrocycles were identified as hits and, after optimisation, robotnikinin emerged. Robotnikinin has been used to interrogate the Sonic Hedgehog signalling pathway (see Section 1.4). Small molecule microarrays have also been used to discovery ligands for human immunoglo-bulin G (IgG),[46] the transcription factor Hap3p,[28] and calmodulin.[45]

Valuable small molecule tools may also be obtained by structure-guided ligand design. The interaction between the lysine 382-acetylated p53 and the bromodomain of the coactivator CBP is essential for p53-induced transcription of the cell cycle inhibitor p21 in response to DNA damage. A structure-guided approach was used to discover small molecule inhibitors of the CBP/p53 interaction.[48] Thus, the structure of the bromodomain of CBP (in complex with a peptide mimic of lysine 382-acetylated p53) enabled the design of a focused small molecule library to target the CBP/p53 interaction: NMR spectroscopy was then used to identify specific small molecules that bound to the bromo-domain of CBP. The active molecules were shown to inhibit CBP/p53 association: the small molecules led to destabilisation of p53, and, thus, deac-tivation of its transcriptional activity in response to DNA damage. In this investigation, the chemical genetic approach enabled the functional con-sequence of a protein-protein interaction to be determined in a cellular context.

### 1.3.1.1   General Reverse Chemical Genetic Approaches for Targeting Specific Members of Families of Macromolecules

A disadvantage of the chemical genetic approach is that it usually lacks gen-erality: in general, for each macromolecule to be targeted, the discovery of a tailored ligand is required. However, more general approaches have been developed to target protein kinases (Section 1.3.1.1.1) and duplex DNA sequences (Section 1.3.1.1.2).

**1.3.1.1.1   Inhibitor-sensitive Protein Kinase Mutants.**   Shokat has pioneered a remarkable approach that allows specific protein kinases to be targeted with small molecule ligands.[49] The design of highly specific ATP-competitive inhibitors of protein kinases is hampered by the large number of proteins with ATP binding sites (including ~600 protein kinases). The key to Sho-kat's approach lies in engineering a protein kinase that is uniquely sensitive to a "bulky" kinase inhibitor. The strategy of the approach is illustrated in Figure 1.4. The structure of the protein kinase Hck in complex with the ligand PP1 is shown in Panel B. By replacing the *p*-tolyl substituent of PP1 with a bulkier substituent (such as a 1-naphthyl or a 1-naphthylmethyl sub-stituent), the ligand is no longer able to inhibit Hck (or indeed any other protein kinase) because of an unfavourable interaction with the gatekeeper residue (T338 in Hck). However, mutating the gatekeeper residue to a smal-ler residue allows the bulkier ligand to be (uniquely) accommodated by the mutant protein kinase.

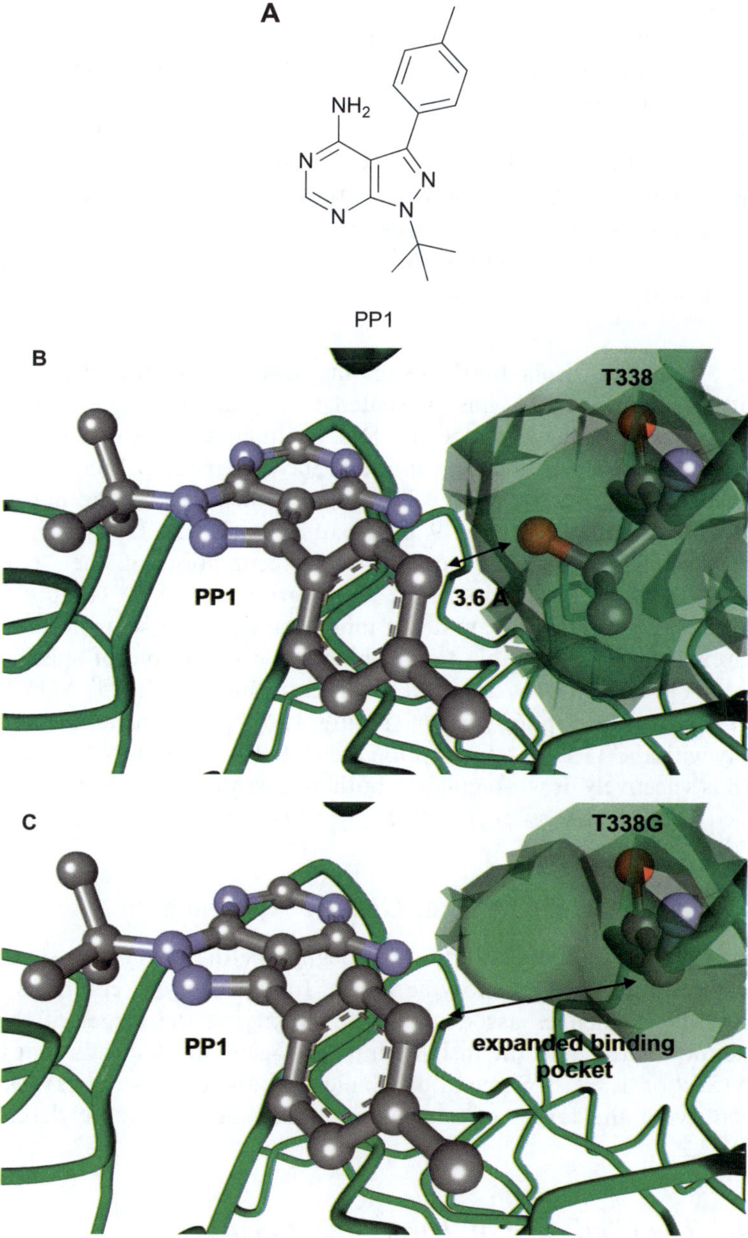



**Figure 1.4** Development of inhibitor-sensitive protein kinases illustrated using the protein kinase Hck. Panel A. Chemical structure of the ligand PP1. Panel B. Crystal structure of PP1 in complex with the protein kinase Hck. The *p*-tolyl substituent of PP1 is in close proximity to the side chain of the gatekeeper residue T338. Panel C: Model of the structure of the Hck·PP1 complex in which the side chain of the gatekeeper residue 338 has been removed *in silico*. The expanded binding pocket in the T338G mutant is able to accommodate bulkier analogues of PP1.

The approach has been demonstrated to be rather general, and has been used to interrogate many biological mechanisms that are regulated by protein kinases.[50] An early investigation focused on the role of the cyclic-dependent kinase Cdc28 (CDK1) in the regulation of the cell cycle in yeast. Specific inhibition of the inhibitor-sensitive mutant of Cdc28 caused a pre-mitotic cell-cycle arrest. Remarkably, the observation was in stark contrast to the G1 arrest that is typically observed in temperature-sensitive *cdc28* mutants.[51] Thus, in this case, the insights gleaned from a conditional (temperature-sensitive) genetic approach were different to those observed using a conditional chemical genetic approach.

**1.3.1.1.2 Chemical Tools for Investigating DNA Function.** The structural regularity of duplex DNA has presented a special opportunity to identify small molecules that target a specific DNA sequence. Dervan has developed rules that allow a a specific polyamide to be designed to bind to any pre-determined DNA sequence.[52] The ligands were inspired by the natural product, distamycin, that binds to 9 base-pairs of AT-rich DNA. The ligands recognise specific DNA sequences through recognition of the "edges" of intact Watson–Crick base pairs in the minor groove of DNA (Figure 1.5).[53]

The ability to design polyamide ligands to target any sequence of duplex DNA is exceptionally valuable for probing the cellular role of specific DNA sequences. A polyamide was designed to target the binding site (5′-AGTACT-3′) of the TFIIIA transcription factor (Figure 1.6).[54] The polyamide competed effectively with the TFIIIA transcription factor for its target DNA sequence: the ligand thus selectively down-regulated, both *in vitro* and *in vivo*, the transcription of a gene (the 5S RNA gene) controlled by TFIIIA (relative to a control gene).

## 1.3.2  The Forward Chemical Genetic Approach

The forward chemical genetic approach begins the discovery of a small molecule modulator using a phenotypic assay. In this section, emphasis will be placed on the challenges associated with identifying the target of the small molecule modulator, and the insights into biological mechanisms that may be accrued (Section 1.3.2.2). Some of the challenges (and opportunities) associated with identifying the target of a small molecule modulator are described in Section 1.3.2.1.

### *1.3.2.1  The Challenge of Target Identification*

A major challenge associated with the forward chemical genetic approach is the identification of the target of the small molecule probe.[11] Affinity chromatography is a common approach to target identification. This approach requires the attachment of the small molecule probe to a solid support: the site of the attachment of the linker must be chosen carefully to avoid the ablation of the probe's biological activity. Trapoxin, is a cyclic tetrapeptide whose

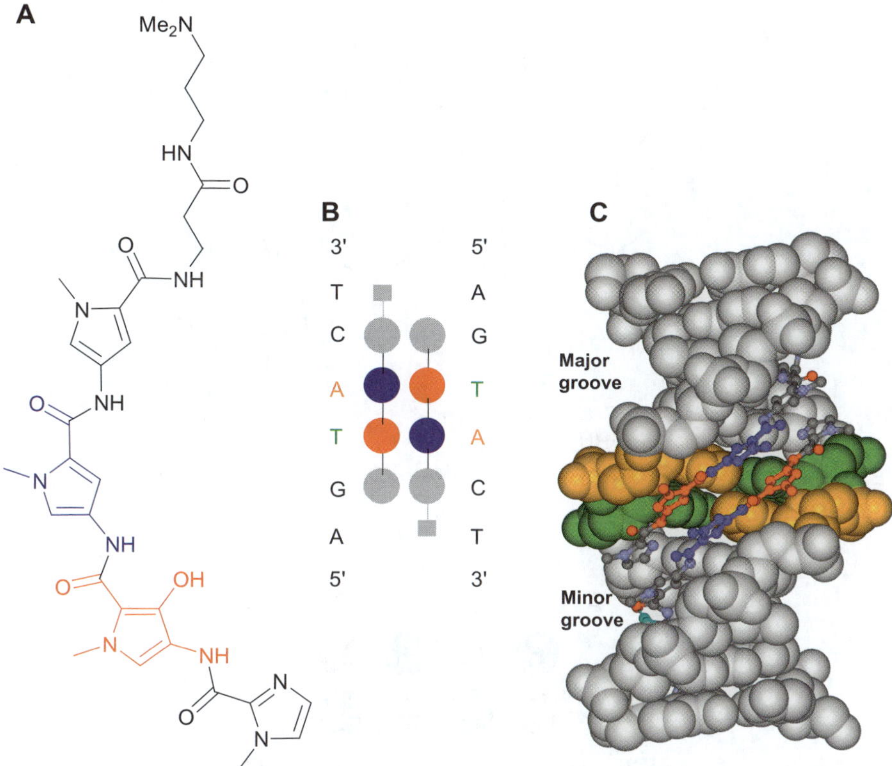

**Figure 1.5**  Sequence-specific recognition of duplex DNA by polyamide ligands. Panel A: Structure of a polyamide ligand designed to target a sequence of duplex DNA (5′-AGTACT-3′). Panel B: Schematic diagram of the binding of two molecules of a polyamide ligand to a sequence of duplex DNA. Panel C: Structure of the complex between duplex DNA and a polyamide ligand. The polyamide ligand occupies the minor groove, with sequence recognition through contacts with the edges of the intact Watson–Crick base pairs.

irreversible mode of action facilitated the identification of its cellular target. Attachment of trapoxin to a solid support gave the K-trap affinity matrix: epoxide-opening led to immobilisation of the target protein, histone deacetylase.[55]

The development of an effective affinity matrix requires functionalisation of the small molecule probe without affecting its biological activity. This usually requires detailed probing of structure–activity relationships which can be time consuming. To address this problem, "tagged" chemical libraries have been developed in which each of the library members is already functionalised with a linker that would allow attachment to a solid support. A specific tagged triazine, **24** (R = H), was found to modulate brain and eye morphogenesis in zebrafish (Figure 1.7).[56] Affinity chromatography, using a matrix obtained by immobilisation of **24** (R = H) to agarose beads, yielded three 40S ribosomal

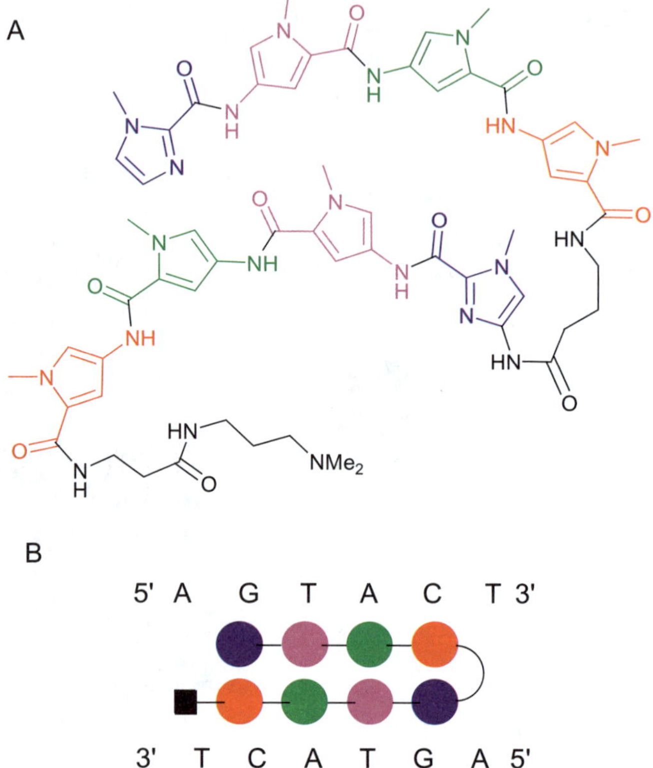



**Figure 1.6**   Design of a polyamide ligand to target the binding site (5′-AGTACT-3′) of the TFIIIA transcription factor. Panel A: Structure of the polyamide ligand. Panel B. Schematic diagram of the binding of the ligand to its target DNA sequence.

subunit proteins from a protein extract. Interestingly, mutations in 40S ribosomal subunit proteins have also been found to modulate zebrafish development. It was suggested that the ligand **24** (R = H) may interfere with the function of a protein complex that includes the S5, S13, S18 and L28 ribosomal proteins, and plays an important role in brain and eye development in the early zebrafish embryo.

Mixed isotope photoaffinity reagents have been developed to facilitate target identification by mass spectrometry. To demonstrate the approach, cyclosporine was both biotinylated and functionalised with a mixed isotope photoactivatable benzophenone probe (Figure 1.8).[57] Irradiation of the probe in the presence of a mixture of proteins led to selective cross-linking to cyclophilin A. The biotin tag enabled easy separation of the cross-linked protein from other proteins in the mixture. Subsequently, after tryptic digestion, the mixed isotope tag facilitated the identification, by mass spectrometry, of the peptides that were cross-linked to the functionalised probe.

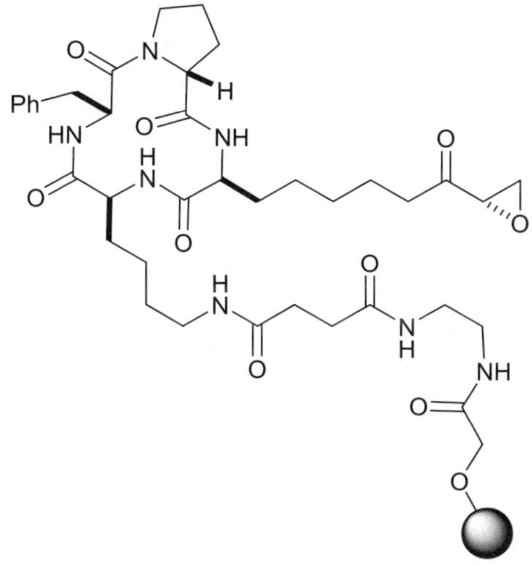

**Figure 1.7** Affinity matrices that allowed discovery of the molecular targets of trapoxin and the triazine **24**, R = H).

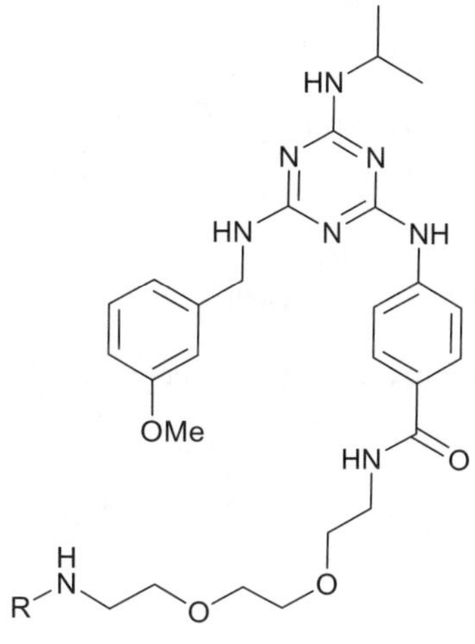

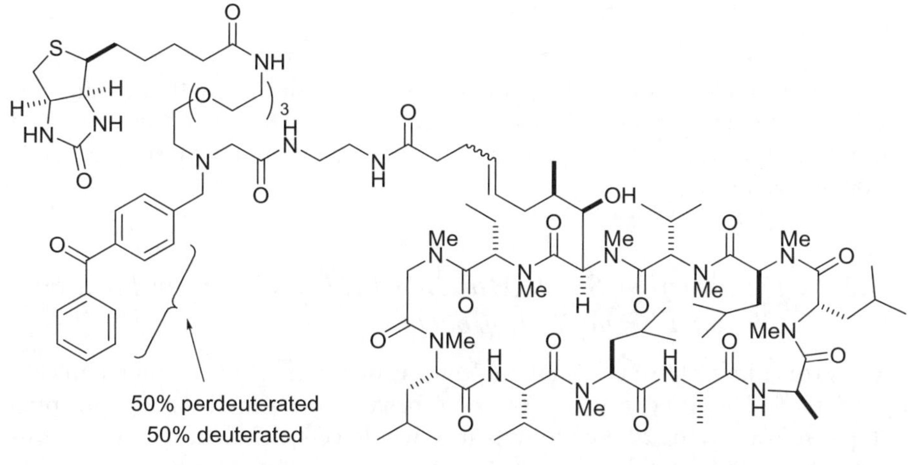

**Figure 1.8** A mixed isotope photoaffinity reagent in which cyclosporine has been functionalised with both biotin and a photoactivatable benzophenone probe.

Protein chips[58] can be valuable tools in target identification, and have been used in chemical genetic studies of the target of rapamycin (TOR) signalling pathway.[59] Small molecule inhibitors of rapamycin (such as SMIR3 and SMIR4; Figure 1.9) suppress rapamycin's effect in the yeast *Saccharomyces cerevisiae*. A chip bearing thousands of yeast proteins was interrogated with

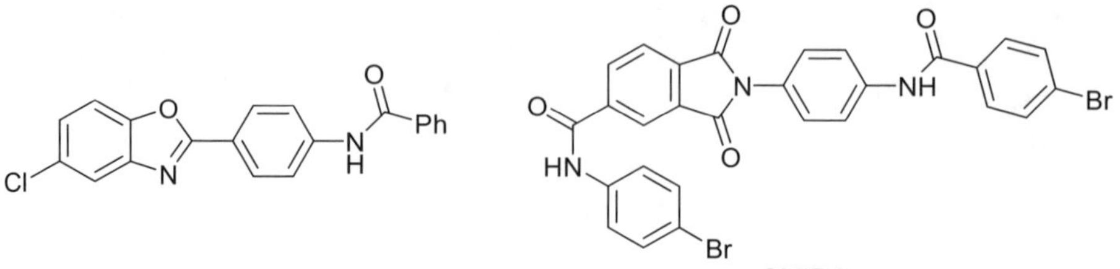

**Figure 1.9** Small molecule inhibitors of rapamycin.

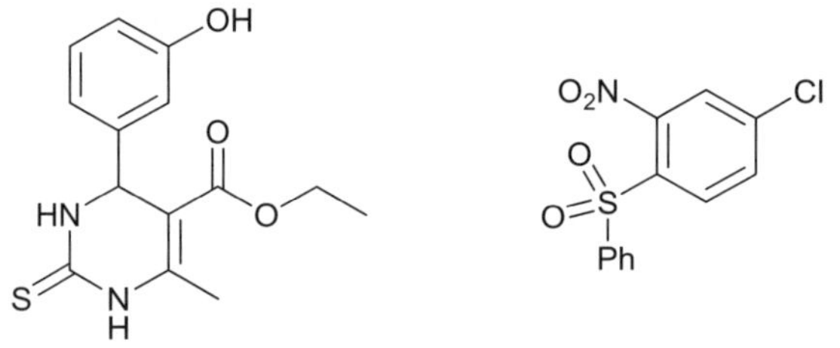

**Figure 1.10** Structures of monastrol and BTB-1 which inhibit the motor proteins Eg5 and Kif18A, respectively.

biotinylated analogues of SMIR3 and SMIR4; Cy3-labelled streptavidin was then used to detect the bound small molecules on the chip. The study revealed putative intracellular target proteins that may be components of the TOR signalling network.

## 1.3.2.2 *Discovery of Small Molecule Probes Using the Forward Chemical Genetic Approach*

Monastrol (Figure 1.10) is a small molecule probe that arrests mammalian cells in mitosis.[60] The discovery of monastrol began with a high-throughput phenotypic screen of 16 320 compounds in a whole cell mitotic arrest assay; this assay yielded 139 active molecules. A number of small molecules were already known to target tubulin, and, therefore, the compounds that target tubulin were eliminated using an *in vitro* tubulin polymerisation assay. The remaining 82 small molecules were assayed for their effect on microtubule, actin and chromatin distribution within cells. Five compounds, including monastrol, were found to affect only mitosis.

In BS-C-1 cells treated with monastrol, the bipolar mitotic spindle was replaced by a monastrol microtubule array surrounded by a ring of chromosomes. The discovery of the target of monastrol was suggested by a very similar phenotypic effect that was induced by specific antibodies that targeted the

BimC kinesin Eg5, a motor protein required for spindle bipolarity. In addition, temperature-sensitive mutations in the BimC family of kinesins do not form bipolar spindles at the restrictive temperature. Monastrol was then shown to inhibit the motility of Eg5 *in vitro*. Monastrol is a useful tool for studying mitotic mechanisms because it complements other small molecules that target tubulin and BTB-1 (Figure 1.10),[61] a small molecule inhibitor of the mitotic motor protein Kif18A (discovered using a reverse chemical genetic approach). In addition, the (conditional) use of monastrol complements the use of temperature-sensitive mutants in a conditional genetic approach.

Purmorphamine (Figure 1.11) was discovered in a high-throughput assay to identify small molecules that induce osteogenesis in multipotent mesenchymal progenitor cells.[62] A library of around 50 000 compounds with heteroaromatic scaffolds was screened for expression of the osteogenesis marker gene ALP whose expression is correlated with osteogenesis. The morphology of cells treated with purmorphamine changes from fibroblast (long and spindle shaped) to osteocyte (small and round). A reporter gene assay for a bone-specific transcription factor was used to confirm that the ligand induced C3H10T1/2 cells to differentiate into osteoblasts. Purmorphamine was subsequently shown to directly target Smoothened, a critical component of the Sonic Hedgehog signalling pathway.[63] Purmorphamine has been used in conjunction with other small molecule tools to interrogate the Sonic Hedgehog signalling pathway (see Section 1.4).

## 1.4   Case Study: Interrogation of the Sonic Hedgehog Pathway Using Complementary Small Molecule Probes

A chemical genetic approach has been used to interrogate the Sonic Hedgehog signalling pathway in which the protein Shh is essential for proper embryonic development (Figure 1.11A).[47,62–64] A feature of these studies is that a range of complementary small molecule tools have been exploited to modulate independently the functions of relevant proteins (Figure 1.11B): the natural product, cyclopamine; robotnikinin, a ligand discovered using a reverse chemical genetic approach (Section 1.3.1); and purmorphamine, a ligand discovered using a forward chemical genetic approach (Section 1.3.2.2).

ShhN, the 20 kDa *N*-terminal fragment of Shh, functions by binding to its transmembrane receptor, Patched (Ptc1). This interaction prevents Patched's inhibition of Smoothened (Smo), thus allowing Smoothened to activate processes within the cell (Figure 1.11B).

Robotnikinin is a ligand for ShhN, and was found to block the Shh signalling pathway in cell lines, keratinocytes and a synthetic model of human skin. Robotnikinin functions by binding to ShhN, preventing its interaction with Patched (Figure 1.11C). Thus, in the presence of robotnikinin, Patched binds to, and represses, Smoothened. Robotnikin, therefore, leads indirectly to the repression of Smoothened, and blocking of the Shh signalling pathway.

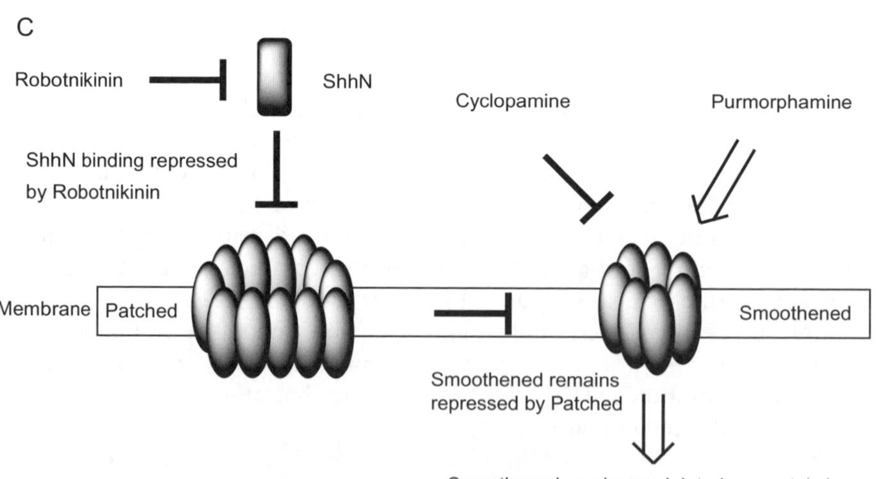

A
robotnikinin
purmorphamine
cyclopamine
B
ShhN
ShhN represses Patched
activating Smoothened
Membrane
Patched
Smoothened
Smoothened activates processes
within cell
C
Robotnikinin
ShhN
Cyclopamine
Purmorphamine
ShhN binding repressed
by Robotnikinin
Membrane
Patched
Smoothened
Smoothened remains
repressed by Patched
Smoothened can be modulated separately by
cyclopamine (suppressing) or Purmorphamine
(activating)

The Shh pathway activity could be rescued by the small molecule agonist of Smoothened, purmorphamine.[63]

The chemical genetic approach allowed the interplay between proteins (ShhN, Patched and Smoothened) to be studied using a pair of small molecule tools (robotnikinin and purmorphamine). The study complemented previous work which demonstrated that cyclopamine[64] – a natural product that antagonises Smoothened – could reverse the effects of mutations in Patched. In both cases, the effects of modulating Patched – either indirectly, by targeting ShhN with Robotnikinin, or by mutation – could be reversed by antagonising Smoothened.

# 1.5  Summary and Outlook

The field of chemical genetics has raised tremendous challenges that continue to be addressed using chemistry. First, the approach requires libraries of diverse small molecules. New diversity-oriented synthetic approaches have been developed to allow broad tracts of biologically relevant chemical space to be explored (see Scheme 1.2). However, chemists' ability to explore chemical space in a systematic way is still limited: in the future, screening collections that cover biologically relevant chemical space in a more even manner will need to be assembled. Second, chemical approaches to small molecule screening continue to be developed. For example, small molecule microarrays (Section 1.3.1) allow large libraries of small molecules to be screened in a format that is accessible to many academic groups. Third, analytical chemistry can be used to address the challenge of identifying the protein that is targeted by a small molecule modulator (Section 1.3.2.1).

The chemical genetic approach does, however, usually, lack generality: for each target protein, a tailored small molecule ligand is required. A generalised approach is sometimes possible, and it is likely that general approaches will be developed for other specific classes of macromolecule. In addition, emerging ultra-high throughput chemical approaches for the synthesis and screening of

---

**Figure 1.11**  Interrogation of the Shh signalling pathway using a chemical genetic approach. Panel A: Ligands used to modulate the Shh signalling pathway; robotnikinin is a ligand for the 20 kDa N-terminal fragment of Shh, ShhN, whereas purmorphamine and cyclopamine agonise and antagonise Smoothened respectively. Panel B: Summary of the biological mechanism of the Shh signalling pathway. ShhN, the 20 kDa N-terminal fragment of Shh, functions by binding to its transmembrane receptor, Patched (Ptc1). This interaction prevents Patched's inhibition of Smoothened (Smo), thus allowing Smoothened to activate processes within the cell. Panel C: Modulation of the Shh signalling pathway using small molecules. Robotnikinin functions by binding to ShhN, preventing its interaction with Patched. In the presence of robotnikinin, Patched binds to, and represses, Smoothened, blocking the signalling pathway. The small molecule agonist of Smoothened, purmorphamine, may be used to rescue the signalling pathway.

small molecule libraries will accelerate the discovery of useful, specific small molecule probes. Synthesis and screening activities are likely to be more routinely integrated: for example, extremely high-throughput approaches have already been developed to pan for small molecule binders whose structure has been encoded by a DNA tag (see Section 1.2.4).

A major feature of the chemical genetic approach is that it is conditional: the experimenter can choose whether to add a small molecule modulator to a biological system. This seemingly trivial feature can be a major advantage over genetics where conditional approaches, such as temperature sensitive mutants, are less common. The chemical genetic approach can also be used to modulate independently the function of more than one protein: either through exploiting a pair of complementary ligands, or in combination with a classical genetic approach. The power of the chemical genetic approach will thus continue to increase dramatically with the range of useful chemical modulators that are discovered. In this chapter, it has been highlighted how the independent modulation of pairs of proteins can reveal insights into cell cycle control (Section 1.3.1), nutrient signalling (Section 1.3.2.1) and mechanisms underpinning developmental biology (Section 1.4). The approach will thereby yield new insights into a wide range of biological mechanisms: by allowing the interplay between proteins to be investigated and by complementing alternative approaches (such as classical genetics and RNAi) for modulating protein function. Where these mechanisms are therapeutically relevant, chemical genetics will provide a direct bridge to drug discovery programmes, thus, allowing some of the greatest challenges facing global society to be addressed.

# References


1. For reviews of the chemical genetic approach, see: (a) D. P. Walsh and Y. T. Chang, *Chem. Rev.*, 2006, **106**, 2476; (b) D. S. Bellows and M. Tyers, *Science*, 2004, **306**, 67; (c) T. U. Mayer, *Trend. Cell Biol.*, 2003, **13**, 270; (d) R. Breinbauer, *Angew. Chem., Int. Ed.*, 2003, **42**, 1086; (e) B. R. Stockwell, *Nat. Rav. Genet.*, 2000, **1**, 116; (f) S. L. Schreiber, *Nat. Chem. Biol.*, 2005, **1**, 64; (g) D. R. Spring, *Chem. Soc. Rev.*, 2005, **34**, 472.
2. (a) C. M. Crews and U. Splittgerber, *Trends Biochem. Sci.*, 1999, **24**, 317; (b) X. F. Zhen and T. F. Chan, *Drug Discovery Today*, 2002, 7, 197; (c) B. R. Stockwell, *Nat. Rev. Genet.*, 2000, **1**, 116.
3. X. S. Zheng, T. F. Chan and H. H. Zhou, *Chem. Biol.*, 2004, **11**, 609.
4. (a) F. G. Kuruvilla, A. F. Shamji, S. M. Sternson, P. J. Hergenrother and S. L. Schreiber, *Nature*, 2002, **416**, 653; (b) A. C. Bishop, J. A. Ubersax, D. T. Petsch, D. P. Matheos, N. S. Gray, J. Blethrow, E. Shimizu, J. Z. Tsien, P. G. Schultz, M. D. Rose, J. L. Wood, D. O. Morgan and K. M. Shokat, *Nature*, 2000, **407**, 395.
5. (a) H. E. Blackwell, L. Pérez, R. A. Stavenger, J. A. Tallarico, E. C. Eatough, M. A. Foley and S. L. Schreiber, *Chem. Biol.*, 2001, **8**, 1167; (b) P. A. Clemons, A. N. Koehler, B. K. Wagner, T. G. Sprigings,

D. R. Spring, R. W. King, S. L. Schreiber and M. A. Foley, *Chem. Biol.*, 2001, **8**, 1183.

6. (a) M. Frank-Kamenetsky, X. M. Zhang, S. Bottega, O. Guicherit, H. Wichterle, H. Dudek, D. Bumcrot, F. Y. Wang, S. Jones, J. Shulok, L. L. Rubin and J. A. Porter, *J. Biol.*, 2002, **1**, 10; (b) R. Weigert, A. Colanzi, A. Mironov, R. Buccione, C. Cericola, M. G. Sciulli, G. Santini, S. Flati, A. Fusella, J. G. Donaldson, M. D. Girolamo, D. Corda, M. A. De Matteis and A. Luini, *J Biol. Chem.*, 1997, **272**, 14200.

7. For progress in this area, see: (a) C. S. Branda and S. M. Dymecki, *Dev. Cell*, 2004, **6**, 7; (b) K. V. Anderson and P. W. Ingham, *Nat. Genet.*, 2003, **33**(Suppl), 285; (c) M. H. Vitaterna, D. P. King, A. M. Chang, J. M. Kornhauser, P. L. Lowrey, J. D. MacDonald, W. F. Dove, L. H. Pinto, F. W. Turek and J. S. Takahashi, *Science*, 1994, **264**, 719.

8. (a) T. U. Mayer, T. M. Kapoor, S. J. Haggarty, R. W. King, S. L. Schreiber and T. J. Mitchison, *Science*, 1999, **286**, 971; (b) A. F. Straight, A. Cheung, J. Limouze, I. Chen, N. J. Westwood, J. R. Sellers and T. J. Mitchison, *Science*, 2003, **299**, 1743.

9. T. Nishimoto, *Methods Enzymol.*, 1997, **283**, 292.

10. H. Siomi and M. C. Siomi, *Nature*, 2009, **457**, 396.

11. L. Burdine and T. Kodadek, *Chem. Biol.*, 2004, **11**, 593.

12. E. O. Peristein, D. M. Ruderfer, G. Ramachandran, S. J. Haggerty, L. Kruglyak and S. L. Schreiber, *Chem. Biol.*, 2006, **13**, 319.

13. J. H. Zhang, T. D. Chung and K. R. Oldenburg, *J. Biomol. Screen.*, 1999, **4**, 67.

14. B. Gunter, C. Brideau, B. Pikounis and A. Liaw, *J. Biomol. Screen.*, 2003, **8**, 624.

15. J. Inglese, C. E. Shamu and R. K. Guy, *Nat. Chem. Biol.*, 2007, **3**, 438.

16. (a) P. J. Hergenrother, *Curr. Opin. Chem. Biol.*, 2006, **10**, 213; (b) J. Soderholm, M. Uehara-Bingen, K. Weis and R. Heald, *Nature Chem. Biol.*, 2006, **2**, 55.

17. (a) M. A. Koch, L.-O. Wittenberg, S. Basu, D. A. Jeyaraj, E. Gourzoulidou, K. Reinecke, A. Odermatt and H. Waldmann, *Proc. Natl. Acad. Sci. U. S. A.*, 2005, **102**, 17272; (b) M. A. Koch, A. Schuffenhaur, M. Scheck, S. Wetzel, M. Casulata, A. Odermatt, P. Ertl and H. Waldmann, *Proc. Natl. Acad. Sci. U. S. A.*, 2004, **101**, 16721.

18. C. M. Dobson, *Nature*, 2004, **432**, 824.

19. A. A. Shelat and R. K. Guy, *Nature Chem. Biol.*, 2007, **3**, 442.

20. (a) A. Nören-Müller, I. Reis Correa Jr., H. Prinz, C. Rosenbaum, K. Saxena, H. Schwalbe, D. Vestweber, G. Cagna, S. Schunk, O. Schwarz, H. Schwiene and H. Waldmann, *Proc. Natl. Acad. Sci. U. S. A.*, 2006, **103**, 10606; (b) K. Grabowski, K.-H. Baringhaus and G. Schneider, *Nat. Prod. Rep.*, 2008, **25**, 892; (c) K. Hübel, T. Leâmann and H. Waldmann, *Chem. Soc. Rev.*, 2008, **37**, 1361.

21. (a) S. Wetzel, K. Klein, S. Renner, D. Rauh, T. I. Oprea, P. Mutzel and H. Waldmann, *Nat. Chem. Biol.*, 2009, **5**, 581; (b) S. Renner, W. A. L. van Otterlo, M. D. Seoane, S. Möcklinghoff, B. Hofmann, S. Wetzel,

A. Schuffenhauer, P. Ertl, T. I. Oprea, D. Steinhilber, L. Brunsveld, D. Rauh and H. Waldmann, *Nat. Chem. Biol.*, 2009, **5**, 585.

22. (a) H. Kubinyi, *Nat. Rev. Drug Discovery*, 2004, **2**, 665; (b) A. L. Hopkins, C. R. Groom and A. Alex, *Drug Discovery Today*, 2004, **9**, 430; (c) E. R. Zartler and M. J. Sharpio, *Curr. Opin. Chem. Biol.*, 2005, **9**, 366; (d) R. A. E. Carr, M. Congreve, C. W. Murray and D. C. Rees, *Drug Discovery Today*, 2005, **10**, 987; (e) D. A. Erlanson, R. S. Mcdowell and T. O'Brien, *J. Med. Chem.*, 2004, **47**, 3463; (f) M. M. Hann, A. R. Leach and G. Harper, *J. Chem. Inf. Comput. Sci.*, 2001, **47**, 856.

23. A. H. Lipkus, Q. Yuan, K. A. Lucas, S. A. Funk, W. F. Bartelt III, R. J. Schenck and A. J. Trippe, *J. Org. Chem*, 2008, **73**, 4443.

24. (a) M. D. Burke and S. L. Schreiber, *Angew. Chem. Int. Ed.*, 2004, **43**, 46; (b) T. E. Nielsen and S. L. Schreiber, *Angew. Chem. Int. Ed.*, 2008, **47**, 48; (c) C. Cordier, D. Morton and S. Murrison, *A. Nelson and C. O'Leary-Steele, Nat. Prod. Rep.*, 2008, **25**, 719.

25. H. E. Pelish, N. J. Westwood, Y. Feng, T. Kirchhausen and M. D. Shair, *J. Am. Chem. Soc.*, 2001, **123**, 6740.

26. S. M. Sternson, J. B. Louca, J. C. Wong and S. L. Schreiber, *J. Am. Chem. Soc.*, 2001, **123**, 1740.

27. (a) H. E. Blackwell, L. Pérez, R. A. Stavenger, J. A. Tallarico, E. C. Eatough, M. A. Foley and S. L. Schreiber, *Chem. Biol.*, 2001, **8**, 1167; (b) P. A. Clemons, A. N. Koehler, B. K. Wagner, T. G. Sprigings, D. R. Spring, R. W. King, S. L. Schreiber and M. A. Foley, *Chem. Biol.*, 2001, **8**, 1183; (c) S. J. Haggerty, K. M. Koeller, J. C. Wong, C. M. Grozinger and S. L. Schreiber, *Proc. Natl. Acad. Sci. U. S. A.*, 2003, **100**, 4389; (d) J. C. Wong, R. Hong and S. L. Schreiber, *J. Am. Chem. Soc.*, 2003, **125**, 5586.

28. A. N. Koehler, A. F. Shamji and S. L. Schreiber, *J. Am. Chem. Soc.*, 2003, **125**, 8420.

29. M. D. Burke, E. M. Berger and S. L. Schreiber, *J. Am. Chem. Soc.*, 2004, **126**, 14095.

30. H. Oguri and S. L. Schreiber, *Org. Lett.*, 2005, **7**, 47.

31. S. Dandapani, M. Duduta, J. S. Panek and J. A. Porco Jr., *Org. Lett.*, 2007, **9**, 3849.

32. N. Kumagai, G. Muncipinto and S. L. Schreiber, *Angew. Chem., Int. Ed.*, 2006, **45**, 3635.

33. (a) T. E. Nielsen and M. Meldal, *J. Org. Chem.*, 2004, **69**, 3765; (b) T. E. Nielsen and M. Meldal, *J. Comb. Chem.*, 2005, **7**, 599; (c) T. E. Nielsen, S. L. Quement and M. Meldal, *Org. Lett.*, 2005, **7**, 3601.

34. D. A. Spiegel, F. C. Schroeder, J. R. Duvall and S. L. Schreiber, *J. Am. Chem. Soc.*, 2006, **128**, 14766.

35. D. Morton, S. Leach, C. Cordier, S. Warriner and A. Nelson, *Angew. Chem., Int. Ed.*, 2009, **48**, 104.

36. (a) S. L. Schreiber, *Nature*, 2009, **457**, 153; (b) H. Waldmann, *Nature Chem. Biol.*, 2009, **5**, 76; (c) W. R. J. D. Galloway, M. Diáz-Gavilán,

A. Isidro-Llobet and D. R. Spring, *Angew. Chem. Int. Ed.*, 2009, **48**, 1194.

37. J. D. Sunderhaus, C. Dockendorff and S. F. Martin, *Org. Lett.*, 2007, **9**, 4223.
38. J. M. Mitchell and J. T. Shaw, *Angew. Chem. Int. Ed.*, 2006, **45**, 1722.
39. E. Comer, E. Rohan, L. Deng and J. A. Porco, *Org. Lett.*, 2007, **9**, 2123.
40. (a) D. Strübing, H. Neumann, S. Klaus, S. Hübner and M. Beller, *Tetrahedron*, 2005, **61**, 11333; (b) D. Strübing, H. Neumann, S. Hübner, S. Klaus and M. Beller, *Tetrahedron*, 2005, **61**, 11345; (c) K. M. Brummond and B. Mitasev, *Org. Lett.*, 2004, **6**, 2245; (d) K. M. Brummond and D. Chen, *Org. Lett.*, 2005, **7**, 3473.
41. (a) G. P. Smith, *Science*, 1985, **228**, 1315; (b) C. F. Barbas III, *Curr. Opin. Biotechnol.*, 1993, **4**, 526; (c) D. S. Wilson, A. D. Keefe and J. W. Szostak, *Proc. Natl. Acad. Sci. U. S. A.*, 2001, **98**, 3750; (d) A. Frankel, S. Li, S. R. Starck and R. W. Roberts, *Curr. Opin. Struct. Biol.*, 2003, **13**, 506; (e) A. D. Ellington and J. W. Szostak, *Nature*, 1990, **346**, 818; (f) C. Tuerk and L. Gold, *Science*, 1990, **249**, 505.
42. C. Heinis, T. Rutherford, S. Freund and G. Winter, *Nat. Chem. Biol.*, 2009, **5**, 502.
43. M. A. Clark, R. A. Acharya, C. C. Arico-Muendel, S. L. Belyanskaya, D. R. Benjamín, N. R. Carlson, P. A. Centrella, C. H. Chiu, S. P. Creaser, J. W. Cuozzo, C. P. Davie, Y. Ding, G. J. Franklin, K. D. Franzen, M. L. Geffer, S. P. Hale, N. J. V. Hansen, D. I. Israel, J. Jiang, M. J. Kavarana, M. S. Kelley, C. S. Kollmann, F. Li, K. Lind, S. Mataruse, P. F. Medeiros, J. A. Messer, P. Myers, H. O'Keefe, M. C. Oliff, C. E. Rise, A. L. Satz, S. R. Skinner, J. L. Svendsen, L. Tang, K. von Vloten, R. W. Wagner, G. Yao, B. Zhao and B. A. Morgan, *Nat. Chem. Biol.*, 2009, **5**, 647.
44. (a) S. J. Wrenn, R. M. Weisinger, D. R. Halpin and P. B. Harbury, *J. Am. Chem. Soc.*, 2007, **129**, 13137; (b) Z. J. Gartner, M. W. Kanan and D. R. Liu, *J. Am. Chem. Soc.*, 2002, **124**, 10304; (c) C. T. Calderone and D. R. Liu, *Angew. Chem. Int. Ed.*, 2005, **44**, 7383; (d) B. N. Tse, T. M. Snyder, Y. Shen and D. R. Liu, *J. Am. Chem. Soc.*, 2008, **130**, 15611.
45. (a) D. Barnes-Seeman, S. Bum Park, A. N. Koehler and S. L. Schreiber, *Angew. Chem. Int. Ed.*, 2003, **42**, 2376; (b) A. J. Vegas, J. E. Bradner, W. Tang, O. M. McPherson, E. F. Greenberg, A. N. Kohler and S. L. Schreiber, *Angew. Chem. Int. Ed.*, 2007, **46**, 7960; (c) N. Pohl, *Angew. Chem. Int. Ed.*, 2008, **47**, 3868.
46. M. Uttamchandani, D. P. Walsh, S. M. Khersonsky, X. Huang, S. Q. Yao and Y.-T. Chang, *J. Comb. Chem.*, 2004, **6**, 862.
47. B. Z. Stanton, L. F. Peng, N. Maloof, K. Nakai, X. Wang, J. L. Duffner, K. M. Taveras, J. M. Hyman, S. W. Lee, A. N. Koehler, J. K. Chen, J. L. Fox, A. Mandinova and S. L. Schreiber, *Nat. Chem. Biol.*, 2009, **5**, 155.
48. X. X. Sachchidanand, L. Resnick-Silverman, S. Yan, S. Mutjaba, W. Liu, L. Zeng, J. J. Manfredi and M. M. Zhou, *Chem. Biol.*, 2006, **13**, 81.

49. A. C. Bishop, J. A. Ubersax, D. T. Petsch, D. P. Matheos, N. S. Gray, J. Blethrow, E. Shimizu, J. Z. Tsien, P. G. Schultz, M. D. Rose, J. L. Wood, D. O. Morgan and K. M. Shokat, *Nature*, 2000, **407**, 395.
50. A. C. Bishop, K. Shah, Y. Liu, L. Witucki, C. Kung and K. M. Shokat, *Curr. Biol.*, 1998, **8**, 257.
51. S. I. Reed, *Genetics*, 1980, **95**, 561.
52. (a) W. S. Wade, M. Mrksich and P. B. Dervan, *J. Am. Chem. Soc.*, 1992, **114**, 8784; (b) M. Mrksich, W. S. Wade, T. J. Dwyer, B. H. Geirstanger, D. E. Wemmer and P. B. Dervan, *Proc. Natl. Acad. Sci. U. S. A.*, 1992, **89**, 7586.
53. (a) C. L. Kielkopf, E. E. Baird, P. B. Dervan and D. C. Rees, *Nat. Struct. Biol.*, 1998, **5**, 104; (b) C. L. Kielkopf, S. White, J. W. Szewczyk, J. M. Turner, E. E. Baird, P. B. Dervan and D. C. Rees, *Science*, 1998, **282**, 111.
54. J. M. Gottesfeld, L. Neely, J. W. Trauger, E. E. Baird and P. B. Dervan, *Nature*, 1997, **387**, 202.
55. J. Taunton, J. L. Collins and S. L. Schreiber, *J. Am. Chem. Soc.*, 1996, **118**, 10412.
56. S. M. Khersonsky, D.-W. Jung, T.-W. Kang, D. P. Walsh, H.-S. Moon, H. Jo, E. M. Jacobsen, V. Shetty, T. A. Neubert and Y.-T. Chang, *J. Am. Chem. Soc.*, 2003, **125**, 11804.
57. S. M. Lamos, C. J. Krusemark, C. J. McGee, M. Scalf, L. M. Smith and P. J. Belshaw, *Angew. Chem. Int. Ed.*, 2006, **45**, 4329.
58. H. Zhu and M. Snyder, *Curr. Opin. Chem. Biol.*, 2003, **7**, 55.
59. J. Huang, H. Zhu, S. J. Haggarty, D. R. Spring, H. Hwang, F. Jin, M. Snyder and S. L. Schreiber, *Proc. Natl. Acad. Sci. U. S. A.*, 2004, **101**, 16594.
60. T. U. Mayer, T. M. Kapoor, S. J. Haggarty, R. W. King, S. L. Schreiber and T. J. Mitchison, *Science*, 1999, **286**, 971.
61. M. Catarinella, T. Grüner, T. Strittmatter, A. Marx and T. U. Mayer, *Angew. Chem. Int. Ed.*, 2009, **48**, 9072.
62. X. Wu, S. Ding, Q. Ding, N. S. Gray and P. G. Schultz, *J. Am. Chem. Soc.*, 2002, **124**, 14520.
63. S. Sinha and J. Chen, *Nat. Chem. Biol.*, 2006, **2**, 29.
64. J. Talpale, J. K. Chen, M. K. Cooper, B. Wang, R. K. Mann, L. Milenkovic, M. P. Scott and P. A. Beachy, *Nature*, 2000, **406**, 1005.

# Applications for Activity-based Probes in Drug Discovery

L.E. EDGINGTON[a] AND M. BOGYO[a,b,*]

[a] Department of Pathology, Stanford University School of Medicine, 300 Pasteur Drive, Stanford, CA 94305, USA; [b] Department of Microbiology and Immunology, Stanford University School of Medicine, 300 Pasteur Drive, Stanford, CA 94305, USA

## 2.1 Background

### 2.1.1 Introduction

The path to a new therapeutic drug is long and difficult and involves many stages including validation of a target, design and selection of a lead compound and finally development of the lead into a drug. This chapter will discuss a relatively new technology that makes use of small molecules termed activity-based probes (ABPs). These probes bind in the active site of a target enzyme or class of enzymes in an activity-dependent fashion. Thus probe labeling serves as an indirect readout of enzyme activity, allowing the dynamic regulation of the target enzyme to be monitored using a number of biochemical and cell biological methods. In addition, labeled targets can be directly isolated by affinity methods, thereby allowing identification of potentially valuable drug targets based solely on their ability to bind a small molecule. Finally, because of the high degree of selectivity of ABPs for a given target protein class, they can be used for studies of drug binding and efficacy in complex cellular mixtures, intact

RSC Drug Discovery Series No. 5
New Frontiers in Chemical Biology: Enabling Drug Discovery
Edited by Mark E. Bunnage

Published by the Royal Society of Chemistry, www.rsc.org

cells and even in whole animals. These attributes of ABPs make them extremely valuable reagents for use at multiple points in the drug discovery process.


## 2.1.2 Activity-based Probes

### 2.1.2.1 The Need for Chemical Probes

While the genomics revolution of the past decade has had a dramatic impact on research science, it is clear that analysis of DNA or RNA content alone is not sufficient to understand cell biology and disease. For example, most classes of enzymes are regulated by a complex set of post-translational mechanisms that make simple assessment of protein abundance or gene expression of limited value for understanding enzyme function. For the drug discovery process, it is essential to understand how enzymatic proteins are regulated in disease pathology in order to be able to predict how modulation of activity with a small molecule drug is likely to impact therapy outcome. Since many of the primary so-called "druggable" classes of targets are enzymes, the development of new methods to dynamically monitor enzyme activity has great potential to impact the drug discovery process.

Activity-based probes are one such technology that has made significant advances in the past decade (for additional reviews see Cravatt *et al.*,[1] Evans and Cravatt,[2] Fonovic and Bogyo[3] and Schmitinger *et al.*[4]). This technology is centered around the development of small molecule probes that bind to a target enzyme using a mechanism that requires enzymatic activity. Thus, an ABP binds to its target only when it is active and labeling can therefore be used as a way to assess levels of activity in a dynamic way. Depending on the desired application for an ABP, these reagents can often be built using previously validated chemistries and knowledge about a particular target protein of interest. In the case where a target is not already established, it is possible to generate diverse sets of probes and use these reagents to find previously poorly characterized enzymes that may play important roles in disease progression. Once a highly selective probe and target pair has been identified, it is possible to use the probe not only to monitor the normal physiologically relevant regulation of the target but also to monitor inhibition by small molecule drug leads. Thus, a suitably designed ABP can aid not only in the identification of novel targets, but also can be applied to the later stages of the drug discovery process to assess the overall efficacy and selectivity of lead compounds.


### 2.1.2.2 Anatomy of the Chemical Probe

In their most basic form, activity-based probes consist of three distinct functional elements (Figure 2.1): a reactive group for covalent attachment to the enzyme, a linker region that can modulate reactivity and specificity of the reactive group, and a tag for identification and purification of modified

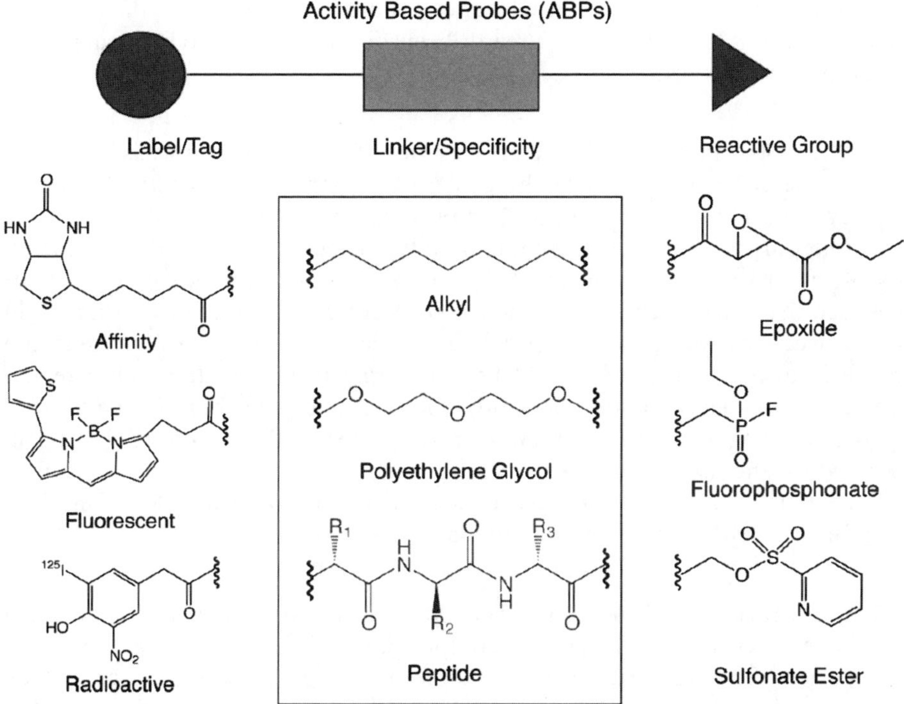



**Figure 2.1**  Structure of a chemical probe. A chemical probe has three basic components: a reactive group for covalent attachment to the enzyme of interest; a linker region to provide spacing and specificity; and a tag to allow for identification and/or purification. Specific examples of each are shown.

enzymes. Each of these elements must be chosen based on the type of application required.

**2.1.2.2.1  Structure of the Reactive Group.** The most significant challenge in the design of ABPs is the selection of a reactive group that provides the necessary covalent modification of a target protein. The reactive group must have sufficient reactivity to allow modification of the target enzyme while not reacting with other non-specific proteins inside the cell. Most of the reactive groups currently in use in ABPs have been designed based on covalent, mechanism-based inhibitors of various enzyme families. Of the many new classes of ABPs that have been developed in the past decade, the majority of the reported agents have been designed to target proteases.[5–13] This is at least partially due to the wide range of covalent reactive groups that have been designed by medicinal chemists as a means to inhibit proteases (for an extensive review see Powers *et al.*[14]). Proteases are also one of the primary families of enzymes that are currently the focus of a number of drug discovery efforts in the pharmaceutical industry. Therefore, there is the potential for ABPs that have already been developed to find immediate use in drug discovery projects.

**2.1.2.2.2 Structure of the Linker Region.** The linker region is generally used to describe the region of the probe that connects the reactive group to the tag used for identification and/or purification. While the linker can have multiple purposes, it often contains binding elements to control the selectivity of the probe for a given target. In addition, it must provide enough space between the reactive group and the tag to prevent steric hindrance that could block access of the reactive group or accessibility of the tag for the purpose of purification. In the case of ABPs that target proteases, this linker region often contains basic peptide or peptide-like elements that bind in the various substrate recognition elements on a target protease. Recent efforts have also focused on the use of the linker to release the resulting labeled proteins after affinity purification. There have been several reports of cleavable linkers that can facilitate such purification methods.[15–17]

**2.1.2.2.3 Structure of the Tag.** The choice of tag for a given activity-based probe depends heavily on the desired application for the probe (for an extensive review see Sadaghiani *et al.*[18]). Some of the more common tags include biotin, radioisotopes and fluorescent tags. Biotin is often used because it is simple, cheap and can be used both for detection by western blot approaches and for direct purification by affinity chromatography. Fluorescent and radioactive tags are generally used for imaging applications (see Section 2.2.3) and fluorescent tags also allow for biochemical analysis of labeled proteins in SDS-PAGE gels using simple laser scanning methods that are much faster and easier than standard western blotting. Fluorescent tags also have the added advantage of allowing direct, microscopic imaging of targets that have been modified by an ABP. Thus the spatial and temporal regulation of enzymes can be monitored *in situ*[19] and *in vivo*.[20] Finally, the use of dyes that emit near infrared fluorescent light allows the use of probes for whole body, non-invasive imaging in living organisms (for more information see Section 2.2.3).

## 2.1.2.3 Classes of Activity-based Probes

The past decade and a half has seen extensive growth in the development of new ABPs. While there still remains many classes of enzymatic proteins for which no ABPs currently exist, the rapid development of new synthesis and screening methods coupled with advances in analytical methods that allow rapid identification of labeled targets has greatly expanded the list of validated probes. Initially, the majority of efforts in probe design were focused on proteases and hydrolases. This is due to the fact that these enzymes use a nucleophilic amino acid to mediate a direct attack on a substrate. This fact, coupled with a wealth of published inhibitors that form covalent bonds with target proteins, has accelerated the development of activity-based probes for proteases. However, the past few years have seen the development of probes for many other classes of enzymes as well as some non-enzymatic receptor proteins. This section will outline some of the major advances in probe design for each of the primary target classes.

**2.1.2.3.1 Proteases.** By far the largest body of work on ABPs has been focused on protease targets. In particular, there have been a number of probes developed for the cysteine proteases that have found widespread use in biological studies of disease[5,7,10,11,21–25] (Figure 2.2). While the concept of covalent inhibition of a protease is not new, the idea of using covalent inhibitors to label proteases with tags that allow isolation and biochemical monitoring of the target enzymes is relatively recent. Some of the earliest examples of ABPs include probes for caspases,[9,13] cathepsins,[6,23] the proteasome[6] and serine hydrolases.[12] All of these probes were originally designed with a specific target protease in mind and made use of either knowledge of substrates or a well-characterized selective inhibitor as a starting point.

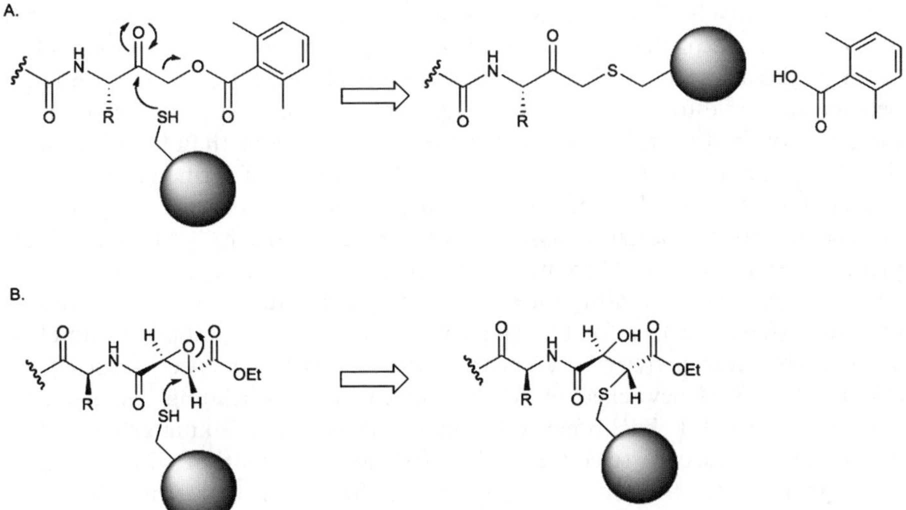

**Figure 2.2** Cysteine protease ABPs. Examples of two of the most commonly used classes of probes that target cysteine proteases. (A) A peptide acyloxymethyl ketone (AOMK) reacts with the active site thiol to produce a stable thio-ether bond with loss of the acyl leaving group. This scaffold has been used to target a number of cysteine protease families including caspases, cathepsins and legumain.[5,8,10,11,21] (B) A peptide epoxide based on the natural product E-64 reacts with the active site cysteine to form a stable adduct upon epoxide ring opening. This class of compounds has been used exclusively to target cysteine cathepsins.[7,23]

Interestingly, several of these original probes continue to find new applications in a wide range of biology. The general cathepsin probe DCG-04,[23] for example, has been used to study cathepsin function in a large number of biological systems. For example, this epoxide-containing probe has been applied to functional studies of the roles of papain family proteases in processes such as tumor progression,[20] angiogenesis,[20] cataract formation,[26] pro-hormone processing,[27] malarial infections,[28] bacterial growth,[29] and plant response to pathogens.[30] In addition, a number of probes containing an acyloxymethyl ketone (AOMK) reactive group have been developed.[5,8,10,21,22] This reactive functional group has proven to be highly selective when used in complex mixtures and even *in vivo*. Thus, it is currently the electrophile of choice for use in imaging probes for proteases (see Section 2.2.3). Probes with the AOMK reactive electrophile have been successfully designed to target cathepsins,[21,22] legumain,[31] caspases[5,8,10] and separase.[32]

There have also been a number of elegant examples of the use of ABPs to target serine proteases and serine hydrolases[12,33,34] (Figure 2.3). Because of the broad reactivity of the first serine hydrolyase probe containing a fluorophosphinate (FP) electrophile, FP-Biotin[12] has been applied in a diverse range of applications and very recently has facilitated the identification of previously uncharacterized target enzymes that have direct links to human diseases such as cancer.[35–39] In addition, there have been a number probes described that can be used to selectively target serine proteases.[33] All of these protease-directed probes make use of the less reactive diphenyl phosphonate (DPP) electrophile to target the active site serine. By attaching a peptide scaffold to the probes, it becomes possible to avoid labeling of general hydrolases and lipases that are the target of the FP probes. Thus, while much less reactive than the FP probes, the DPP probes can be used for more selective studies of serine proteases.

Metalloproteases (MPs), like serine and cysteine proteases, play key roles in peptide hormone processing, tissue remodeling, and cancer.[40–42] However, this protease family uses a tightly bound water molecule to initiate attack of substrate, thereby circumventing the acyl enzyme intermediate. As a result, design of activity-based probes for this class of proteases is substantially more challenging. However, activity-based probes targeting MPs have been reported[43,44] (Figure 2.4). This new class of ABPs contains a zinc-chelating hydroxamate coupled to a peptide backbone containing a photo-cross-linking group. These probes can be used to selectively label MPs after irradiation with UV light. The high affinity of the hydroxamic acid group for the active site zinc allows the use of low probe concentrations thereby producing low background of probe labeling. While this is a big leap forward for MP-specific probes, the need for UV light to facilitate probe binding limits their use to *ex vivo* applications.


**2.1.2.3.2  Kinases and Phosphatases.**  Due to the rapid growth in interest in kinases as drug targets in conditions such as cancer, there has been a push to develop new methods to study kinase function. In addition, the kinase family is large and it is necessary to understand the overall selectivity of a given

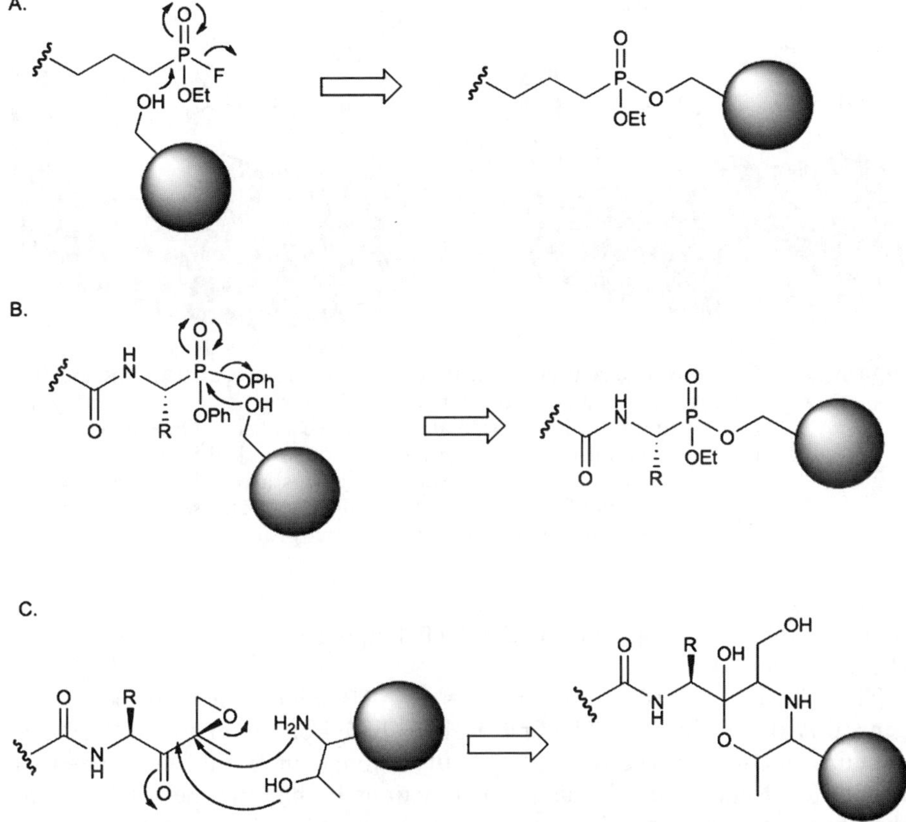

**Figure 2.3**  Serine protease and hydrolase ABPs. (A) Reaction of a general serine hydrolase probe containing a fluorophosphonate (FP) reactive electrophile. This class of probes has been used extensively to label various classes of serine hydrolases including proteases, esterases, lipases and others.[12] (B) The peptide diphenyl phosphonate (DPP) reacts with the serine nucleophile in the active site of serine proteases. This probe is much less reactive than the FP class of probes but is more selective towards serine proteases over other types of serine hydrolases.[33] (C) The natural product epoxomicin contains a keto-epoxide that selectively reacts with the catalytic N-terminal threonine of the proteasome β-subunit. This reaction results in the formation of a stable six-membered ring. This class of electrophile has been used in probes of the proteasome.[101]

kinase drug lead. This information becomes even more important for kinase inhibitors since most are designed to bind in the highly conserved ATP binding pocket of these enzymes. Because virtually all kinases as well as other ATP binding enzymes have similar ATP binding sites, there is a significant potential for unwanted off-target reactivity. Thus, a number of groups have focused on the design of probes for kinases. These probes have focused either on scaffolds based on the structure of ATP or on synthetic small

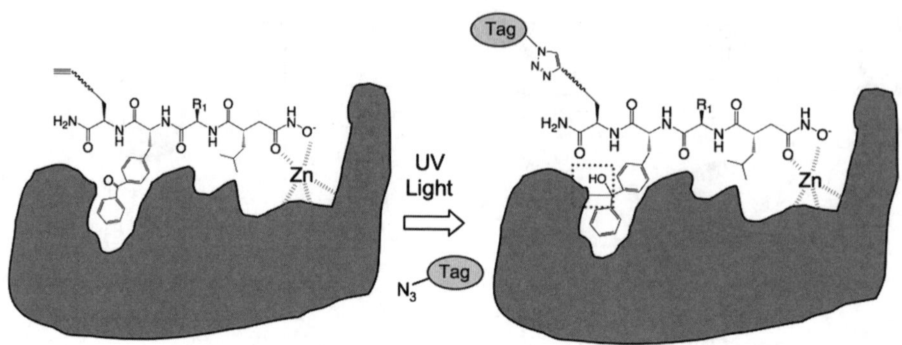



**Figure 2.4**  Mechanism of action of ABPs for metallo proteases. Current metallo-protease probes bind in the active site of a target enzyme using a hydroxamic acid group that coordinates the active site zinc. Once bound in the active site the probe can be permanently cross-linked into the active site using a photo-cross-linking group that forms a reactive oxygen radical upon irradiation of the probe with UV light.[43,44] This secondary reaction allows permanent modification of the target protease.

molecule inhibitors that bind in the ATP binding pocket of a target kinase (Figure 2.5).

There have been two primary reports of ATP analogs that can be used as general ABPs for kinases. The first probe is based on an analog of ATP that contains an *O*-biotinoyl group linked to the terminal nucleotide phosphate *via* a reactive acyl phosphate.[45] This probe functions by binding the ATP site and then facilitating the transfer of the *O*-acyl biotin group to a conserved lysine found in most kinases. Since this lysine residue forms close contacts with the γ-phosphate, it is ideally situated to form a covalent bond resulting in tagging of the target kinase. Amazingly, this general probe was shown to label nearly 75% of the known kinases and it is therefore a very useful reagent for applications to determine the overall inhibitor specificity of kinase drug leads using proteomic methods (see Section 2.2.1). A second probe is a biotin labeled analog of 5′-fluorosulfonylbenzoyl 5′-adenosine (FSBA).[46,47] This probe has a similar mode of action as the acyl phosphate probe in that it covalently binds to the conserved lysine residues near the ATP binding pocket. However, unlike the acyl phosphate probe that transfers only the acyl biotin group, this probe remains intact when modifying the conserved lysine residue.

The other major class of kinase probes is based on the structures of synthetic kinase inhibitors. The advantage of these probes is that it is possible to generate probes with a much narrower profile of targets. For example, the use of the natural fungal metabolite Wortmannin as a starting point yields a probe that is selective for a subset of protein and lipid kinases.[48,49] Other probes based on inhibitor scaffolds have made use of photoaffinity tags,[50,51] acrylamide electrophiles,[52] and a fluoromethyl ketone electrophile.[53,54] The compounds that use photoaffinity tags can be used to target virtually any kinase since the

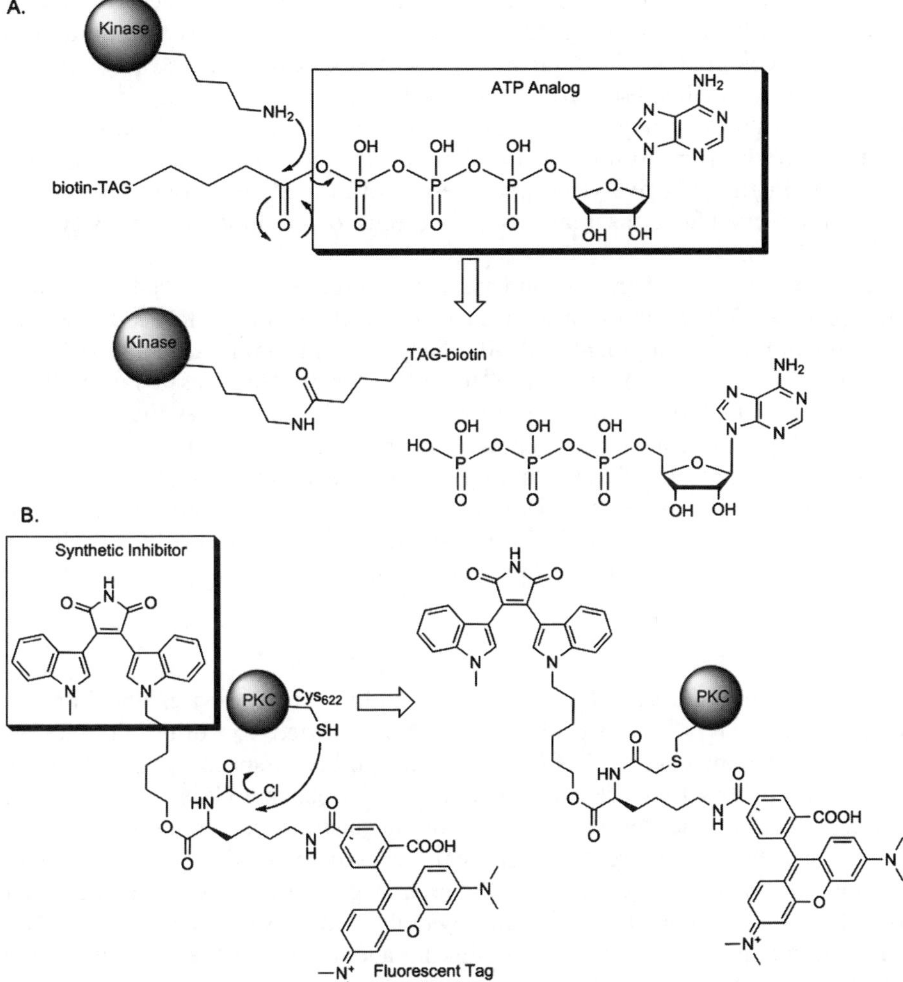



**Figure 2.5** ABPs for kinase targets. Examples of two different probe classes that have been designed to target kinases. (A) Structure and mechanism of action of an acyl phosphate ATP mimetic. This probe binds in the conserved ATP binding pocket of kinases and other ATP dependent enzymes and forms a permanent covalent linkage to the kinase target by acylation of a lysine residue near the γ-phosphate of ATP. This general probe was found to label nearly 75% of all known kinases.[45] (B) Structure and mechanism of action of the protein kinase C (PKC) probe AX4697.[91] This probe uses a quinazoline-based inhibitor to facilitate selective binding to PKC. It contains a reactive chloroacteamide to covalently modify a distal cysteine residue near the ATP binding pocket.

linkage of the probe to the kinase is mediated by UV light. The probes with electrophiles have been designed to specifically target kinases that contain a cysteine nucelophile in the ATP site. These probes can be tuned to be highly selective for a small subset of kinases based on the need for a reactive cysteine coupled with structural elements of the inhibitor that only bind kinases with a small "gate-keeper" residue. Thus, kinase probes have made significant advances in the past five years and these probes are now proving useful for drug discovery efforts as reagents for screens of compound selectivity (see Section 2.2.1).

The flip side of protein phosphorylation by kinases is controlled by protein phosphatases. Phosphatases are another family of enzymes whose study would benefit from the development of ABPs. They make ideal targets for ABP design as their catalytic site includes a cysteine residue that acts as a nucleophilic thiolate for attack of a substrate. Previously, suicide substrates that contain a masked electrophile have been reported as probes for phosphatases.[55,56] These reagents, while potentially exciting, have yet to be shown effective for labeling of endogenous phosphatases in biologically relevant samples. Additionally, the design of cell permeable probes for phosphatases is likely to be difficult due to the need for a highly charged phosphate mimetic.

**2.1.2.3.3   Other Classes of Enzymes.**   In addition to proteases and kinases, there have been a number of new probes reported for other families of enzymes. This includes probes that target a number of specific target proteins as well as probes that are more general for an enzyme class. For example, recent efforts in the design of probes of the histone deacetylases (HDACs) have generated general reagents that can be used for profiling of the activity of this class of metallo enzymes.[57,58] The first generation probe was designed based on the clinical candidate suberoylanilide hydroxamic acid (SAHA) by simple attachment of a photo-cross-linking agent. Due to the prominent role of these enzymes in the regulation of gene expression in cancer, a number of companies have initiated clinical trials with HDAC inhibitors. Thus, HDAC probes are particularly valuable for drug discovery efforts. More recently, a probe for HDAC enzymes was designed based on a lead series of pimelic diphenylamide inhibitors.[59] By converting the lead compound into a probe by the addition of a benzophenone linked to an alkyne tag, it was possible to identify the primary target of the lead compound as HDAC3 and define a link between this enzyme and Friedreich's Ataxia Gene Silencing.

The past 5 years have also seen a rapid increase in the number of new probe classes. This has included the development of probes that target non-enzymatic proteins such as GABA receptors[60] and acetylcholine receptors.[61] In addition, there have been recent reports of probes that target other diverse enzyme classes including sulfatases,[62] deaminases,[35] dimethylaminohydrolases,[63] lipases,[64,65] esterases,[65,66] nitrilases,[67] cytochrome P450s,[68] gylcosidases,[69] gylcanases[70] and beta galactosidase.[71] This ever-expanding list of enzyme families and receptors that can be targeted by ABPs suggests that the field continues to advance.

In addition to the directed proteomic profiling strategies outlined above, Cravatt and co workers have pioneered an approach using non-directed probes to target multiple enzyme classes concurrently (Figure 2.6). In the first example of this approach, a library of reactive sulfonate esters was used to profile various complex proteomes, including extracts from cancer cell lines.[35,72,73] In addition to labeling class I aldehyde dehydrogenase (ALDH-1), a number of the probes in the libraries labeled mechanistically different enzyme classes that had not been previously identified or characterized in activity-based studies. These sulfonate probes were later shown to react with a diverse range of nucleophiles in an enzyme active site including cysteine, aspartic acid, glutamic acid and tryptophan.[74] Interestingly, a more recent study has made use of the same type of sulfonate ester electrophile to facilitate covalent transfer of a probe from a reversible binding ligand to protein target.[75] These studies also demonstrated that the sulfonate ester electrophile is capable of reacting with histadine as well. Using a similar strategy, Cravatt and co-workers also developed a library of probes containing a reactive chloroacetamide group and showed that they could select individual probes from the library that labeled a set of targets that are differentially expressed in obese mice.[76]

## 2.1.2.4 Tagging and Detection Methods

Because ABPs are used for a number of different applications ranging from *in vivo* imaging to affinity purification of labeled target proteins, there are a number of different labels that are generally used on an ABP. This section will focus on the primary types of tags used in ABPs. This includes isotope tags for biochemical and imaging applications, affinity tags for biochemical purification, stable isotope tags for mass spectrometry quantification and fluorescent tags for cell biological and whole body imaging applications. For a more extensive review see Sadaghiani *et al.*[18]

### 2.1.2.4.1 Isotope Tags

*2.1.2.4.1.1 Radioisotopes.* Radioactive isotopes are commonly used in various aspects of biology. They can also be used as tags on activity-based probes. The most commonly used isotope is $^{125}$I, which has been incorporated into many classes of ABPs. This isotope can easily be introduced by simple iodination methods designed for proteins and peptides (for specific protocols for iodination of ABPs see Bogyo *et al.*[77]). In addition, some probes have been designed to incorporate $^{3}$H as it can be added without significant alteration of the probe structure (for example see Fenteany *et al.*[78]). However, the use of tritium is generally not optimal since its specific activity is extremely low, thereby requiring long exposure times to analyze labeling patterns.

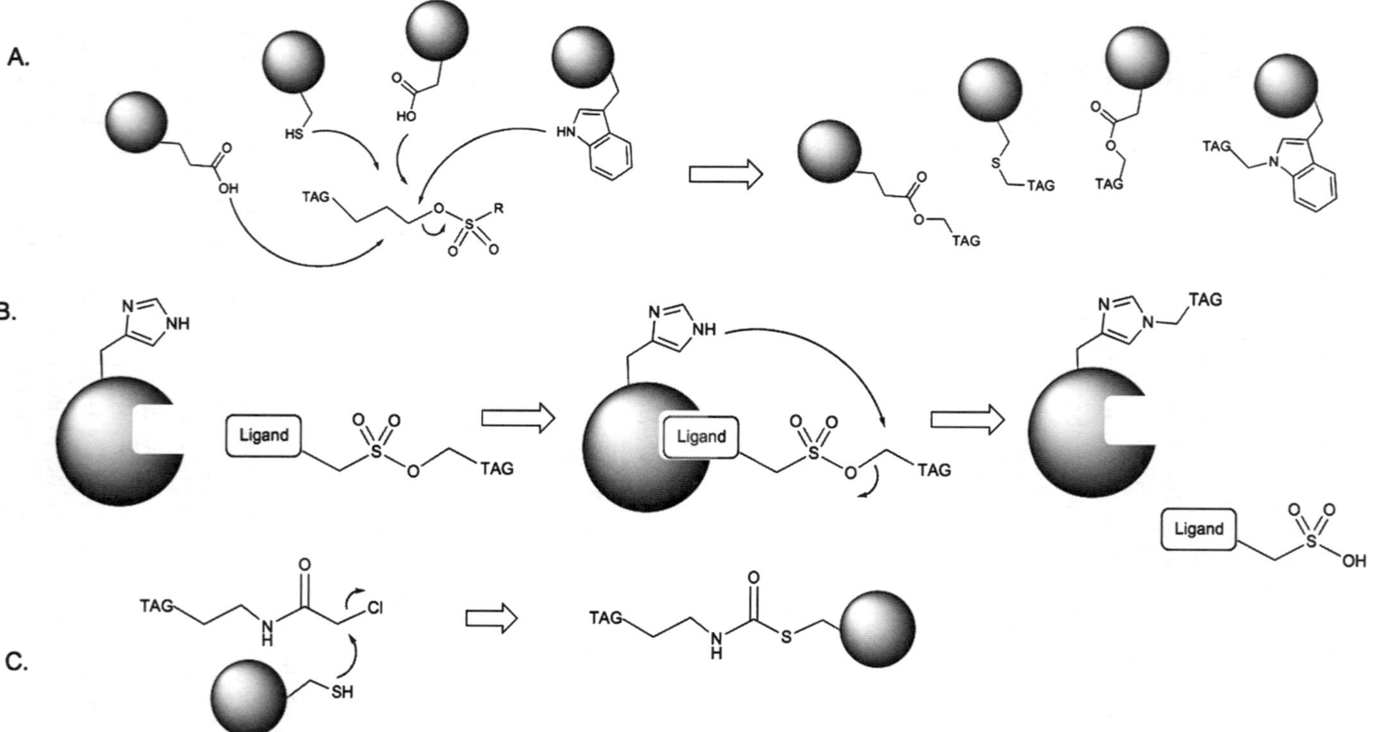



**Figure 2.6**  ABPs that target multiple classes of enzyme targets. (A) Reactivity of a general sulfonate ester probe. This probe has been shown to react with mechanistically distinct targets and is capable of modifying cysteine, aspartic acid, glutamic acid and tryptophan residues in the active site.[72,74] (B) The sulfonate ester reactive group has also been used to generate probes that transfer a tag to the target enzyme after binding of a high affinity ligand.[75] In this example, a high affinity ligand is linked to the sulfonate ester and tag. Binding of the ligand induces the transfer of the tag to a nearby histidine. (C) General probes containing the chloro-acetamide electrophile have been used to label cysteine containing enzymes such as cysteine proteases.[76] This electrophile has been shown to be highly selective for cysteine residues.[74]

There are a number of positive and negative sides to the use of small radioisotopes as tags for ABPs. The small size of iodine makes it possible to generate an ABP based on a well-characterized inhibitor without needing to change the structure significantly by the addition of a bulky tag. However, this advantage is somewhat less important with the advent of multi functional tags that allow attachment of the tag after the probe has bound its target. The other major benefit of radioactive tags is that they can be used in very low concentrations and therefore yield high signal to noise in complex proteomes. These labeling patterns tend to be easier to analyze and provide less background noise. The major drawback of using an isotope label is that there is no direct way to affinity purify the labeled target. Thus, probes must be converted back to an affinity tagged version. However, if the target identities have already been established, it is possible to use the radiolabeled ABPs in screens of compound selectivity using SDS-PAGE analysis in a competition assay (see Section 2.2.1).

*2.1.2.4.1.2 Stable Isotopes.*   As an alternative to the use of radioisotopes, it is possible to use stable isotopes on ABPs for applications to mass spectrometry. Isotope tags have recently been used for a number of quantitative applications in mass spectrometry. This includes the use of isotope coded affinity tags such as ICAT,[79] iTRAC[80] and metabolic isotope labeling SILAC[81] and AQUA.[82] These heavy and light reporters can be distinguished by differences in mass using mass spectrometry. For the technique of ICAT, a general cysteine reactive probe is used in heavy and light labeled form to covalently modify cysteine containing proteins in two samples of different origin. After combination and digestion, relative abundance can be quantified by measuring the abundance of heavy and light labeled peptides by mass spectrometry. An isotope coded activity-based probe has been reported by Overkleeft and co-workers, who synthesized DCG-04, a general papain probe, in a light and heavy version.[83] Alternatively, an ABP can be combined with stable isotope labeling of proteins to yield a highly quantitative method for enzyme activity monitoring.[84] It is also possible that spectral counting methods can be used for relative quantification without the need for isotope labels.[85] In fact, Cravatt and co-workers have demonstrated the use of non-isotopically labeled ABPs to quantify changes in enzyme activity levels by direct mass spectrometry methods, suggesting that the use of such stable isotope labels in ABPs may only be required for monitoring subtle changes in enzyme activity.[86]

**2.1.2.4.2 Affinity Tags.**   One of the primary applications for activity-based probes is the direct isolation of labeled target proteins. Therefore, many of the commonly used activity-based probes contain tags for enrichment of labeled proteins using affinity resins. The most commonly used affinity tag is biotin, which binds to avadin resins with diffusion-limited kinetics. This high affinity is particularly important when probes are used to label low abundance targets. The direct attachment of biotin to an ABP is often the most efficient method of tagging; however, the main drawback is the overall lack of cell permeability of the biotin tag. Thus, biotin-labeled probes are mainly used for labeling applications in cell and tissue extracts rather than for *in vivo* or cell based labeling studies. To get around this problem of cell permeability, a number of groups have developed orthogonal tagging methods in which the biotin or fluorescent tag is added after the probe has already modified the target enzyme (see tandem labeling methods). Thus, it is possible to use ABPs that have small, cell permeable tags that are suitable for subsequent chemical ligation with a biotin affinity tag.

As an alternative to biotin, it is also possible to use short peptide tags such as a HA or FLAG tag on ABP. This approach has been mainly used for larger protein-based probes such as those designed to target ubiquitin-specific proteases.[87–90] These probes are synthesized as recombinant proteins and the reactive electrophile is added to the C-terminus after the protein is purified. This approach allows one to genetically encode the tag into the primary peptide scaffold. Another recent development has been the use of antibodies that are specific for a small organic fluorophore such as TAMRA or BODIPY.[91,92] This allows the use of a relatively small, cell permeable fluorophore for imaging and biochemical applications and then subsequently for direct immuno-precipitation of labeled protein targets.

Another example of an affinity tag for ABPs is short stretches of peptide nucleic acids (PNAs). This method was used to create probes in which the structure of the probe is "coded" by the PNA tag. The tag also serves as a way to purify the probe labeled targets through direct hybridization to a DNA containing chip.[93,94] Finally, there have been a number of examples of the use of small molecule probes that have been directly attached to a solid support. In this case, the tag is the resin bead. While this is not optimal for covalent probes as it would require a method to cleave the probe from the resin, it has proven to be valuable for use in isolation of kinase targets using reversible probe bound resins.[95,96]

**2.1.2.4.3 Fluorophores.** The biggest advance in probe technology has been the addition of a wide range of fluorescent tags on ABPs. While it is obvious that fluorescent tags can facilitate imaging applications, these tags are also valuable for direct biochemical profiling studies. Since the fluorescent tags can be quantitatively detected in SDS-PAGE gels using a simple flatbed laser scanner, it has become possible to label target enzymes in intact cells or even whole organisms and then rapidly analyze the profile of labeled proteins by SDS-PAGE followed by gel scanning. This also allows quantitative profiling without the need to blot gels or even remove the gel from the gel plates. This advance has greatly improved workflow. In addition, many of the small organic dyes are highly fluorescent, cell permeable and have proven to be as sensitive as $^{125}$I-labeled probes. In addition, there are a wide variety of small organic fluorophores that can be purchased from commercial sources.

By using dyes that emit light in the near infrared (NIR) region of the spectrum, it is possible to generate probes that can be used for non-invasive *in vivo* imaging applications (see Section 2.2.3).

A wide variety of fluorophores have now been used for the development of activity-based probes. The most commonly used classes include fluorescein and rhodamine,[97] dansyl,[98] NBD (nitrobenz-2-oxa-1,3-diazole),[65] BODIPY[19] and the Cy-dyes.[22,99] The first class is relatively inexpensive, but suffers from rapid photobleaching, making them less suitable for most imaging applications. BODIPY and Cy-dyes display high absorption coefficients, high quantum yields, narrow absorption peaks and relatively large Stokes' shifts. Furthermore, these fluorescent tags are hydrophobic and freely penetrate cell membranes. These combined properties make them suitable for a variety of biological applications. However, the commercially available activated ester forms of these fluorophores are extremely expensive and therefore have prevented large scale production of fluorescent ABPs.

#### 2.1.2.4.4  Tandem Labeling Strategies.

Another significant advance in probe development has been the generation of tandem labeling methods. This allows a probe to be synthesized with a relatively small tag that can serve as a site for a chemo-selective ligation reaction at a later point in time (Figure 2.7). Ideally, the probe will be cell permeable and can be used in cells and in whole organisms. This so-called tandem labeling process has been developed recently and has been successfully demonstrated for two different ligation chemistries.

The first method makes use of a modified Staudinger reduction in which a probe containing an azide functional group is reduced and subsequently

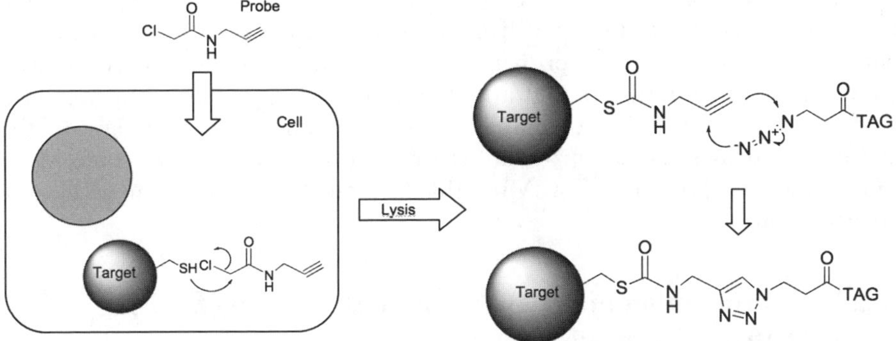

**Figure 2.7**  Tandem labeling methods. In this example, a general chloroacetamide probe containing an alkyne tag is used to label a target enzyme inside the cell. Once the probe labels the target, the cell is lysed and the probe labeling is visualized by a secondary CLICK reaction with an azide containing tag. The result is the formation of a stable triazole between the tag and the probe. Using this method it is possible to label targets with small, cell-permeable probes and then attach a label after cells have been lysed.[120]

reacted to form a stable amide with an appropriately derivatized phosphene carrying a reporter tag. This strategy was first developed by Bertozzi[100] to tag modified carbohydrates on the cell surface and also proved to be useful for ABPs. In particular, this two-step labeling strategy has been employed for detection of active proteasomes in intact cells using an azide-containing probe that was then used for labeling with a biotin-derivitized phosphine.[101] Interestingly, more recent work by the same group has demonstrated that direct *in vivo* labeling of proteasomes is possible using cell permeable fluorescent probes thus obviating the need for the tandem labeling reaction.[98]

A second and more common way for tandem labeling makes use of the so-called "click" chemistry, in which the $2 + 3$ cycloaddition of an alkyne and azide functional group is facilitated by addition of a copper catalyst. Originally developed by Sharpless and co-workers, this strategy was modified by Cravatt and co-workers for use in activity-based proteomics.[102] The two reaction partners are both fully biocompatible and cyclizations can be initiated in aqueous solutions that even contain strong denaturants such as SDS and urea. Thus, CLICK probes have become the method of choice for tandem labeling applications and a number of probes containing either an azide or alkyne tag have been recently reported.[44,86,99] Overall, both of these tandem labeling methods have had a significant impact on the field of activity-based proteomics and will likely be extensively used in future generations of ABPs.


## 2.2 Applications of Activity-based Probes to Drug Discovery

Perhaps the greatest potential impact for activity-based probes is in the area of drug discovery. Since ABPs can be used to assess the activity of a given target enzyme in the context of complex proteomic samples and even in whole organisms, they can be used effectively to assess such important parameters as efficacy, pharmacodynamic properties and overall target selectivity of drug leads. These are parameters that are often difficult to measure using standard techniques yet the ability to make these measurements is often key to the success of a drug discovery program. In this section we will outline some of the most valuable applications of ABPs that are most relevant to the drug discovery progress.


### 2.2.1 Identification and Validation of Drug Targets Using Activity-based Probes

The drug discovery process begins with selection of a "validated" target. Once this target is selected, efforts can begin to identify small molecule lead compounds for advancement into clinical trials. In many cases, the targets of small molecule drugs are enzymes and the small molecule lead acts as an inhibitor of its enzymatic activity. Before the process of lead identification can begin, one must be certain the target enzyme is relevant for the given disease indication and that it can be selectively inhibited to produce a positive therapeutic outcome. ABPs are extremely valuable reagents for the early validation phases of the drug discovery process. Using broad-spectrum probes, it is possible to begin to assess the repertoire of related target enzymes in a given disease model (Figure 2.8). The relevance of a specific enzyme or family of enzymes can then be assessed by monitoring changes in activity levels during disease pathogenesis. Finally, by using selective small molecule inhibitors coupled with an ABP, it is possible to correlate inhibition of specific targets with effects on disease progression in order to validate specific targets for drug discovery efforts.

There have been a number of specific examples of this type of application of ABPs over the past few years. One prime example is the application of a general cysteine protease ABP to validate the cysteine cathepsins in the regulation of cancer pathogenesis.[20] This study demonstrated that several cysteine cathepsins were up-regulated during multiple stages of tumorigenesis in a mouse model of pancreatic cancer.[20] By using a fluorescently labeled, broad-spectrum ABP, it was possible to monitor changes in cathepsin activity and furthermore to determine which cell types were producing these enzyme in the growing tumors. By coupling the use of a small molecule inhibitor with the ABP, it was also possible to assess the extent of inhibition of the target cathepsins and subsequently the effects of target inhibition on tumor growth, angiogenesis and tumor invasiveness. Thus, it was possible to directly correlate cathepsin activities with specific disease pathologies. These results suggest that the cysteine cathepsins are valuable targets for the development of new anti-cancer drugs and that ABPs are powerful reagents for validating specific cathepsin targets *in vivo*.

In several more recent examples, a general ABP for the serine hydrolyase family of enzymes was used to identify enzymes whose increase in activity is correlated with various aspects of cancer pathogenesis.[38,39] In both studies a general fluorophosphonate probe (FP) was used to profile total cell extracts from cancer cells. In the first study, the probe identified a monoacylglycerol lipase that is over-expressed specifically in invasive cancer cell lines. This enzyme was then validated by forced expression in non-invasive cell lines resulting in the transformation of these lines to more aggressive cancers.[38] This study demonstrates that by identifying enzymes whose activities are linked to a given pathology (*i.e.* invasiveness) it is possible to identify new targets that are likely to be key regulators of disease. In the second example, the same probe was used to profile primary human ductal adenocarcinomas. The probe identified the retinoblastoma binding protein 9 (RBBP9) as a previously uncharacterized serine hydrolyase that promotes anchorage independent growth and pancreatic carcinogenesis *in vivo*. This study is a particularly interesting example because the target that was identified by the probe was not known to have serine hydrolyase activity and thus was not considered to be a viable drug target. By using an activity-based probe, it was possible to demonstrate that, not only did this protein have enzymatic activity, but that this activity was required in order to regulate tumor growth. Thus, inhibitors of this activity are likely to have therapeutic benefits for cancer treatment.

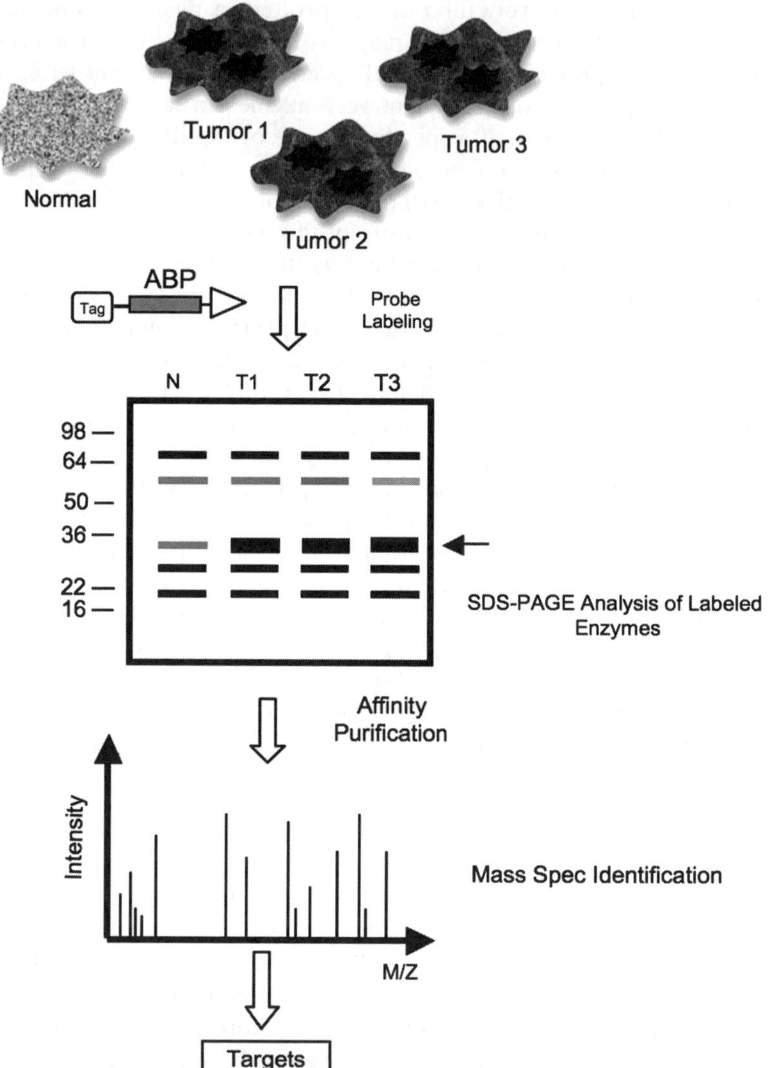



**Figure 2.8**  Using ABPs to identify novel drug targets associated with human disease. A general ABP can be used to profile the levels of active enzymes in normal and disease tissues. In this example, a general probe is used to label normal tissues and tissues derived from tumors. The resulting labeled enzymes can then be visualized by SDS-PAGE followed by scanning for the tag on the probe (often a fluorescent tag). Since labeling requires activity, the intensity of labeled proteins is an indication of the levels of active enzymes in the tissues. In this example, there is one protein that has higher levels of activity specifically in tumor tissues (shown with arrow). This protein can be identify by affinity purification of probe labeled proteins followed by mass spectrometry based identification of the target. Upon further analysis of the biological role of this enzyme in disease progression, it can be validated as a target for drug discovery efforts.

There have been many other examples of the use of ABPs to study the basic biology of enzymes. For example, ABPs have been used to study the function of proteases in the parasite pathogen that is the causative agent of malaria.[28,103] By using activity-based probes, a number of both host and parasite proteases that are key regulators of processes such as host cell invasion and host cell rupture have been identified. Once validated by ABPs as essential players in parasite pathogenesis, they become viable targets for development of new drug leads.

## 2.2.2 Use of Activity-based Probes for Drug Lead Identification and Assessment of Selectivity

As outlined above, ABPs are often valuable reagents for the identification and validation of a potential drug lead. Another major advantage of using an ABP for selection of a target is that probe labeling can be used as an assay to assess both potency and selectivity of inhibitor leads (Figure 2.9). Since the probe binds in the active site of the target enzyme, it can be used as an indirect measure of inhibition by a small molecule. Thus a competition method involves treating a target enzyme with a range of concentrations of a lead compound followed by labeling with the ABP. Since inhibition by the lead compound blocks labeling, it is possible to quantify the amount of inhibition by measuring residual labeling by the probe. This assay typically makes use of SDS-PAGE as

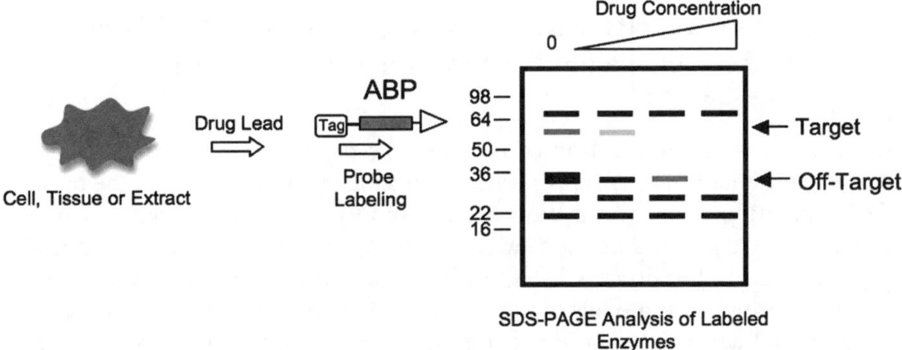


**Figure 2.9**  Profiling the selectivity of a drug lead using ABPs, which are valuable tools for assessing the selectivity of a given drug lead for a target enzyme of interest. This method is often referred to as a competition assay. In this assay, a tissue extract or intact cell is treated with a range of concentrations of a drug lead. The sample is then treated with a broad-spectrum probe that labels multiple related enzyme targets. Samples are analyzed by SDS-PAGE and labeled proteins are detected by scanning the gel to detect the probe tag. When the drug target binds in the active site of a target, it blocks labeling by the probe and signal is lost. In this example the drug lead is selective for the target but also shows off-target activity at higher drug concentrations.

a readout of residual probe labeling. There are several major benefits of this approach. First, the assay can be performed without any *a priori* knowledge of the native enzyme substrates. Second, if the competition assay is performed in complex mixtures that contain other related off target enzymes, the final SDS-PAGE analysis provides information regarding overall selectivity of the lead compound. Finally, compound potency and selectivity can be assessed without the need to purify and isolate an active form of the desired target enzyme since screening can be performed in crude cell extracts, in intact cells or even in whole organisms (see Section 2.2.1). It should also be noted that the ABP competition assay can be used to assess inhibition of a target enzyme by both reversible and irreversible inhibitors, thus making it suitable for most common classes of drug leads.[28,104,105]

One of the main drawbacks of using an ABP competition assay to assess inhibitor selectivity and potency is that the assay is generally not high throughput and therefore is not suitable for use as a primary screen of a large library of possible lead compounds. However, if the probe is used with purified enzyme or if the probe is highly selective for a given target enzyme, it is possible to switch to a readout that can be adapted to a HTS format. In a recent example, a competition assay was developed by measuring changes in fluorescence polarization of a fluorescent ABP upon target binding. Using this method it was possible to screen a large library of diverse small molecule and identify inhibitor leads of multiple enzymes including RBBP9, an enzyme for which no known substrates have been reported.[106]

### 2.2.3  Use of Activity-based Probes for Assessment of *In vivo* Pharmacodynamic Properties and Efficacy of Lead Compounds

Another beneficial application of ABPs in the drug discovery process is their use for assessment of *in vivo* properties of drug leads. In most cases, the process of identification and optimization of an inhibitor of a given target enzyme is straightforward and makes use of a simple *in vitro* assays using the purified target enzyme. This process usually results in a lead compound that is then advanced into animal models of disease to confirm therapeutic value. However, it is often difficult to predict how compounds will behave *in vivo* and therefore, results from studies in animal models of a disease can be difficult to interpret. For example, if a lead compound fails to produce the desired effect in the disease model, it is often difficult to determine if this is due to a poor choice of target enzymes or the lack of ability of the lead compound to reach the target enzyme in the tissue of interest (*i.e.* within a tumor). Furthermore, the compound may have altered specificity once introduced *in vivo* as the result of accumulation in distinct tissues and cells within the organism. ABPs can provide valuable information about the overall *in vivo* selectivity, potency and pharmacodynamic properties of a lead inhibitor (Figure 2.10). This information is generally obtained by treating animals with a candidate compound

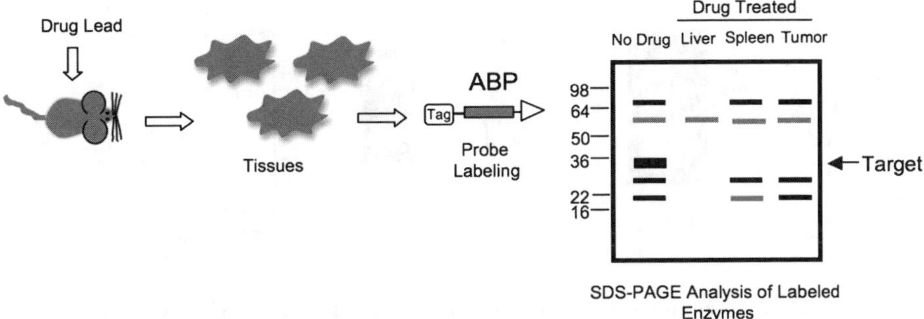


**Figure 2.10**  Using ABPs to monitor pharmacodynamics and *in vivo* selectivity of drug lead. ABPs can be used to perform *in vivo* competition analysis to determine the selectivity and tissue distribution of a drug lead. In this example, a mouse is treated with a drug lead. Tissues from primary organs and disease tissues are isolated and lysates labeled with a general ABP. Analysis of labeled proteins by SDS-PAGE shows how much of each target enzyme has been inhibited in each tissue. The drug shows highly selective inhibition of single target in the tumor tissue but show slight cross-reactivity in the spleen and strong cross-reactivity in the liver. This data suggests that the drug may clear through the liver and accumulate in multiple organs leading to a loss of selectivity in those organs.

followed by *in vivo* administration of an ABP and assessment of residual labeling in tissues by SDS-PAGE. If probes are sufficiently selective and stable *in vivo,* it is possible to perform such *in vivo* competition studies and obtain valuable information about the distribution, potency and selectivity of a drug lead.

In a recent example of this approach, a small library of epoxide-based inhibitors was screened for selectivity against the cysteine cathepsins.[107] Compounds that were identified as being selective for a given cathepsin were then injected into mice and overall potency and selectivity in various organs were assessed by labeling of tissue extracts with a general radiolabeled probe [125]I-DCG-04. These results indicated that, while the overall selectivity that was observed *in vitro* was also observed *in vivo*, the inhibitors accumulated in certain locations resulting in loss of overall selectivity. These results demonstrate the power of using ABPs to determine how compounds behave *in vivo* and to better understand the targets of the compounds *in vivo*. This information is essential to understanding how therapeutic effects correlate with inhibition of specific targets. It also can provide critical information about the value of further optimization of compound selectivity.

### 2.2.4  *In vivo* Imaging Applications

One of the more recent applications for ABPs has been to apply them for imaging applications *in vivo* (Figure 2.11). While ABPs that carry a cell

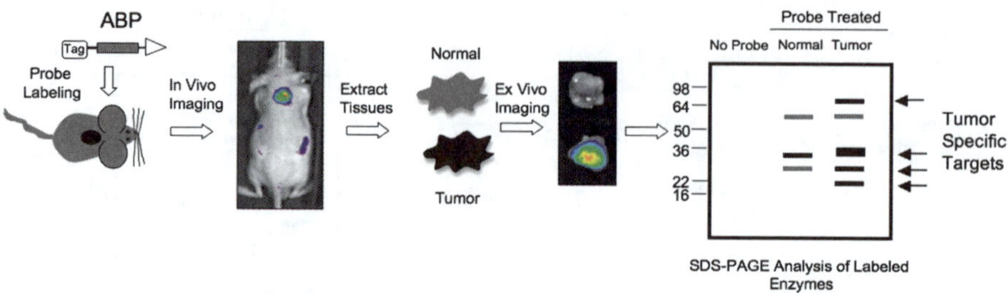



**Figure 2.11** Applications for ABPs in non-invasive imaging. ABPs containing near infrared fluorophores (NIRFs) can be used to image probe labeling of target enzymes using non-invasive, whole body imaging methods. In this example, a mouse bearing a small tumor on its back is treated with a NIRF-ABP. After the probe is allowed to clear the animal it is imaged using a non-invasive imaging system. Probe accumulates in the tumor as the result of elevated levels of active target enzymes. Tumor and normal tissue is then removed, imaged *ex vivo* and analyzed by SDS-PAGE. This allows specific signals observed in the animal to be correlated with tumor specific targets that are labeled by the probe. This method can be used in early detection of disease tissue but can also be used for real-time, non-invasive measurement of target inhibition using the *in vivo* competition assay outlined in Figure 2.10.

permeable tag have been in use since the early days of probe development, they have only recently found applications in whole body, non-invasive imaging applications.[8,11,22] Specific advances include the use of near infrared (NIR) fluorophores that emit light that can be visualized through skin and other tissues. Thus, it is possible to apply NIRF-labeled probes *in vivo* and then assess levels of enzyme activity in the whole organism. This section will outline some of the most recent advances in the use of ABPs for *in vivo* imaging applications with a focus on applications that are relevant to the drug discovery process.

### 2.2.4.1 *Non-invasive Monitoring of Drug Efficacy*

With the development of new classes of NIRF-labeled ABPs has come a number of possible applications for this technology. First and foremost, probes that label targets that are highly up-regulated in disease tissues such as tumors can be used to assess critical parameters of pathogenesis such as tumor location, growth rate, *etc.* Recent examples of probes that target cysteine cathepsins and legumain demonstrate the utility of detecting tumor tissues using non-invasive methods.[11,22] Because of the potential roles of cathepsins and legumain in a number of conditions that involve inflammation, these probes are likely to find use for diagnostic purposes in the early detection and monitoring of diseases such as atherosclerosis.

In addition, these recent studies have demonstrated the potential to make a fluorescently quenched version of an ABP that produces signal only after it has bound to the target enzyme.[22] These quenched probes therefore can be used for real-time imaging applications in whole cells and also provide a more rapid readout of specific signal when used for *in vivo* imaging. Quenched ABPs also have potential value for topical applications where the unbound probe cannot be washed away prior to imaging.

NIRF-labeled ABPs can also be used to non-invasively assess the efficacy of a small molecule drug lead. In a recent example with a cysteine cathepsin ABP, the effects of inhibition of target proteases using a general inhibitor were assessed by direct non-invasive imaging methods.[22] Since the probes bind covalently, it was possible to measure effects of the drug using non-invasive methods and then collect tumor tissues and confirm that *in vivo* signals correlated with target labeling by SDS-PAGE analysis. Therefore, it was possible to interpret the non-invasive imaging data with confidence due to link between the imaging data and the *ex vivo* biochemical analysis.

In another example of the use of non-invasive imaging of probe labeling, Edgington *et al.* used NIRF-labeled ABPs that label caspases to monitor the rates of apoptosis *in vivo*.[8] In this study, the ABP was used to assess the efficacy of a clinical antibody that induces apoptosis in tumors by ligation of death receptors on the tumor cell surface. This study demonstrated that the NIRF-ABP could be used to monitor changes in the rate of apoptosis of tumors using non-invasive imaging methods. Furthermore, accumulation of probe in apoptotic tissues could be linked to the amount of labeled caspase proteases that regulate the process of cell death. These results provide evidence that ABPs could be used to continuously monitor the effects of chemotherapy agents on tumors. This could allow assessment of patient response without the need to perform invasive analysis such as biopsy. Furthermore, such imaging-based monitoring of tumor response could greatly reduce the time required to make decisions about treatment strategies, thus reducing the amount of time patients are given drugs that are not effective against their disease.

## 2.2.4.2  *Activity-based Probes versus Substrates*

ABPs are finding increasing use for imaging applications. However, it should be noted that there are alternative methods for monitoring enzyme activity. The most commonly used methods for monitoring enzyme activity have focused on the use of reporter substrates as probes. These types of substrate probes can range from small molecules and short peptides to large proteins containing multiple reporter tags. Over the past decade, significant efforts have focused on protease reporters for use in imaging applications. Unlike ABPs that covalently bind and inhibit target proteases, reporter substrates are usually small peptide or peptide-like molecules that, when processed by the protease, produce an optical signal.

Although proteases were originally thought to completely degrade proteins in order to maintain homeostasis of overall proteins levels in the cell, it is now clear that proteases perform limited proteolysis of substrates at defined cleavage sites. This allows proteases to regulate structure, function and localization of substrates. Although the ability to cleave a specific site on a protein substrate can be controlled by a number of factors including secondary structure and localization of target and protease, in many cases, substrate cleavage is controlled by the primary amino acid sequences surrounding the scissile amide bond. Therefore, it is possible to use optimal recognition sequences to generate fluorescent substrate probes, whose spectral properties change in the presence of active proteases. Several examples of this type of probe exist. The simplest and perhaps most widely used fluorogenic substrate probes consist of a peptide sequence attached at the C-terminus to a fluorophore, such as AFC (7-amino-4-trifluoromethyl coumarin). In the presence of the active protease, the AFC is cleaved from the peptide, leading to a detectable shift in its fluorescent spectrum.

Other examples of substrate probes include peptide recognition sequences that are flanked by a fluorophore and quencher pair, such as Cy5 and QSY21. In these probes, the quencher absorbs the photons emitted by the fluorophore, preventing fluorescence until the quencher is released by cleavage of the substrate. Specific recent examples include probes for caspase proteases as markers of apoptosis.[108–110] It is also possible to replace the fluorophore/quencher pair with a fluorescent resonance energy transfer (FRET) pair. Because FRET can be performed using genetically encodable fluorecent proteins, a number of examples of protease responsive reporters have been developed using this approach.[111–113]

Another important class of substrate-based probes for proteases uses two or more fluorophores, that are self-quenched when in close proximity.[114] Multiple fluorophores can be linked to graft polymers containing peptide substrate sequences. When these linkers are cleaved by the protease, free fluorescent monomer can be detected. This class of probes has been widely used to study the activity of the cysteine cathepsin family of proteases across many diverse disease models.

While the natural substrate sequence often serves as a starting point for probe development, these sites may be engineered to enhance potency and sensitivity towards the protease target of choice using positional scanning libraries. Proteolytic cleavage sites are denoted by P4-P3-P2-P1/P1′-P2′, with the scissile bond occurring between P1 and P1′ sites. A significant number of substrate libraries use a positional scanning approach in which pools of fluorogenic peptide substrates are generated with one position, P1 for example, held constant as a fixed amino acid, while the other positions are mixtures of all possible natural amino acids. Each pool contains a different fixed amino acid, and the pools that yield the highest fluorescence in the presence of the enzyme identify the most optimal residues for that site. Subsequently, if multiple positions are scanned, the results can be combined to produce optimal substrate recognition sequences for various proteases.[104,115,116] This approach is also useful for finding sequences that are less potent towards off-target proteases, allowing for the development of more selective reagents. However, the use of positional scanning methods has its drawbacks since it fails to provide information on the cooperativity of multiple substrate positions for binding in the active site of the protease target. Therefore, some have chosen to focus on smaller libraries of individual protease substrates.[117,118] Once optimal substrate scaffolds are identified, these structures can be used to generate substrate based probes or they can be converted to covalent inhibitors for use as ABPs.

The addition of a number of new substrate-based probes to the imaging field has helped to advance progress in the use of molecular imaging agents. While substrates and ABPs make use of two different methods to assess activity, both methods have their strengths and weaknesses. The major strength of the ABP is that it binds covalently to the target protease resulting in retention of the label at the site of enzyme activity. In addition, the covalent modification allows the subsequent biochemical analysis of labeled targets as described above. Finally, the ABPs tend to be small molecules that readily diffuse into cells. Substrate probes, on the other hand, are often large peptide or polymer-based scaffolds that have long retention times in blood but have slow diffusion into tissues. On the flip side, substrates have the potential to provide signal amplification due to the fact that the target enzyme is not inhibited by the probe and can therefore process multiple substrate molecules. Interestingly, a recent study comparing large polymer-based substrates to ABPs suggests that this amplification process may not be a significant component of signal intensity.[119] Thus, both methods have their advantages for imaging and these advantages and disadvantages depend on the type of probe scaffold being used.

## 2.3   Outlook

The field of activity-based proteomics is progressing at a rapid rate. The past decade has seen a significant growth in the number of new ABPs reported and also a growth in the possible applications for ABPs. This chapter outlined the growth in both of these areas with a specific focus on applications for drug discovery. If the field continues to progress on the same trajectory, it is likely that ABPs will find applications in virtually all aspects of basic biology research and drug discovery. This chapter has outlined some of the major benefits of using ABPs in drug discovery efforts. ABPs have the potential to be applied to each and every step of the discovery process beginning with the identification and validation of new targets to use in clinical studies of drug efficacy. This chapter provides specific examples of how ABPs have been applied to target identification, HTS screening for lead compounds, medicinal chemistry efforts and selectivity profiling, *in vivo* pharmacodynamic/selectivity studies and finally *in vivo* efficacy assessment. While most of the examples presented here have focused on probes for proteases, it is clear that as the number of new probe families continues to expand, there will be examples of ABPs being applied to a more diverse range of enzyme targets. Looking toward the future for activity-based probes, it is likely that additional applications in drug discovery will be developed. In particular, the use of ABPs for monitoring clinical endpoints and for direct assessment of drug dynamics in human patients will be the next big step forward. Most of these applications are limited by the lack of probes that have advanced through FDA approval for use in humans. Similarly, there are a large number of potentially valuable clinical imaging applications for ABPs. However, these will also require significant effort and financial resources to obtain regulatory approval for use as systemic agents in humans. A number of small start-up companies are already beginning to develop optical imaging systems for applications during surgery and for early stage diagnostic monitoring. These companies will likely rely on new developments in optical contrast agents, such as ABPs, to make a significant impact on medicine. Hopefully, an interest in contrast agents such as ABPs will help advance the more late stage applications of ABPs in the drug discovery process.

## Acknowledgements

The authors would like to thank Benjamin Cravatt for helpful discussions and Aaron Puri and Fangfang Yin for assistance with graphics for the figures. This work was funded by NIH grants R01 EB005011 and R01 AI078947.

## References


1. B. F. Cravatt, A. T. Wright and J. W. Kozarich, *Annu. Rev. Biochem.*, 2008, **77**, 383.
2. M. J. Evans and B. F. Cravatt, *Chem. Rev.*, 2006, **106**, 3279.
3. M. Fonovic and M. Bogyo, *Expert Rev. Proteomics*, 2008, **5**, 721.
4. H. Schmidinger, A. Hermetter and R. Birner-Gruenberger, *Amino Acids*, 2006, **30**, 333.
5. A. B. Berger, M. D. Witte, J. B. Denault, A. M. Sadaghiani, K. M. Sexton, G. S. Salvesen and M. Bogyo, *Mol. Cell*, 2006, **23**, 509.
6. M. Bogyo, J. S. McMaster, M. Gaczynska, D. Tortorella, A. L. Goldberg and H. Ploegh, *Proc. Natl. Acad. Sci. U. S. A.*, 1997, **94**, 6629.
7. M. Bogyo, S. Verhelst, V. Bellingard-Dubouchaud, S. Toba and D. Greenbaum, *Chem. Biol.*, 2000, **7**, 27.
8. L. E. Edgington, A. B. Berger, G. Blum, V. E. Albrow, M. G. Paulick, N. Lineberry and M. Bogyo, *Nat. Med.*, 2009, **15**, 967.
9. L. Faleiro, R. Kobayashi, H. Fearnhead and Y. Lazebnik, *EMBO J.*, 1997, **16**, 2271.
10. D. Kato, K. M. Boatright, A. B. Berger, T. Nazif, G. Blum, C. Ryan, K. A. Chehade, G. S. Salvesen and M. Bogyo, *Nat. Chem. Biol.*, 2005, **1**, 33.
11. J. Lee and M. Bogyo, *ACS Chem. Biol.*, 2010, **19**, 233.
12. Y. Liu, M. P. Patricelli and B. F. Cravatt, *Proc. Natl. Acad. Sci. U. S. A.*, 1999, **96**, 14694.
13. N. A. Thornberry, E. P. Peterson, J. J. Zhao, A. D. Howard, P. R. Griffin and K. T. Chapman, *Biochemistry*, 1994, **33**, 3934.
14. J. C. Powers, J. L. Asgian, O. D. Ekici and K. E. James, *Chem. Rev.*, 2002, **102**, 4639.

15. M. Fonovic, S. H. Verhelst, M. T. Sorum and M. Bogyo, *Mol. Cell Proteomics*, 2007, **6**, 1761.
16. S. H. Verhelst, M. Fonovic and M. Bogyo, *Angew. Chem. Int. Ed. Engl.*, 2007, **46**, 1284.
17. E. Weerapana, A. E. Speers and B. F. Cravatt, *Nat. Protoc.*, 2007, **2**, 1414.
18. A. M. Sadaghiani, S. H. Verhelst and M. Bogyo, *Curr. Opin. Chem. Biol.*, 2007, **11**, 20.
19. D. Greenbaum, A. Baruch, L. Hayrapetian, Z. Darula, A. Burlingame, K. F. Medzihradszky and M. Bogyo, *Mol. Cell Proteomics*, 2002, **1**, 60.
20. J. A. Joyce, A. Baruch, K. Chehade, N. Meyer-Morse, E. Giraudo, F. Y. Tsai, D. C. Greenbaum, J. H. Hager, M. Bogyo and D. Hanahan, *Cancer Cell*, 2004, **5**, 443.
21. G. Blum, S. R. Mullins, K. Keren, M. Fonovic, C. Jedeszko, M. J. Rice, B. F. Sloane and M. Bogyo, *Nat. Chem. Biol.*, 2005, **1**, 203.
22. G. Blum, G. von Degenfeld, M. J. Merchant, H. M. Blau and M. Bogyo, *Nat. Chem. Biol.*, 2007, **3**, 668.
23. D. Greenbaum, K. F. Medzihradszky, A. Burlingame and M. Bogyo, *Chem. Biol.*, 2000, **7**, 569.
24. K. B. Sexton, D. Kato, A. B. Berger, M. Fonovic, S. H. Verhelst and M. Bogyo, *Cell Death Differ.*, 2007, **14**, 727.
25. F. Yuan, S. H. Verhelst, G. Blum, L. M. Coussens and M. Bogyo, *J. Am. Chem. Soc.*, 2006, **128**, 5616.
26. A. Baruch, D. Greenbaum, E. T. Levy, P. A. Nielsen, N. B. Gilula, N. M. Kumar and M. Bogyo, *J. Biol. Chem.*, 2001, **276**, 28999.
27. S. Yasothornsrikul, D. Greenbaum, K. F. Medzihradszky, T. Toneff, R. Bundey, R. Miller, B. Schilling, I. Petermann, J. Dehnert, A. Logvinova, P. Goldsmith, J. M. Neveu, W. S. Lane, B. Gibson, T. Reinheckel, C. Peters, M. Bogyo and V. Hook, *Proc. Natl. Acad. Sci. U. S. A.*, 2003, **100**, 9590.
28. D. C. Greenbaum, A. Baruch, M. Grainger, Z. Bozdech, K. F. Medzihradszky, J. Engel, J. DeRisi, A. A. Holder and M. Bogyo, *Science*, 2002, **298**, 2002.
29. A. Oleksy, E. Golonka, A. Banbula, G. Szmyd, J. Moon, M. Kubica, D. Greenbaum, M. Bogyo, T. J. Foster, J. Travis and J. Potempa, *Biol. Chem.*, 2004, **385**, 525.
30. R. A. van der Hoorn, M. A. Leeuwenburgh, M. Bogyo, M. H. Joosten and S. C. Peck, *Plant Physiol.*, 2004, **135**, 1170.
31. K. B. Sexton, M. D. Witte, G. Blum and M. Bogyo, *Bioorg. Med. Chem. Lett.*, 2007, **17**, 649.
32. F. Uhlmann, D. Wernic, M. A. Poupart, E. V. Koonin and K. Nasmyth, *Cell*, 2000, **103**, 375.
33. Z. Pan, D. A. Jeffery, K. Chehade, J. Beltman, J. M. Clark, P. Grothaus, M. Bogyo and A. Baruch, *Bioorg. Med. Chem. Lett.*, 2006, **16**, 2882.
34. E. Sabido, T. Tarrago, S. Niessen, B. F. Cravatt and E. Giralt, *ChemBioChem*, 2009, **10**, 2361.

35. G. C. Adam, E. J. Sorensen and B. F. Cravatt, *Mol. Cell Proteomics*, 2002, **1**, 828.
36. K. P. Chiang, S. Niessen, A. Saghatelian and B. F. Cravatt, *Chem. Biol.*, 2006, **13**, 1041.
37. N. Jessani, Y. Liu, M. Humphrey and B. F. Cravatt, *Proc. Natl. Acad. Sci. U. S. A.*, 2002, **99**, 10335.
38. D. K. Nomura, J. Z. Long, S. Niessen, H. S. Hoover, S. W. Ng and B. F. Cravatt, *Cell*, **140**, 49.
39. D. J. Shields, S. Niessen, E. A. Murphy, A. Mielgo, J. S. Desgrosellier, S. K. Lau, L. A. Barnes, J. Lesperance, M. Bouvet, D. Tarin, B. F. Cravatt and D. A. Cheresh, *Proc. Natl. Acad. Sci. U. S. A.*, 2009, **107**, 2189.
40. C. Chang and Z. Werb, *Trends Cell Biol.*, 2001, **11**, S37.
41. C. M. Overall and C. Lopez-Otin, *Nat. Rev. Cancer*, 2002, **2**, 657.
42. X. S. Puente, L. M. Sanchez, C. M. Overall and C. Lopez-Otin, *Nat. Rev. Genet.*, 2003, **4**, 544.
43. A. Saghatelian, N. Jessani, A. Joseph, M. Humphrey and B. F. Cravatt, *Proc. Natl. Acad. Sci. U. S. A.*, 2004, **101**, 10000.
44. S. A. Sieber, S. Niessen, H. S. Hoover and B. F. Cravatt, *Nat. Chem. Biol.*, 2006, **2**, 274.
45. M. P. Patricelli, A. K. Szardenings, M. Liyanage, T. K. Nomanbhoy, M. Wu, H. Weissig, A. Aban, D. Chun, S. Tanner and J. W. Kozarich, *Biochemistry*, 2007, **46**, 350.
46. S. J. Ratcliffe, T. Yi and S. S. Khandekar, *J. Biomol. Screen.*, 2007, **12**, 126.
47. S. S. Khandekar, B. Feng, T. Yi, S. Chen, N. Laping and N. Bramson, *J. Biomol. Screen.*, 2005, **10**, 447.
48. M. C. Yee, S. C. Fas, M. M. Stohlmeyer, T. J. Wandless and K. A. Cimprich, *J. Biol Chem*, 2005, **280**, 29053.
49. Y. Liu, K. R. Shreder, W. Gai, S. Corral, D. K. Ferris and J. S. Rosenblum, *Chem. Biol.*, 2005, **12**, 99.
50. K. R. Shreder, M. S. Wong, T. Nomanbhoy, P. S. Leventhal and S. R. Fuller, *Org. Lett.*, 2004, **6**, 3715.
51. P. R. Young, M. M. McLaughlin, S. Kumar, S. Kassis, M. L. Doyle, D. McNulty, T. F. Gallagher, S. Fisher, P. C. McDonnell, S. A. Carr, M. J. Huddleston, G. Seibel, T. G. Porter, G. P. Livi, J. L. Adams and J. C. Lee, *J. Biol. Chem.*, 1997, **272**, 12116.
52. J. A. Blair, D. Rauh, C. Kung, C. H. Yun, Q. W. Fan, H. Rode, C. Zhang, M. J. Eck, W. A. Weiss and K. M. Shokat, *Nat. Chem. Biol.*, 2007, **3**, 229.
53. M. S. Cohen, C. Zhang, K. M. Shokat and J. Taunton, *Science*, 2005, **308**, 1318.
54. M. S. Cohen, H. Hadjivassiliou and J. Taunton, *Nat. Chem. Biol.*, 2007, **3**, 156.
55. S. Kumar, B. Zhou, F. Liang, W. Q. Wang, Z. Huang and Z. Y. Zhang, *Proc. Natl. Acad. Sci. U. S. A.*, 2004, **101**, 7943.

56. Q. Zhu, X. Huang, G. Y. J. Chen and S. Q. Yao, *Tetrahedron Lett.*, 2004, **45**, 707.

57. C. M. Salisbury and B. F. Cravatt, *Proc. Natl. Acad. Sci. U. S. A.*, 2007, **104**, 1171.

58. C. M. Salisbury and B. F. Cravatt, *J. Am. Chem. Soc.*, 2008, **130**, 2184.

59. C. Xu, E. Soragni, C. J. Chou, D. Herman, H. L. Plasterer, J. R. Rusche and J. M. Gottesfeld, *Chem. Biol.*, 2009, **16**, 980.

60. X. Li, J. H. Cao, Y. Li, P. Rondard, Y. Zhang, P. Yi, J. F. Liu and F. J. Nan, *J. Med. Chem.*, 2008, **51**, 3057.

61. M. Tantama, W. C. Lin and S. Licht, *J. Am. Chem. Soc.*, 2008, **130**, 15766.

62. C. P. Lu, C. T. Ren, S. H. Wu, C. Y. Chu and L. C. Lo, *ChemBioChem*, 2007, **8**, 2187.

63. Y. Wang, S. Hu and W. Fast, *J. Am. Chem. Soc.*, 2009, **131**, 15096.

64. M. Schicher, I. Jesse and R. Birner-Gruenberger, *Methods Mol. Biol.*, 2009, **580**, 251.

65. H. Schmidinger, R. Birner-Gruenberger, G. Riesenhuber, R. Saf, H. Susani-Etzerodt and A. Hermetter, *ChemBioChem*, 2005, **6**, 1776.

66. D. Kidd, Y. Liu and B. F. Cravatt, *Biochemistry*, 2001, **40**, 4005.

67. K. T. Barglow and B. F. Cravatt, *Angew. Chem. Int. Ed. Engl.*, 2006, **45**, 7408.

68. A. T. Wright, J. D. Song and B. F. Cravatt, *J. Am. Chem. Soc.*, 2009, **131**, 10692.

69. K. A. Stubbs, A. Scaffidi, A. W. Debowski, B. L. Mark, R. V. Stick and D. J. Vocadlo, *J. Am. Chem. Soc.*, 2008, **130**, 327.

70. S. Misaghi, G. A. Korbel, B. Kessler, E. Spooner and H. L. Ploegh, *Cell Death Differ.*, 2006, **13**, 163.

71. T. Komatsu, K. Kikuchi, H. Takakusa, K. Hanaoka, T. Ueno, M. Kamiya, Y. Urano and T. Nagano, *J. Am. Chem. Soc.*, 2006, **128**, 15946.

72. G. C. Adam, B. F. Cravatt and E. J. Sorensen, *Chem. Biol.*, 2001, **8**, 81.

73. G. C. Adam, E. J. Sorensen and B. F. Cravatt, *Nat. Biotechnol.*, 2002, **20**, 805.

74. E. Weerapana, G. M. Simon and B. F. Cravatt, *Nat. Chem. Biol.*, 2008, **4**, 405.

75. S. Tsukiji, M. Miyagawa, Y. Takaoka, T. Tamura and I. Hamachi, *Nat. Chem. Biol.*, 2009, **5**, 341.

76. K. T. Barglow and B. F. Cravatt, *Chem. Biol.*, 2004, **11**, 1523.

77. M. Bogyo, A. Baruch, D. A. Jeffery, D. Greenbaum, A. Borodovsky, H. Ovaa and B. Kessler, *Curr. Protoc. Protein Sci.*, 2004, Chapter 21, Unit 21 17.

78. G. Fenteany, R. F. Standaert, W. S. Lane, S. Choi, E. J. Corey and S. L. Schreiber, *Science*, 1995, **268**, 726.

79. S. P. Gygi, B. Rist, S. A. Gerber, F. Turecek, M. H. Gelb and R. Aebersold, *Nat. Biotechnol.*, 1999, **17**, 994.

80. L. R. Zieske, *J. Exp. Bot.*, 2006, **57**, 1501.

81. S. E. Ong, B. Blagoev, I. Kratchmarova, D. B. Kristensen, H. Steen, A. Pandey and M. Mann, *Mol. Cell Proteomics*, 2002, **1**, 376.
82. D. S. Kirkpatrick, S. A. Gerber and S. P. Gygi, *Methods*, 2005, **35**, 265.
83. P. F. van Swieten, R. Maehr, A. M. van den Nieuwendijk, B. M. Kessler, M. Reich, C. S. Wong, H. Kalbacher, M. A. Leeuwenburgh, C. Driessen, G. A. van der Marel, H. L. Ploegh and H. S. Overkleeft, *Bioorg. Med. Chem. Lett.*, 2004, **14**, 3131.
84. P. A. Everley, C. A. Gartner, W. Haas, A. Saghatelian, J. E. Elias, B. F. Cravatt, B. R. Zetter and S. P. Gygi, *Mol. Cell Proteomics*, 2007, **6**, 1771.
85. W. M. Old, K. Meyer-Arendt, L. Aveline-Wolf, K. G. Pierce, A. Mendoza, J. R. Sevinsky, K. A. Resing and N. G. Ahn, *Mol. Cell Proteomics*, 2005, **4**, 1487.
86. A. E. Speers and B. F. Cravatt, *J. Am. Chem. Soc.*, 2005, **127**, 10018.
87. A. Borodovsky, B. M. Kessler, R. Casagrande, H. S. Overkleeft, K. D. Wilkinson and H. L. Ploegh, *EMBO J.*, 2001, **20**, 5187.
88. A. Borodovsky, H. Ovaa, N. Kolli, T. Gan-Erdene, K. D. Wilkinson, H. L. Ploegh and B. M. Kessler, *Chem. Biol.*, 2002, **9**, 1149.
89. J. Hemelaar, P. J. Galardy, A. Borodovsky, B. M. Kessler, H. L. Ploegh and H. Ovaa, *J. Proteome Res.*, 2004, **3**, 268.
90. K. R. Love, R. K. Pandya, E. Spooner and H. L. Ploegh, *ACS Chem. Biol.*, 2009, **4**, 275.
91. Y. Liu, J. Wu, H. Weissig, J. M. Betancort, W. Z. Gai, P. S. Leventhal, M. P. Patricelli, B. Samii, A. K. Szardenings, K. R. Shreder and J. W. Kozarich, *Bioorg. Med. Chem. Lett.*, 2008, **18**, 5955.
92. E. S. Okerberg, J. Wu, B. Zhang, B. Samii, K. Blackford, D. T. Winn, K. R. Shreder, J. J. Burbaum and M. P. Patricelli, *Proc. Natl. Acad. Sci. U. S. A.*, 2005, **102**, 4996.
93. N. Winssinger, R. Damoiseaux, D. C. Tully, B. H. Geierstanger, K. Burdick and J. L. Harris, *Chem. Biol.*, 2004, **11**, 1351.
94. N. Winssinger and J. L. Harris, *Expert Rev. Proteomics*, 2005, **2**, 937.
95. D. Hesek, M. Toth, V. Krchnak, R. Fridman and S. Mobashery, *J. Org. Chem.*, 2006, **71**, 5848.
96. D. Hesek, M. Toth, S. O. Meroueh, S. Brown, H. Zhao, W. Sakr, R. Fridman and S. Mobashery, *Chem. Biol.*, 2006, **13**, 379.
97. M. P. Patricelli, D. K. Giang, L. M. Stamp and J. J. Burbaum, *Proteomics*, 2001, **1**, 1067.
98. C. R. Berkers, M. Verdoes, E. Lichtman, E. Fiebiger, B. M. Kessler, K. C. Anderson, H. L. Ploegh, H. Ovaa and P. J. Galardy, *Nat. Methods*, 2005, **2**, 357.
99. E. W. Chan, S. Chattopadhaya, R. C. Panicker, X. Huang and S. Q. Yao, *J. Am. Chem. Soc.*, 2004, **126**, 14435.
100. E. Saxon and C. R. Bertozzi, *Science*, 2000, **287**, 2007.
101. H. Ovaa, P. F. van Swieten, B. M. Kessler, M. A. Leeuwenburgh, E. Fiebiger, A. M. van den Nieuwendijk, P. J. Galardy, G. A. van der Marel, H. L. Ploegh and H. S. Overkleeft, *Angew. Chem. Int. Ed. Engl.*, 2003, **42**, 3626.

102. A. E. Speers, G. C. Adam and B. F. Cravatt, *J. Am. Chem. Soc.*, 2003, **125**, 4686.
103. S. Arastu-Kapur, E. L. Ponder, U. P. Fonovic, S. Yeoh, F. Yuan, M. Fonovic, M. Grainger, C. I. Phillips, J. C. Powers and M. Bogyo, *Nat. Chem. Biol.*, 2008, **4**, 203.
104. D. C. Greenbaum, W. D. Arnold, F. Lu, L. Hayrapetian, A. Baruch, J. Krumrine, S. Toba, K. Chehade, D. Bromme, I. D. Kuntz and M. Bogyo, *Chem. Biol.*, 2002, **9**, 1085.
105. D. Leung, C. Hardouin, D. L. Boger and B. F. Cravatt, *Nat. Biotechnol.*, 2003, **21**, 687.
106. D. A. Bachovchin, S. J. Brown, H. Rosen and B. F. Cravatt, *Nat. Biotechnol.*, 2009, **27**, 387.
107. A. M. Sadaghiani, S. H. Verhelst, V. Gocheva, K. Hill, E. Majerova, S. Stinson, J. A. Joyce and M. Bogyo, *Chem. Biol.*, 2007, **14**, 499.
108. E. M. Barnett, X. Zhang, D. Maxwell, Q. Chang and D. Piwnica-Worms, *Proc. Natl. Acad. Sci. U. S. A.*, 2009, **106**, 9391.
109. K. Bullok and D. Piwnica-Worms, *J. Med. Chem.*, 2005, **48**, 5404.
110. D. Maxwell, Q. Chang, X. Zhang, E. M. Barnett and D. Piwnica-Worms, *Bioconjugate Chem.*, 2009, **20**, 702.
111. P. L. Bardet, G. Kolahgar, A. Mynett, I. Miguel-Aliaga, J. Briscoe, P. Meier and J. P. Vincent, *Proc. Natl. Acad. Sci. U. S. A.*, 2008, **105**, 13901.
112. A. A. Gerencser, K. A. Mark, A. E. Hubbard, A. S. Divakaruni, Z. Mehrabian, D. G. Nicholls and B. M. Polster, *J. Neurochem.*, 2009, **110**, 990.
113. C. T. Hellwig, B. F. Kohler, A. K. Lehtivarjo, H. Dussmann, M. J. Courtney, J. H. Prehn and M. Rehm, *J. Biol. Chem.*, 2008, **283**, 21676.
114. R. Weissleder, C. H. Tung, U. Mahmood and A. Bogdanov Jr., *Nat. Biotechnol.*, 1999, **17**, 375.
115. B. J. Backes, J. L. Harris, F. Leonetti, C. S. Craik and J. A. Ellman, *Nat. Biotechnol.*, 2000, **18**, 187.
116. J. L. Harris, B. J. Backes, F. Leonetti, S. Mahrus, J. A. Ellman and C. S. Craik, *Proc. Natl. Acad. Sci. U. S. A.*, 2000, **97**, 7754.
117. A. W. Patterson, W. J. Wood and J. A. Ellman, *Nat. Protoc.*, 2007, **2**, 424.
118. W. J. Wood, A. W. Patterson, H. Tsuruoka, R. K. Jain and J. A. Ellman, *J. Am. Chem. Soc.*, 2005, **127**, 15521.
119. G. Blum, R. M. Weimer, L. E. Edgington, W. Adams and M. Bogyo, *PLoS One*, 2009, **4**, e6374.
120. A. E. Speers and B. F. Cravatt, *Chem. Biol.*, 2004, **11**, 535.

# Targeted Intracellular Protein Degradation as a Potential Therapeutic Strategy

A.R. SCHNEEKLOTH[a] AND C.M. CREWS[a,b,c,*]

[a] Department of Chemistry, Yale University, PO Box 208107, New Haven, CT 06520-8107, USA; [b] Department of Molecular, Cellular, and Developmental Biology, Yale University, PO Box 208103, New Haven, CT 06520-8103, USA; [c] Department of Pharmacology, Yale University, PO Box 208103, New Haven, CT 06520-8103, USA

## 3.1   Background

### 3.1.1   Introduction

Investigating the function and importance of specific intracellular proteins is critical for our understanding of both normal cellular activities and the pathology of disease. Traditional genetic approaches involve the identification of a phenotype and its relevant proteins, and then the manipulation of the gene sequences encoding those proteins. The newly emerging field of chemical genetics offers a complementary approach, in which small molecules perturb the function of proteins in order to understand their significance with regards to a particular phenotype. Though the traditional approach will no doubt continue to be of great importance in the study of cellular function, chemical genetics provides several advantages over cumbersome genetic manipulation techniques. In addition, small molecule-induced perturbation of a system often provides a more

RSC Drug Discovery Series No. 5
New Frontiers in Chemical Biology: Enabling Drug Discovery
Edited by Mark E. Bunnage

Published by the Royal Society of Chemistry, www.rsc.org

realistic context for experiments, in which the genetic foundation of a cell has not been altered. While there have been many different chemical genetic methods developed for a variety of applications, we discuss here a method known as PROteolysis TArgeting Chimeric molecules, or PROTACs, developed to target a specific protein for post-translational degradation.

Despite much research effort in the years since the human genome was sequenced, much remains to be learned about the protein products of those genes. The continual development of methods to modulate protein levels will enable the further exploration of their function, and enhance our understanding of the complex interactions within the cell. The PROTACs method, first described in 2001, utilizes the ubiquitin–proteasome pathway for protein degradation.[1] The PROTAC molecule is a heterobifunctional complex containing a ligand for the target protein connected by a linker to a ligand for an E3 ubiquitin ligase (see Figure 3.1). Upon entering the cell, the PROTAC binds to both the target protein and the E3 ligase, inducing the ubiquitination of the target protein by the enzyme. Following ubiquitination, the target protein is recruited to the 26S proteasome, where it is degraded. This method has the potential to be useful as both a tool for understanding the role of specific proteins in cellular pathways, and also as a therapeutic designed to specifically eliminate disease-causing proteins from the cell. Herein, we describe the basic biological pathways and components utilized in the PROTAC method, give an overview of PROTAC development, and explore the current state of the art.

## 3.1.2 The Ubiquitin–Proteasome System

### 3.1.2.1 Introduction

The integrity and regulation of cellular processes requires the precise control of protein levels in the cell. One of the primary means by which this can be accomplished is by controlling protein half-life through post-translational degradation, which has emerged as a major cellular regulatory mechanism. This degradation can occur through either the lysosome, or through the ubiquitin–proteasome pathway.

Degradation *via* the ubiquitin–proteasome system (UPS) requires the convergence of two separate systems, which work together to provide a highly specific, yet broadly applicable process, enabling the degradation of a wide range of protein substrates. First, the ubiquitin cascade utilizes three different classes of enzymes to mark a specific protein for degradation by the covalent attachment of a polyubiquitin chain. Second, the 26S proteasome recognizes the ubiquitin-tagged proteins and delivers them into the multicatalytic proteolytic core for degradation (Figure 3.2).

Proteasome-mediated protein degradation is required for many aspects of cellular homeostasis. Many misfolded or aged proteins are recycled in this way, keeping the cell free from potentially harmful or useless components. Cyclin degradation *via* this pathway regulates cell cycle progression, and even immune recognition is dependent on proteasomal degradation for antigenic peptide

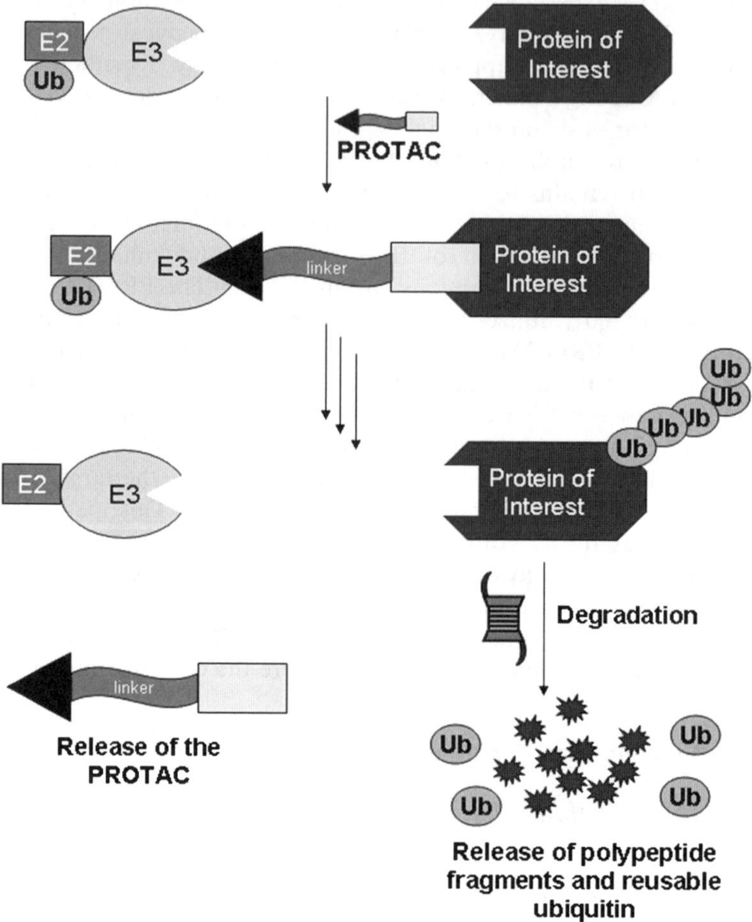


**Figure 3.1**   Mechanism of proteolysis targeting chimeric molecules (PROTACs). Cell lysates or whole cells are treated with a PROTAC, which contains a binding domain for an E3 ligase (black triangle) and a binding domain for a protein of interest (pale gray rectangle) connected by a linker. PROTAC treatment promotes the close proximity of the E3 ligase and the target protein, resulting in the synthesis of a polyubiquitin chain on the target protein. The target protein is consequently recognized and degraded by the proteasome, while the PROTAC is free to bind another protein of interest.

processing. Thus, this critical regulatory system is of significant interest, both to expand our knowledge of cellular processes, and as a target for therapeutic intervention.

## 3.1.2.2   The Ubiquitin Cascade

The ubiquitin–proteasome pathway is a finely tuned system, coordinating the function of multiple diverse components to enable the highly specified

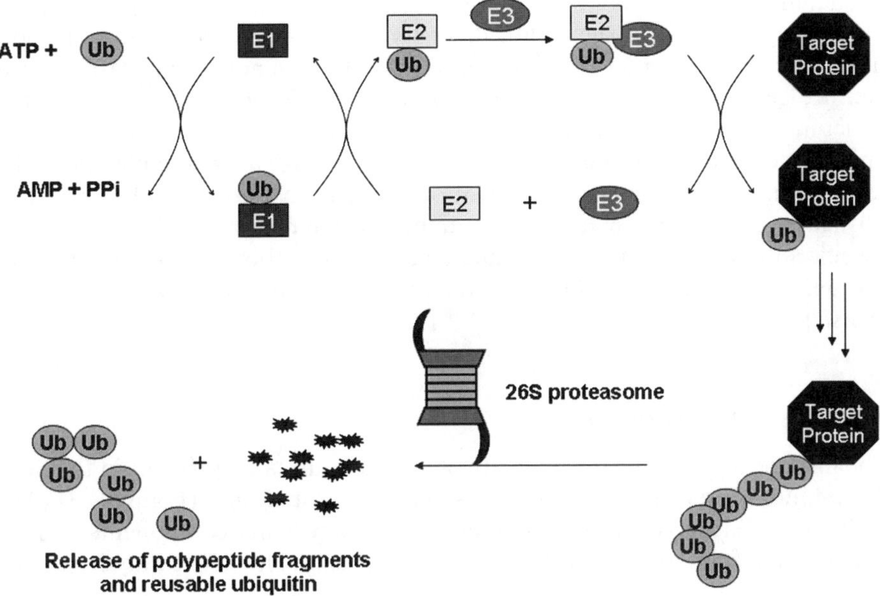


**Figure 3.2**  The ubiquitin–proteasome system (UPS). Ubiquitin is activated in an ATP-dependent manner by a ubiquitin activating enzyme (E1), leading to its covalent attachment to the E1 through a high-energy thioester bond. The ubiquitin is subsequently transferred *via* a transacylation reaction to a reactive thiol on a ubiquitin conjugating enzyme (E2). Depending on the nature of the E2 enzyme, it will then either transfer ubiquitin directly to an ubiquitin ligase (E3), or form a ternary complex with an E3. The E3 ligase facilitates substrate recognition, culminating in the synthesis of a polyubiquitin chain on the target protein. The target is then recognized and degraded by the 26S proteasome, with concomitant release of ubiquitin.

degradation of an incredibly diverse range of proteins. The cascade involves the coordination of three different classes of enzymes (E1s, E2s and E3s) to activate and specifically transfer ubiquitin to a target protein. Ubiquitin itself is composed of 76 amino acids, and is highly conserved from yeast to mammals.[2] The terminal residue, Gly-76, is activated in an ATP-dependent manner, allowing ligation with the ε-amino group of an internal lysine residue of the target protein through an isopeptide bond.

In the initial step of the ubiquitin conjugation cascade, the ubiquitin-activating enzyme, E1, activates ubiquitin in an ATP-dependent manner, resulting in the covalent attachment of ubiquitin to the enzyme *via* a high-energy thioester bond. Next, ubiquitin is transferred by a transacylation reaction to a thiol group on the active site Cys residue of an E2 enzyme. Depending on the nature of the E2 enzyme, it will then either transfer ubiquitin directly to an E3 ubiquitin ligase, or form a ternary complex with an E3 ligase, culminating in ubiquitin transfer directly from the E2.

The final components of the ubiquitin cascade are the E3 ubiquitin ligases. These enzymes exist in significant diversity in the cell, reflecting their responsibility for protein target recognition. Many E3s have now been characterized, and are generally found to have one of several different defining E3 domain structures. Some E3 ligases recognize a constitutively active degradation signal on a target protein, while others respond to the covalent modification of a sequence motif, such as phosphorylation or hydroxylation. In addition, some E3 ligases require association with adaptor proteins in order to recognize a specific subset of protein targets, and as mentioned above, some require association with an E2 enzyme in a ternary complex in order to facilitate ubiquitin transfer.[3]

### 3.1.2.3   The Proteasome

The final destination for the majority of polyubiquitinated proteins in the cell is the multicatalytic protease known as the 26S proteasome (Figure 3.3). This high molecular mass enzyme is composed of two primary components: the 20S catalytic particle, and the 19S regulatory particle. The 20S catalytic particle

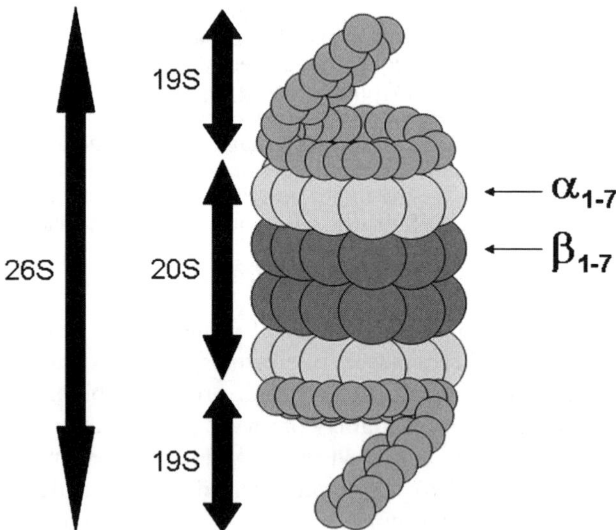


**Figure 3.3**   The proteasome. The proteasome is composed of two primary components: the 20S catalytic particle and the 19S regulatory particle. The 20S catalytic particle is made up of two pair of heptameric rings, $\alpha$ (outer) and $\beta$ (inner), and contains the catalytic sites for protein degradation. The 19S regulatory particle is coupled to the 20S catalytic particle at either end, to make up the 26S proteasome. The 19S regulatory particle is responsible for substrate recognition, and the ATP-dependent unfolding of proteins necessary for entrance into the catalytic core.

forms a barrel shape, with a central chamber through which the target substrate is fed. This hollow particle is comprised of four stacked rings. The two heptameric outer rings are composed of seven α-subunits each, while the two heptameric inner rings are composed of seven β-subunits each. The amino terminal threonines of three of the β-subunits serve as the catalytic nucleophiles for protein degradation.[4]

The 20S proteasome is coupled with a 19S regulatory particle at either one or both α-rings.[5] The 19S regulatory particle is crucial to the specificity of the ubiquitin–proteasome system, as it facilitates the recognition of polyubiquitin-tagged protein substrates, and catalyzes the initiation of protein degradation. The 20S catalytic core, unlike typical proteases, enjoys a broad range of proteolytic specificities, allowing the near complete degradation of targeted protein substrates. While all the protease active sites in eukaryotic proteasomes utilize comparable N-terminal threonine residues for catalysis, the substrate specificity is determined by different substrate-binding channels, containing binding sites for peptides seven to nine amino acids in length.[5]

## 3.1.3 Modulation of Protein Levels in Cells

### 3.1.3.1 Introduction

The ability to reduce or eliminate a protein of interest from the cell is one of the primary ways in which the activity of that protein can be fully understood and explored. While methods have been developed to manipulate protein levels at both the DNA and RNA level, our interest is in the methods that act at the post-translational level to disrupt protein expression, allowing the study of the resulting phenotype.

Many different techniques for the regulation of protein expression at the post-translational level have been developed over the last several years. While preventing protein expression at the DNA or RNA level is appropriate in many situations, the destruction of a protein following its synthesis has multiple advantages. Unlike RNAi, which blocks the translation of specific mRNA sequences and thus does not affect the level of the protein already present in the cell, a post-translational method can effectively target proteins regardless of their cellular half-life. In addition, many methods of post-translational protein inactivation are inducible, and can thus examine the effects of protein loss in a finely tuned temporal manner. This can be particularly important for proteins involved in processes such as cell cycle regulation, or embryonic development, where a fundamental loss of the protein (as would be found with a genetic knockout technique) would preclude a thorough understanding of its function.

The majority of these techniques designed to post-translationally degrade target proteins utilize the cell's primary pathway for intracellular protein degradation: the ubiquitin–proteasome system (UPS). By hijacking the UPS and artificially marking a protein of interest as a target for the pathway, potentially any protein could be subjected to complete degradation in the proteasome. One of the first such methods involved the creation of a fusion protein between an E3 ubiquitin ligase and a substrate recognition domain for the protein of interest.[6,7] Other methods have involved the fusing of the protein of interest directly to a destabilizing domain which promotes proteasome-mediated degradation of the fusion pair.[8–13]

While these techniques have significant utility for the cell-based study of many proteins, the necessity of creating a fusion protein not only makes the system a less realistic representation of the natural environment, but for some protein targets could also disrupt the normal activity of the protein. Thus, the development of an externally administered small molecule-based system for post-translational protein degradation was of significant interest. Such a method would not only be a useful tool in cell biology, but by eliminating the need for biochemical manipulation of the system, it would also have the potential to be optimized as a therapeutic. Ultimately, the continued exploration of a variety of techniques for post-translational protein degradation will no doubt contribute to our understanding of protein function by acting on target proteins that may be inaccessible or intractable at the DNA or mRNA level.

### 3.1.3.2   *PROTACs: A Chemical Genetic Approach to Protein Degradation*

A chemical genetic approach utilizes small molecules to perturb a biological system in order to understand the function and importance of its components. These small molecule probes can take the form of chemical inducers of dimerization (CIDs), consisting of a ligand for a target protein connected to another ligand *via* a linker.[14,15] These types of probes are prepared using traditional synthetic methods, and can induce complexation between two proteins to result in a desired biological effect. One of the advantages of a chemical genetics approach to the study of protein degradation is the requirement for little to no biochemical manipulation of the system. This decreases the chance that the introduction of a fusion protein could produce aberrant results. In addition, the use of small molecules to stimulate protein degradation provides the temporal control and dose-response capabilities to thoroughly dissect the function of almost any type of protein target.

A broadly applicable method for small molecule-mediated post-translational protein degradation, and the focus of this chapter, was developed by Crews and co-workers, and is known as PROTACs (PROteolysis TArgeting Chimeric molecules). The PROTACs method is centered on the design of a hetero-bifunctional molecule that targets a specific protein for degradation *via* the ubiquitin–proteasome pathway by forming a complex between the target protein and an E3 ubiquitin ligase. The PROTAC itself is comprised of a recognition element for the target, a linker, and a recognition element for an E3 ligase. Following the treatment of cells with a PROTAC, the target protein and the E3 ligase are brought into close physical proximity, facilitating the poly-ubiquitination of the target, and its subsequent recognition and degradation by the proteasome (Figure 3.1). As a chemical genetics approach to protein degradation this method does not inherently require any manipulation of the system, and these molecules function by promoting the direct ubiquitination of the substrate.


## 3.2 Development of the PROTACs Approach

### 3.2.1 Initial Proof-of-Concept

Based on earlier work, it was clear that the ubiquitin–proteasome system could be successfully manipulated to promote the degradation of proteins that are normally stable in the cell.[6–13] In 2001, Crews and co-workers first detailed the development of PROteolysis TArgeting Chimeric molecules (PROTACs), a method that provides the ability to induce degradation without biochemical manipulation.[1] In order for this method to work, however, it would be necessary to identify ligands for both the E3 ubiquitin ligase and the protein of interest.

The E3 ligase chosen for the initial experiments was the mammalian F-box protein β-TRCP/E3RS. F-box proteins are the substrate recognition components of a class of heterotetrameric E3 ligases known as Skp1-Cullin-F-box (SCF) complexes.[16] This E3 ligase had previously been shown to bind IκBα, a negative regulator of NFκB, promoting its degradation.[17] This results in the activation of NFκB during the inflammatory response. The binding of SCF$^{\beta\text{-TRCP}}$ to IκBα is mediated by a 10 amino acid region within the protein, DRHDSGLDSM.[17] Upon receipt of inflammatory signals, IκBα kinase (IKK) phosphorylates this peptide motif on both serine residues, promoting the binding of the SCF$^{\beta\text{-TRCP}}$ complex to IκBα, resulting in its ubiquitination. Knowing that this short peptide fragment from IκBα is capable of recruiting the SCF$^{\beta\text{-TRCP}}$ E3 ligase complex, it was chosen for incorporation into the first PROTAC molecule for use in targeting an unrelated protein for ubiquitination and degradation.

Following the identification of an E3 ligase–ligand pair, the next step in the development of the first generation of PROTACs was to choose a target protein for degradation studies. Methionine aminopeptidase-2 (MetAP2) is an enzyme that catalyzes the cleavage of the N-terminal methionine residue from nascent peptide chains. MetAP2 was chosen as the PROTAC target largely because of the existence of a known covalent ligand, the small molecule angiogenesis inhibitor ovalicin (OVA).[18] In addition to the existence of a known ligand, MetAP2 was an ideal PROTAC target because it was not known to be ubiquitinated or a natural substrate for any SCF complex.

The design of PROTAC-1, the MetAP2-targeting PROTAC (**1**, shown in Figure 3.4), incorporates the IκBα phosphopeptide (IPP) as the E3 ligase ligand, connected by a linker to ovalicin, the target protein ligand. It was hypothesized that the phosphopeptide moiety would bind β-TRCP, ovalicin would bind MetAP2, and the PROTAC would essentially hold the two proteins together, allowing the E3 ligase the opportunity to ubiquitinate the target.

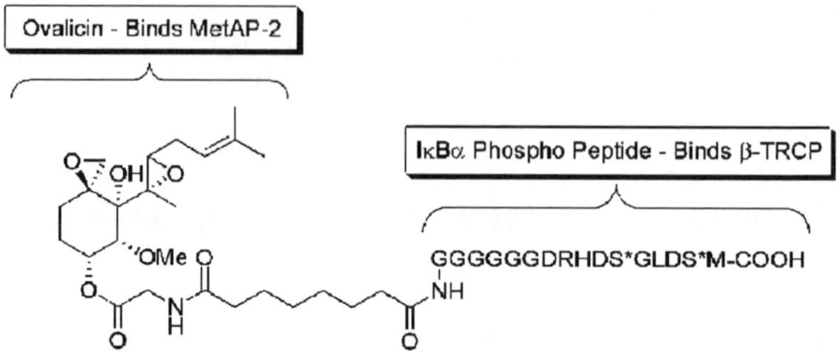

**Figure 3.4**  Structure of PROTAC-1. The small molecule ovalicin, which binds covalently to methionine aminopeptidase 2 (MetAP2), is connected by a linker to the IκBα phosphopeptide, which recruits the E3 ligase β-TRCP. S* indicates a phosphorylated serine residue.

Initial experiments showed by western blot that PROTAC-1 was, in fact, interacting specifically with the purified target protein MetAP2. Further *in vitro* experiments verified that PROTAC-1 can simultaneously bind to MetAP2 and recruit the remaining components of the SCF$^{\beta\text{-TRCP}}$ complex.[1] Having confirmed that both ends of the PROTAC undergo functional interactions, it was necessary to determine whether the E3 ligase was successfully ubiquitinating the target protein. The components of the SCF$^{\beta-\text{TRCP}}$ complex were immunoprecipitated and then supplemented with ATP, MetAP2 pre-bound to PROTAC-1, ubiquitin, and purified E1 and E2 enzymes. Western blot analysis showed multiple increased molecular weight bands corresponding to MetAP2, indicating the presence of ubiquitin chains of various length (Figure 3.5A).[1] In addition, this ubiquitination activity was successfully competed away by free IPP (though not with free OVA, as the MetAP-2-OVA interaction is covalent) (Figure 3.5B).[1] These data taken together suggest that PROTAC-1 is functioning as anticipated, bringing together the E3 ligase and the target protein, and promoting the specific ubiquitination of the target.

Finally, it was important to determine whether MetAP2 could be ubiquitinated in a PROTAC-1-dependent manner with the use of endogenous ubiquitin–proteasome pathway components at normal intracellular concentrations. The MetAP2-PROTAC-1 complex was pre-formed, and then added to a mixture of *Xenopus* egg extract. As shown in Figure 3.6, the MetAP2-PROTAC-1 complex (top band) but not MetAP2 alone (bottom band) was mostly degraded after 30 min. This degradation was attenuated in the presence of proteasome inhibitors LLnL and epoxomicin, proving it to be proteasome-mediated.[1]

These initial experiments were crucial in the development of the PROTAC technique, investigating each individual interaction in the three component E3 ligase–PROTAC–Target protein complex. In order for the PROTAC

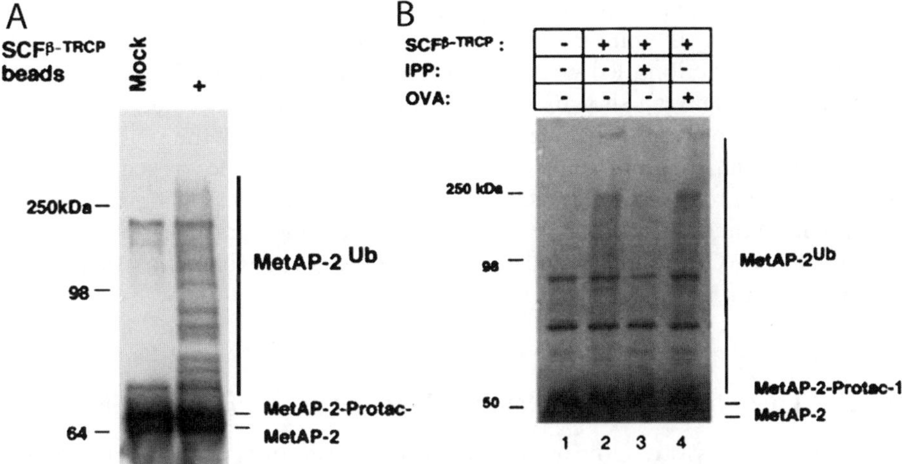

**Figure 3.5** PROTAC-1 mediates MetAP2 ubiquitination by SCF$^{\beta\text{-TRCP}}$. (A) Ubiquitination of full-length (67 kDa) MetAP2. MetAP2-PROTAC-1 mixture was added to either control (Mock) or SCF$^{\beta\text{-TRCP}}$ beads ($+$) supplemented with ATP plus purified E1, E2 and ubiquitin. Reactions were incubated for 1 h at 30 °C, and samples were analyzed by western blot with an anti-MetAP2 antibody. (B) PROTAC-1-dependent ubiquitination of MetAP2 is competitively inhibited by IPP. Conditions the same as A, except that reactions in lanes 3 and 4 were supplemented with 100 μM each IPP or OVA (respectively). Reprinted, with permission, from Sakamoto *et al.*[1] © 2001 National Academy of Sciences, U.S.A.

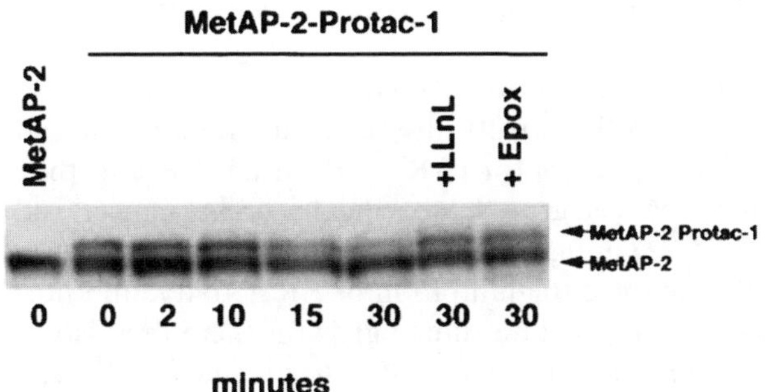

**Figure 3.6** The MetAP2-PROTAC-1 complex is degraded in *Xenopus* extracts. The MetAP2-PROTAC-1 mixture or MetAP2 alone was added to *Xenopus* egg extract. Reactions were incubated for the indicated times at room temperature, terminated by adding SDS-PAGE loading buffer, and evaluated by SDS-PAGE followed by western blotting with an anti-MetAP2 antibody. Where indicated, reactions were further supplemented with 50 μM LLnL or 10 μM epoxomicin (Epox.). Reprinted, with permission, from Sakamoto *et al.*[1] © 2001 National Academy of Sciences, U.S.A.

method to be applicable in an endogenous cellular environment, it was necessary to verify that the binding interactions were taking place as expected, and that these interactions led to the functional consequence of target protein ubiquitination. Having shown this to be the case, it was extremely exciting to see that in the presence of the endogenous components of the ubiquitin proteasome system, that not only would the target protein be tagged with ubiquitin, but that this did, in fact, lead to its degradation in a proteasome-dependent manner. These first experiments set the stage for further exploration of the PROTAC technique, and the next logical step, which was to test the system within a cell.

## 3.2.2   Targeting Tumor-relevant Proteins *In vitro* and *In vivo*

One of the greatest strengths of targeted post-translational protein degradation is the capacity to pursue potentially any protein in the cell. Proteins of particular interest include those that facilitate the development or progression of a particular disease state. Thus, the next step forward in the development of the PROTAC technique focused on the development of PROTACs targeting cancer-promoting proteins for degradation.

Following the initial development of the technique, there were some key questions that still needed to be addressed in order to validate its broad applicability. First, it was necessary to determine whether the method could, in fact, be used to target other proteins besides MetAP2 that might be of therapeutic interest. It was also critical to address the issue of covalent versus non-covalent interactions between the target protein and its ligand. The first generation PROTAC targeting MetAP2 utilized a known covalent ligand, but for the technique to be broadly applicable it must work for non-covalent ligands as well. And finally, it was necessary to assess the activity of the PROTAC within the context of a living cell.

These questions were all addressed with the second generation of PROTACs designed to target two therapeutically relevant proteins: the estrogen receptor (ER) and the androgen receptor (AR).[19] Both the ER and the AR have been shown to promote cancer growth, specifically breast cancer and prostate cancer, respectively.[20,21] Though there are pre-existing treatment mechanisms for breast cancer that involve the inhibition of ER activity, and hormonal therapy can be used to control prostate tumor growth, there are also many situations where these existing treatments show little to no efficacy. Therefore, new drugs that down-regulate AR and ER by novel mechanisms could have a significant impact on the treatment of both breast and prostate cancers. Though both the AR and the ER have previously been shown to be regulated through the ubiquitin–proteasome pathway, a PROTAC should simply increase the rate of turnover utilizing a secondary E3 ligase, leading to an overall decrease in cellular levels of these disease-enhancing proteins.[22,23]

The identification of ligands for targeting the AR and the ER was simplified due to the existence of natural high-affinity non-covalent ligands for both proteins. Dihydrotestosterone (DHT) binds the AR, and estradiol binds the ER. Thus, two new PROTACs were synthesized (**2** and **3**, Figure 3.7) consisting of either DHT or estradiol connected by a short linker to the IκBα-derived 10 amino acid peptide (phosphorylated on both serine residues) DRHDSGLDSM, which recruits the E3 ligase F-box protein β-TRCP.

The estradiol-based PROTAC, PROTAC-2 (**2**), was tested for its ability to mediate ubiquitination and proteasome-dependent degradation of ER *in vitro*. As described for PROTAC-1, it was determined that PROTAC-2 was capable of facilitating the ubiquitination of the ER *in vitro*, at concentrations as low as 0.1 μM. *In vitro* competition experiments in which PROTAC-2 was tested in the presence of excess amounts of either free IκBα phosphopeptide or free estradiol demonstrated that PROTAC activity is negated when binding to either target (the E3 ligase or the ER) is reduced, and that a physical link between the two

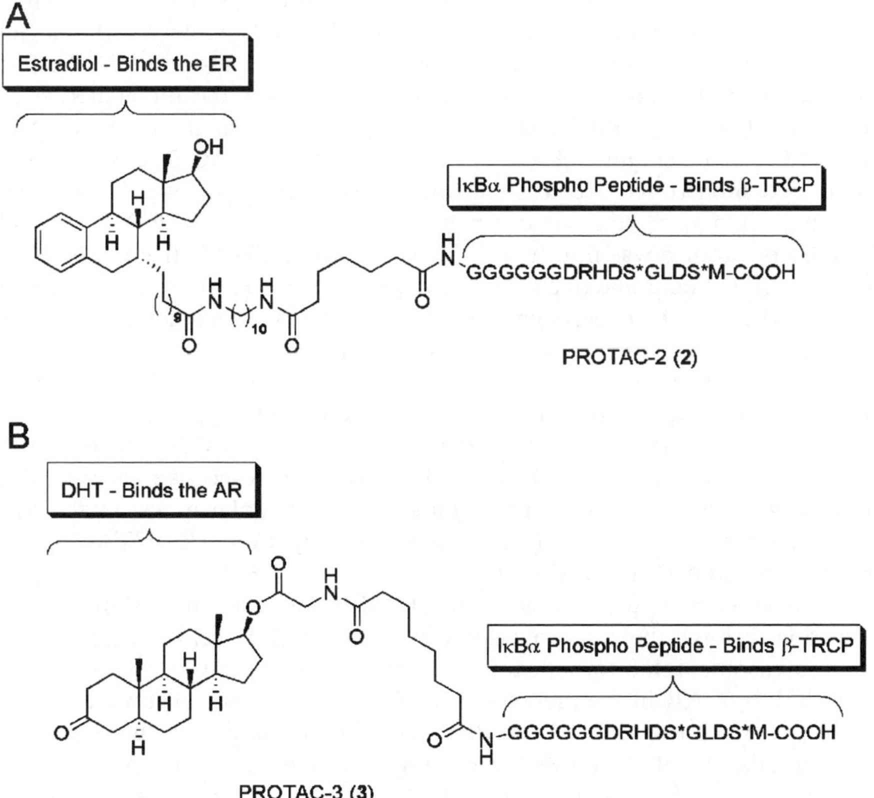



**Figure 3.7**   Structures of PROTAC-2 (**2**) and PROTAC-3 (**3**). Both PROTACs utilize the IκBα phosphopeptide as the E3 ligase-binding motif. S* indicates a phosphorylated serine residue. (A) Estradiol-containing PROTAC (**2**) degrades the estrogen receptor (ER). (B) DHT-containing PROTAC (**3**) degrades the androgen receptor (AR).

binding components is required for PROTAC activity. Finally, it was determined that PROTAC-2 was capable of promoting the degradation of the ER *in vitro* in the presence of purified yeast 26S proteasome.[19] Collectively, these experiments validate the hypothesized mechanism of the PROTAC, emphasizing the requirement of both independent binding events, as well as the necessity of the physical proximity between the E3 ligase and the target facilitated by the PROTAC.

The success of PROTAC-2 in promoting the ubiquitination and degradation of the ER *in vitro* demonstrates that a non-covalent interaction between the target protein and the PROTAC is sufficient for PROTAC activity. However, the important question of PROTAC activity within the context of a living cell still needed to be addressed. For this issue, the androgen receptor-targeting PROTAC (PROTAC-3, **3**) was chosen, due to the availability of a HEK293 cell line stably expressing a GFP-AR conjugate that would allow easy visual observation of PROTAC effects on AR levels in the cell. Due to the phosphate groups on the IκBα peptide the PROTAC was not cell permeable, and thus it was necessary to use microinjection to assess PROTAC activity within a cell.

For these experiments, PROTAC-3 was microinjected into 293[GFP-AR] cells to a final concentration of 1 μM. Initial experiments showed the anticipated loss of GFP-AR following PROTAC-3 treatment, with maximal degradation observed after 1 h (Figure 3.8A).[19] Observation of PROTAC-3 activity on over 200 cells demonstrated a consistent loss of GFP-AR, with over 70% of cells showing minimal, partial, or complete disappearance of GFP-AR after 1 h. This disappearance was not due to the leakage of GFP-AR from cells, as the presence of a rhodamine dye co-injected with the PROTAC was not affected. As with the *in vitro* experiments, it was found that the linkage of the two binding motifs was necessary for PROTAC activity. Co-injection of the free IκBα phosphopeptide and free testosterone (metabolized to DHT in the cell) at 10 μM each did not result in a decrease of the GFP-AR signal (Figure 3.8B).[19] Additionally, it was verified that PROTAC activity was dependent on the IκBα phosphopeptide and DHT binding to their respective targets. Cells co-injected with PROTAC-3 (10 μM) and a 10-fold molar excess (100 μM) of free phosphopeptide or testosterone showed a mitigation of PROTAC activity (Figure 3.8C and D, respectively).[19]

PROTAC-3-mediated degradation of GFP-AR was also shown to be occurring through the 26S proteasome as anticipated. In cells pre-treated with the proteasome inhibitor epoxomicin prior to PROTAC-3 treatment, the degradation of AR in the presence of PROTAC-3 was not observed.[19]

The success of PROTAC-2 and PROTAC-3 in promoting the ubiquitination and degradation of their respective targets both *in vitro* and in cell culture demonstrates the capacity of the PROTAC technique to target a wide variety of proteins, with both covalent and non-covalent ligands. The technique was also shown to function in both a selective and dose-dependent manner to promote ubiquitination and subsequent degradation *via* the 26S proteasome. This work was an important step in the continued exploration and validation of the PROTAC technique, showing it to be both widely applicable with regards to

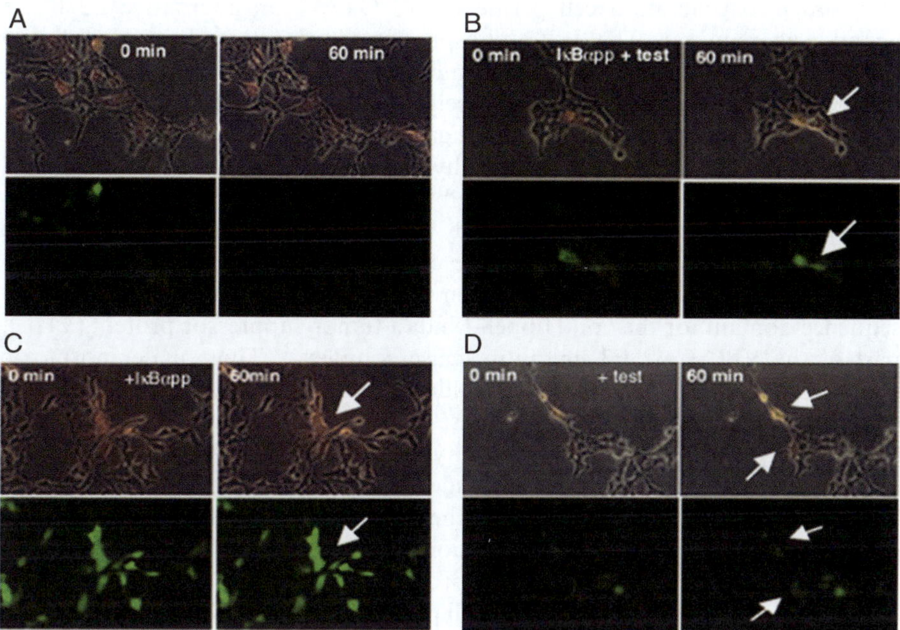

**Figure 3.8**  Microinjection of 10 μM PROTAC-3 leads to GFP-AR degradation in cells. PROTAC-3 was microinjected into 293$^{GFP-AR}$ cells in a solution containing rhodamine dextran (50 μg mL$^{-1}$) to a final concentration of 1 μM. (A) PROTAC-3 induces GFP-AR degradation within 60 min. The top panels show light microscopy images overlaid with images of cells injected with PROTAC-3 as indicated by rhodamine fluorescence (pink). The bottom panels show images of GFP fluorescence. (B) 293$^{GFP-AR}$ cells were microinjected with free IκBα phosphopeptide (IκBαpp) plus free testosterone (test). (C) 293$^{GFP-AR}$ cells were microinjected with PROTAC-3 (10 μM) plus 10-fold molar excess (100 μM) of IκBα phosphopeptide (IκBαpp). (D) 293$^{GFP-AR}$ cells were microinjected with PROTAC-3 (10 μM) plus 10-fold molar excess (100 μM) testosterone. Reprinted, with permission, from Sakamoto *et al.*[19] © The American Society for Biochemistry and Molecular Biology.

target protein selection, as well as functional in a therapeutically-relevant cellular environment.

## 3.2.3  Development of Cell Permeable PROTACs

The next logical step in the development of the PROTAC technique as a viable therapeutic option was the design of a PROTAC scaffolding that would be independently cell permeable, negating the need for microinjection as the method for PROTAC delivery. A cell permeable PROTAC would also lessen the molecular biological manipulations necessary for PROTAC activity, emphasizing the capacity for the technique as both a tool for understanding protein function in the cell and as a potential therapeutic.

In order to generate a cell permeable PROTAC, an alternative E3 ligase ligand was required. The IκBα phosphopeptide functioned successfully to recruit the SCF$^{\beta\text{-TRCP}}$ E3 ligase components, but the negatively charged phosphoserines effectively prevented cell permeability. In the absence of any known small molecule ligands for E3 ligases, a new peptide motif was chosen. Under normoxic conditions, a proline hydroxylase catalyzes the hydroxylation of the protein known as hypoxia inducible factor 1α (HIF1α) at P564 resulting in its recognition and ubiquitination by an E3 ligase.[24] The seven amino acid sequence ALAPYIP from HIF1α (the central proline in the sequence representing the hydroxylated P564) has been shown to be the minimum recognition sequence domain for the von Hippel–Lindau tumor suppressor protein (VHL), part of the VBC-Cul2 E3 ubiquitin ligase complex.[25] Thus, under normoxic conditions, the HIF1α peptide should be constitutively hydroxylated and capable of recruiting the VHL E3 ligase *in vivo*.

In order to ensure the cell permeability of the new PROTAC design, an eight amino acid polyarginine tag was included on the carboxy-terminus of the E3 ligase-targeting peptide sequence of the PROTAC. This significant positive charge serves to facilitate translocation of the PROTAC into cells *via* a mechanism that mimics the HIV Tat protein.[26]

To complete the design of the first cell permeable PROTAC, a protein target–ligand pair developed by ARIAD Pharmaceuticals was chosen. The protein target is a mutated version of the mammalian FK506 binding protein (FKBP12), which is part of a highly conserved family of isomerases involved in protein folding.[27] The mutant FKBP12 contains an F36V mutation, opening up a space in the tertiary structure of the protein where the phenylalanine would normally reside. This engineered "hole" in the protein is exploited with a hydrophobic "bump" on the artificial ligand AP21998 (**4**, Figure 3.9), which is incapable of binding to the wild-type protein.[28] Inclusion of AP21998 as the protein target-binding motif in the PROTAC allows the mutant (F36V)FKBP12 to be targeted orthogonally within the cell.

Thus, the completed design of the cell-permeable (F36V)FKBP12 (henceforth referred to as FKBP) PROTAC, PROTAC-4, includes the target protein ligand AP21998 connected by a short linker to the HIF1α peptide, followed by a polyarginine tag.[29] AP21998 was synthesized as previously described, and coupled to the peptide components to give PROTAC-4 (**5**, Figure 3.10) as the final product.[28,30] Though the AP21998 ligand was synthesized and carried through as a 1:1 mixture of diastereomers at C9, each diastereomer had previously been shown to bind to the target.[28,30] Based on this design, it was hypothesized that PROTAC-4 would enter the cell, be recognized and hydroxylated by a prolyl hydroxylase, and subsequently be bound by both the mutant FKBP target protein and the VHL E3 ligase, promoting ubiquitination and degradation of FKBP.

In order to observe the effects of PROTAC-4 on FKBP, a HeLa cell line stably expressing a GFP-tagged FKBP was generated. Thus, loss of the target protein could be visually assessed by diminished fluorescence. The GFP-FKBP cells were then treated with PROTAC-4, which was simply added to the cell

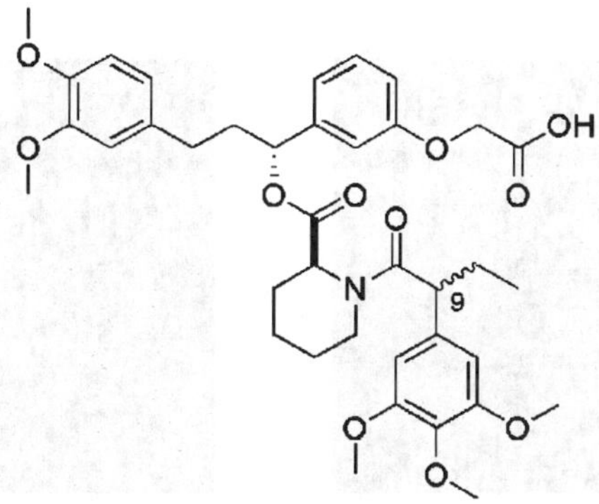

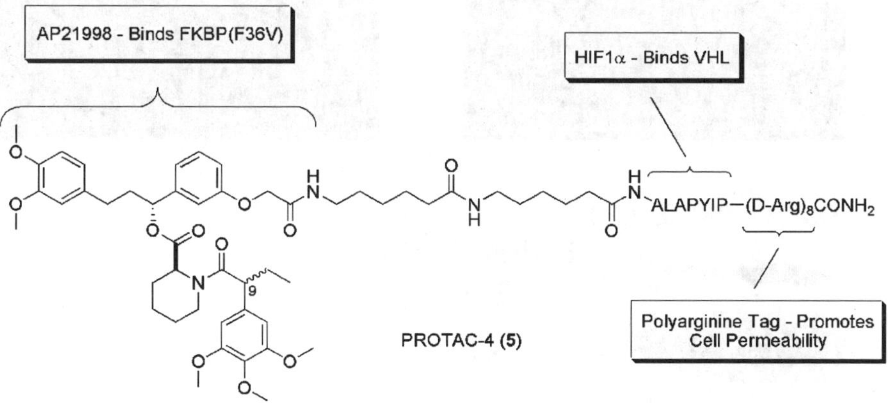



**Figure 3.9** AP21998 (**4**) was synthesized as previously described, as an approximately 1:1 mixture of diastereomers at C9.[28]



**Figure 3.10** Structure of PROTAC-4 (**5**). AP21998 binds the target protein FKBP(F36V), while the HIF1α peptide recruits the von Hippel–Lindau E3 ligase (VHL). A polyarginine tag provides cell permeability.

culture medium. Initial experiments showed that treatment with 25 μM PRO-TAC-4 for 2.5 h led to significant loss of FKBP, while treatment with the DMSO vehicle alone had no effect (Figure 3.11A–D).[29] Western blot analysis of EGFP-FKBP levels confirmed the visual observation of FKBP degradation resulting from PROTAC-4 treatment (Figure 3.11E).[29] Additionally, as with previous PROTACs, cells treated with the uncoupled PROTAC ligands (AP21998 and the HIF1α peptide–polyarginine tag) at 25 μM showed no loss of GFP-FKPB fluorescence.[29] This demonstrates that the two binding domains need to be physically linked for PROTAC efficacy.

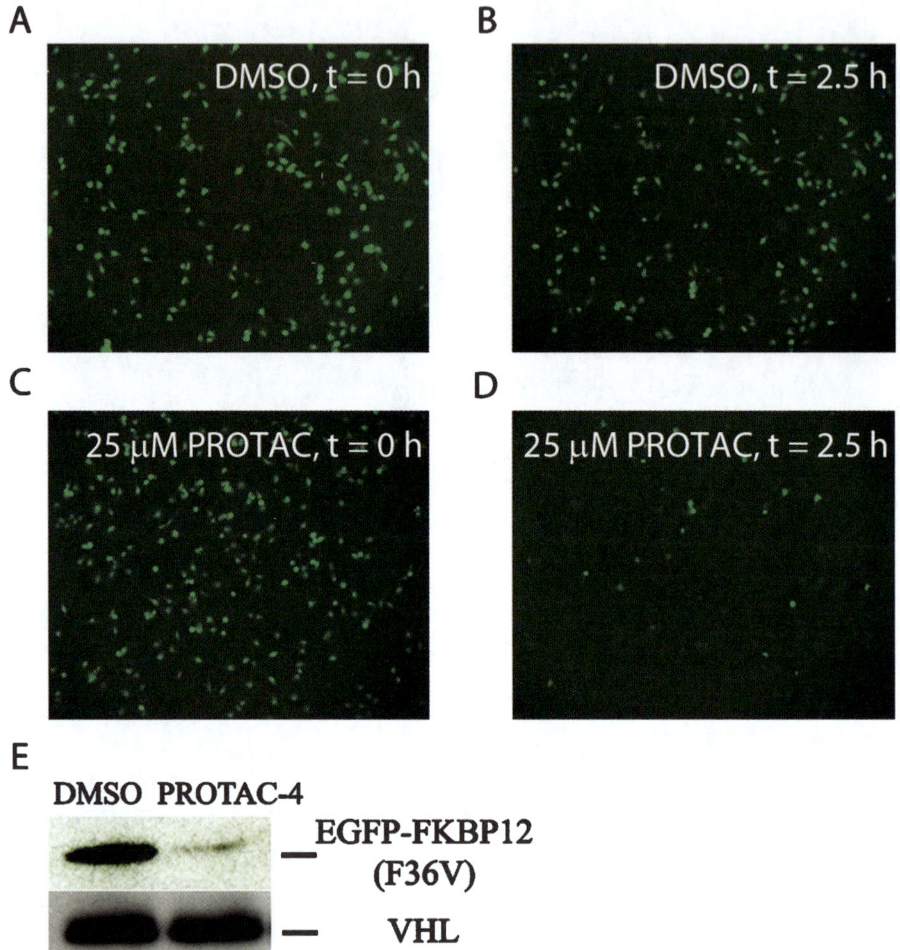



**Figure 3.11**  PROTAC-4 is cell-permeable and mediates EGFP-FKBP degradation. HeLa cells expressing EGFP-FKBP were treated externally with PROTAC-4 diluted in growth media. No change in fluorescence is observed before (A) and 2.5 h after (B) treatment in DMSO control, while a significant change is observed before (C) and 2.5 h after (D) treatment with 25 µM PROTAC-4. (E) Western blot analysis of EGFP-FKBP levels with an anti-GFP antibody confirms loss of EGFP-FKBP in cells treated with 25 µM PROTAC-4 for 2.5 h compared to the vehicle (DMSO) treated cells. Reprinted, with permission, from Schneekloth et al.[29] © 2004 American Chemical Society.

Having validated the design of the cell permeable PROTAC with PROTAC-4, the next step was to test the design on an endogenous target protein. The androgen receptor was once again an attractive target, due to the existence of a natural, high-affinity ligand: DHT. The role played by AR in the growth of

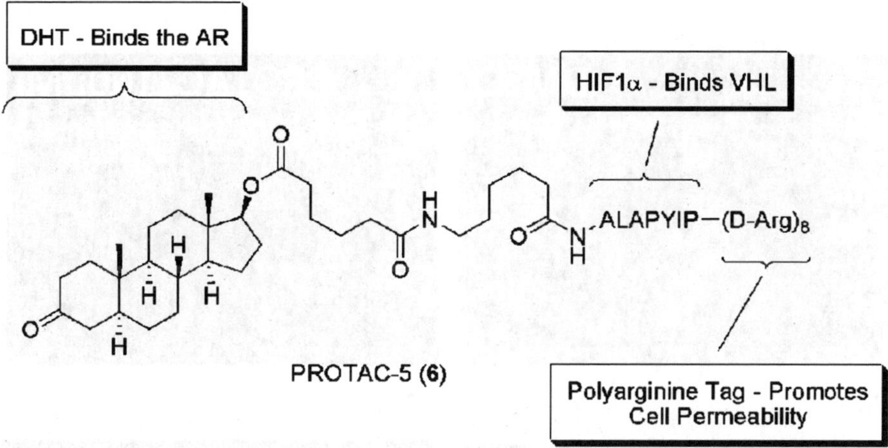


**Figure 3.12**  Structure of PROTAC-5 (**6**). DHT binds the target protein, the androgen receptor (AR), while the HIF1α peptide recruits the von Hippel–Lindau E3 ligase (VHL). A polyarginine tag provides cell permeability.

prostate tumor cells and the fact that inhibition of AR has previously been shown to repress that growth, suggest that a successful, cell-permeable AR-targeting PROTAC could potentially lead to a legitimate therapeutic candidate.[21]

PROTAC-5 (**6**), the cell-permeable AR-targeting PROTAC, is shown in Figure 3.12, and is composed of DHT connected by a linker to the HIF1α peptide with a C-terminal polyarginine tag.[29] As with PROTAC-3, a HEK293 cell line stably expressing a GFP-AR conjugate (293[GFP-AR]) was used to monitor the effects of PROTAC-5. The 293[GFP-AR] cells were treated with varying concentrations of PROTAC-5 in DMSO, or a DMSO vehicle control alone. Within 1 h, a significant decrease in the fluorescence signal was observed with 100, 50 and 25 µM PROTAC-5, while no change was seen with the DMSO control (Figure 3.13A–D).[29] In addition, cells pre-treated with the proteasome inhibitor epoxomicin followed by treatment with PROTAC-5, showed no observable degradation of the GFP-AR conjugate.[29] This suggests that pro-teasome-mediated degradation is behind the observed decrease in AR fluor-escence with PROTAC treatment. The PROTAC-5-mediated loss of AR as well as the inhibition of degradation in the presence of epoxomicin were both verified by western blot experiments in addition to the visual observation of fluorescence.[29]

Finally, additional experiments investigating the mechanism of PROTAC-5-mediated degradation were consistent with results from previous PROTACs. It was demonstrated that competing away PROTAC-5 with free testosterone or free HIF1α peptide inhibited PROTAC activity. In addition, it was verified that neither testosterone nor the HIF1α peptide alone could initiate AR degrada-tion, nor could the two free ligands induce degradation when added

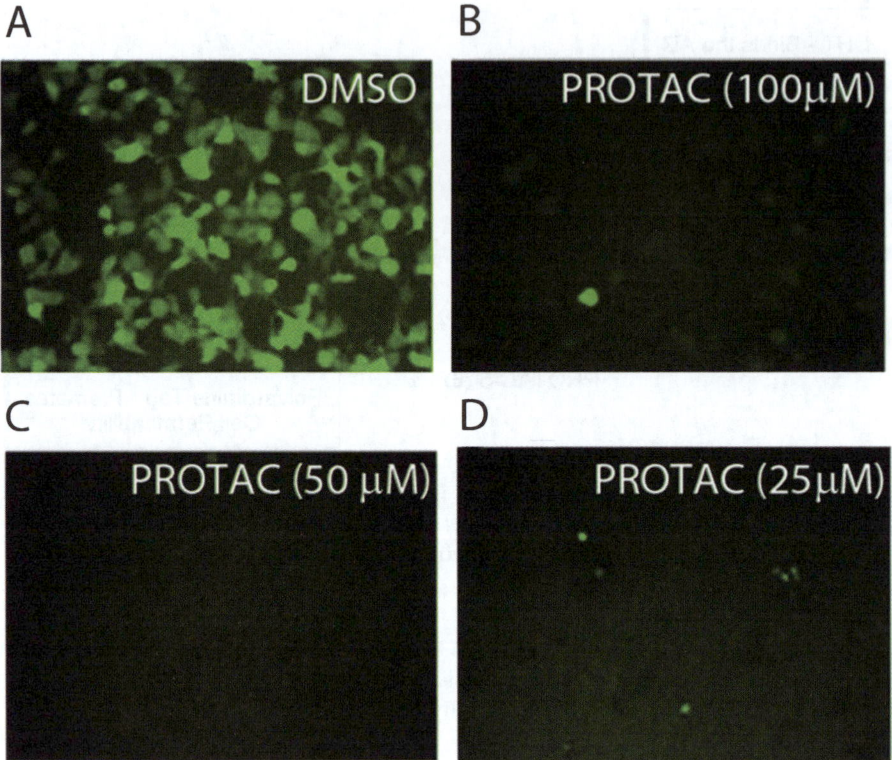


**Figure 3.13**   PROTAC-5 is cell-permeable and mediates GFP-AR degradation. One hour after treatment, 293$^{GFP-AR}$ cells treated with 100 µM (B), 50 µM (C) or 25 µM (D) PROTAC-5 show a significant loss of fluorescence compared to the DMSO-treated control cells (A). Reprinted, with permission, from Schneekloth *et al.*[29] © 2004 American Chemical Society.

simultaneously (but lacking a chemical linkage).[29] Once again, these data validate the proposed mechanism of PROTAC action, in which the PROTAC acts as a bridging molecule, binding both the target protein and the E3 ligase, and stimulating the ubiquitination of the target, leading to its degradation in the 26S proteasome.

Having once again proven the general efficacy of the PROTAC method through development of a cell-permeable PROTAC, the next step in the advancement of this technique must involve the continued improvement of the design of the molecule. If the ultimate goal is to shape the PROTAC technique into a viable therapeutic technology, it must be expanded in terms of both the variety of E3 ligases that can be utilized and the drug-like qualities of the PROTAC itself. Thus, the elimination of the peptide component of the PRO-TAC, and the identification of a small molecule ligand for a novel E3 ligase would be significant improvements.

## 3.2.4   Development of an All-small Molecule PROTAC

One of the primary drawbacks to the design of the PROTAC molecules thus far has been their high molecular weight and peptidic nature. These characteristics, particularly the presence of peptide bonds, complicate the further development of PROTACs as legitimate therapeutic leads. In order to improve the pharmacokinetics, cell permeability, and intracellular stability of the compounds, it was necessary to replace the peptidic E3 ligase ligands with a small-molecule-based ligand. This had not been attempted in the early stages of PROTAC development due to the absence of any known small-molecule ligands for E3 ligases. They do not have any natural small-molecule ligands, precluding the use of an evolutionarily designed binding partner, and this class of proteins has only recently begun to attract attention as a drug target, and thus to date there has only been a minor effort to discover small-molecule inhibitors of E3 ligases.[31–34] This is beginning to change, however, as the role of various E3s in a variety of disease states is constantly being elucidated.[35–38]

One of the first E3 ligases to be identified as a viable drug target is an E3 known as MDM2, which promotes the degradation of the tumor-suppressor protein p53.[39] In 2004, Roche developed a series of small molecule antagonists of MDM2 which were found to block the MDM2–p53 interaction, leading to the activation of the p53 pathway *in vivo*.[40] They called these molecules the Nutlins, and found that the most active analog, Nutlin-3 (**7**, Figure 3.14), bound to MDM2 with an IC$_{50}$ of 90 nM in enantiomerically pure form. In addition to their potential utility as anti-cancer drugs, the Nutlins provided the perfect opportunity to begin developing an all-small molecule PROTAC, due to the fact that they are high affinity ligands for MDM2, an E3 ligase.

By choosing the small molecule Nutlin as a PROTAC E3 ligase-targeting ligand, the technique should see improvement in a variety of areas. First, all cleavage-prone peptide bonds are eliminated from the PROTAC design, which

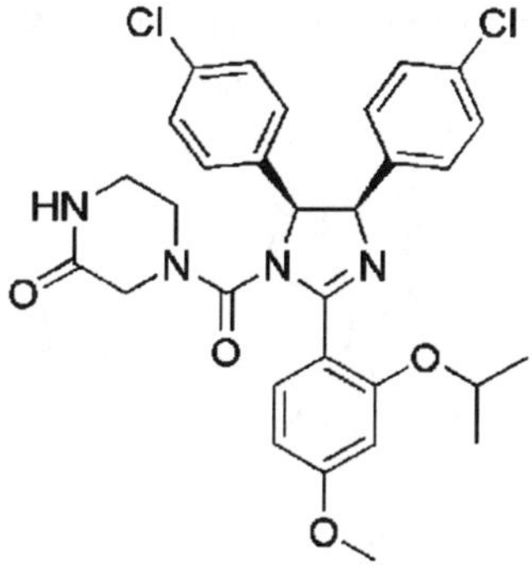

**Figure 3.14**   Structure of Nutlin-3 (**7**).

should greatly improve the half-life of the molecule within cells. Cell permeability should also be improved as a result of the drastically reduced molecular weight and polarity, allowing entrance to the cell by passive diffusion, and eliminating the need for the polyarginine tag. And finally, by targeting a new E3 ligase, the broad applicability of the technique is emphasized. The development of a variety of successful PROTACs utilizing an array of E3 ligase targets not only provides significant flexibility to the technique, but also lays the groundwork for the design of specifically tailored PROTAC molecules for potentially any application.

For the design of the first all-small molecule PROTAC, it was logical to change only one major variable: the E3 ligase. This would allow a more complete understanding of the capacity for this E3 ligase to ubiquitinate a non-natural substrate. Thus, the target protein chosen for this application was, once again, the androgen receptor, as this protein had been successfully targeted for degradation by the PROTAC method in the past. This time, however, a different AR ligand was utilized, as a result of the strongly agonistic effect of the previous AR ligand, DHT. The new ligand chosen for recruiting the target protein is known as the selective androgen receptor modulator (SARM), and acts as a partial agonist towards AR (**8**, Figure 3.15). The SARM ligand $(K_i = 4\,\text{nM})^{41}$ has a similarly high affinity for AR as compared to DHT

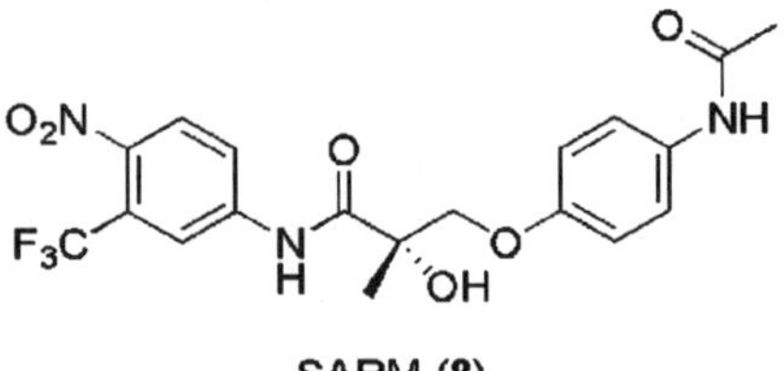

SARM (8)

**Figure 3.15**   Structure of the selective androgen receptor modulator (SARM) (**8**).

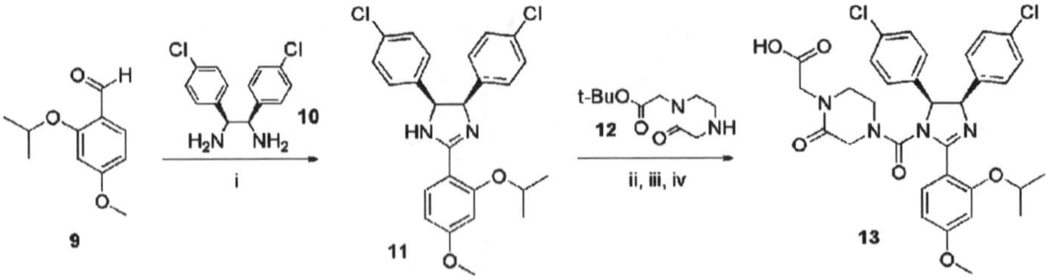

**Scheme 3.1**   Synthesis of Nutlin-3 derivative **13**. Reagents and conditions: (i) **10**, CH$_2$Cl$_2$, 0 °C, 2 h, then NBS, 0 °C, rt, 16 h, 88%; (ii) Triphosgene, Et$_3$N, THF, 0 °C, 2.5 h; (iii) **12**, CH$_2$Cl$_2$, 0 °C, 1.5 h, 96%; (iv) TFA, CH$_2$Cl$_2$, 96%.

($K_i = 0.27\,\text{nM}$),[41] but minimizes the upregulation of AR transcription induced by DHT that counteracts the PROTAC-mediated degradation.

In order to synthesize the all-small molecule PROTAC, it was advantageous to develop a more succinct method for the synthesis of Nutlin. A one-pot synthesis of a derivative of Nutlin-3 (**13**) was developed, as shown in Scheme 3.1.[42] Following the synthesis of **13**, it was verified by fluorescence polarization that this racemic derivative (modified for incorporation into a PROTAC) exhibited a similar ability to interact with MDM2 as the published Nutlin structures. It was observed that the $\text{IC}_{50}$ of **13** was $21.4\,\mu\text{M}$, which is very similar to the reported $\text{IC}_{50}$ for racemic Nutlin-3 of $13.6\,\mu\text{M}$.[40]

The structure of the SARM-Nutlin PROTAC (PROTAC-6) is shown in Figure 3.16.[42] In order to assess the activity of PROTAC-6, HeLa cells transiently transfected with AR were treated with varying concentrations of compound and assessed by western blot at multiple time points. As seen in Figure 3.17, treatment of AR-expressing cells with $10\,\mu\text{M}$ PROTAC-6 led to a significant decrease in AR concentrations after $7\,\text{h}$ compared to the vehicle control.[42] Additionally, pre-treatment with the proteasome inhibitor epoxomicin fully inhibited the PROTAC-mediated degradation (shown in Figure 3.17), proving once again that PROTAC activity is dependent on proteasome activity.[42] This is also important because it verifies that the use of a novel E3 ligase (MDM2) did not affect the mechanism of PROTAC action, thus suggesting that any E3 ligase could potentially be targeted by a PROTAC with great success.

The success of the all-small molecule PROTAC-6 was an important step forward in the continued development of the PROTAC technique. Not only did it show the potential for significant variations in the structural design of the PROTAC, but also the flexibility in the choice of an E3 ligase target. In addition, it showed the successful improvement of PROTAC cell-permeability

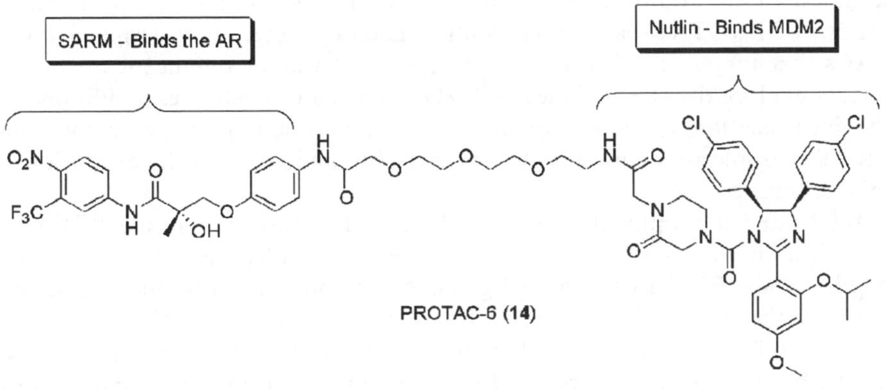

**Figure 3.16** Structure of PROTAC-6 (**14**). The SARM ligand binds the target protein, the androgen receptor (AR), while the Nutlin ligand binds the E3 ligase MDM2.

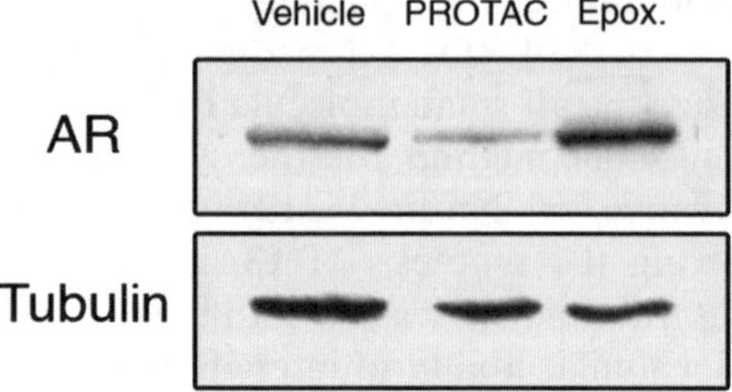


**Figure 3.17**   PROTAC-6 effectively degrades the androgen receptor (AR) *in vivo*. HeLa cells transiently expressing AR were treated for 7 h with either vehicle, 10 μM PROTAC-6, or pre-treated with 10 μM epoxomicin and then 10 μM PROTAC-6 for 7 h. The total level of AR was detected by western blotting using an anti-AR antibody (top). As a loading control, the same membrane was probed using an anti-α-tubulin antibody (bottom). Reprinted from Schneekloth *et al.*[42] © 2008 with permission from Elsevier.

through the removal of the peptide motif. By decreasing the overall molecular weight of the compound, it was found that PROTAC-6 was able to enter cells successfully by passive diffusion, with no need for either microinjection or the attachment of a polyarginine tag. While PROTAC-6 represents a significant improvement in PROTAC design, there are still many avenues left unexplored with regards to this technique, and hopefully the therapeutic potential of small molecule-mediated post-translational protein degradation will inspire others to pursue the many possible applications of the PROTAC technology.

## 3.2.5   Further Exploration of the PROTAC Technique

While significant work has been done to explore and expand the PROTAC technique itself, additional work has also been done to understand the implications of PROTAC-based therapeutics. Having successfully designed PRO-TACs that target proteins involved in cancer, it was of significant interest to further explore the effect of these PROTACs on cancer cell lines. Additionally, the identification of new protein targets and designing the corresponding PROTAC molecules continues to be incredibly important in the expansion of this technique.

While the development of the Nutlin-based all-small molecule PROTAC described in the previous section provided many obvious advantages to a peptide-based E3 ligase targeting ligand, it is still important to understand and optimize the peptide ligands. Providing a wide range of tools for targeting a variety of E3 ligases, each with different degradation kinetics, could prove invaluable in the design of specific PROTACs in the future. Thus, an important improvement to the PROTAC technique was the optimization of the size of the HIF1α-based peptide ligand used to recruit the E3 ligase VHL. This work was done by the Mohan and Kim laboratories, and involved the systematic shortening of the HIF1α octapeptide, with the intention of identifying a shorter sequence that might maintain the same functionality.[43] Interestingly, they observed that the pentapeptide sequence LAPYI (shortened from the original ALAPYIP, with the central P residue undergoing hydroxylation) seemed to produce a more robust PROTAC response than the octapeptide. They also found that the HIF1α peptide-based PROTACs do not require the presence of the polyarginine tag for cell permeability. The shortening of the VHL-targeting peptide sequence from eight residues down to five constitutes an important step forward in the optimization of this PROTAC variation.

Concurrent with their exploration of the minimum sequence requirement for VHL-based PROTAC activity, Mohan and Kim also investigated the anti-angiogenic effects of an estrogen receptor-targeting PROTAC. Angiogenesis involves the growth of new blood vessels from existing vasculature. It is a necessary and important function with respect to normal development and wound healing, but is also a crucial event in the progression and expansion of solid tumors. Thus, methods of inhibiting angiogenesis have the potential to be useful in the treatment of a variety of cancers. The estrogen receptor is thought to play a role in angiogenesis through a variety of mechanisms.[43] Therefore, a PROTAC stimulating the down-regulation of the ER could potentially function in an anti-angiogenic capacity. Mohan and Kim tested the activity of a PROTAC containing the natural ER ligand estradiol, connected by a linker to the HIF1α pentapeptide, LAPYI (estradiol-penta PROTAC (**15**), shown in Figure 3.18).[43] When tested in a three-dimensional endothelial cell sprouting assay (3D-ECSA), which can be used to screen angiogenesis inhibitors, the estradiol-penta PROTAC decreased endothelial cell sprouting by about 80% compared to vehicle-treated controls.[43] These data suggest that an ER-targeting PROTAC has the potential to be therapeutically useful as an inhibitor of angiogenesis, and thus be broadly applicable in the treatment of a variety of cancers.

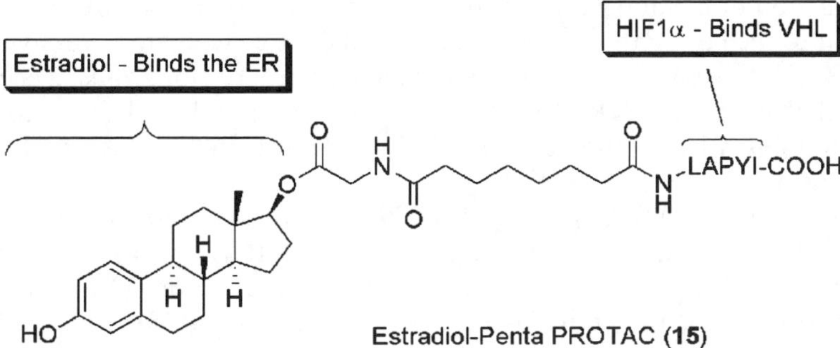


**Figure 3.18** Structure of estradiol-penta PROTAC (**15**). Estradiol binds the target protein, the estrogen receptor (ER), while the HIF1α pentapeptide recruits the von Hippel–Lindau E3 ligase (VHL).

Further work exploring the therapeutic utility of androgen receptor and estrogen receptor-targeting PROTACs in prostate and breast cancers, respectively, has also been pursued by a variety of groups.[44,45] In 2008, Sakamoto and co-workers published their work exploring the effects of an AR-targeting PROTAC on an androgen-dependent cancer cell line, and the effects of an ER-targeting PROTAC on an estrogen-dependent cancer cell line.[44] The lymph node carcinoma of prostate (LNCaP) cell line expresses AR and can only proliferate in medium that contains androgens. They compared the growth inhibitory activity of a DHT-HIF1α PROTAC (PROTAC-A (**16**), shown in Figure 3.19A) in LNCaP cells versus two cell lines that are androgen independent and do not express AR. The results were striking, showing clear inhibition of LNCaP cell growth in the presence of PROTAC-A (IC$_{50}$ at $72\,h = 12.5\,\mu M$), with no observable effect in the androgen independent cell lines.[44] Similarly, they compared the activity of an estradiol-HIF1α PROTAC (PROTAC-B (**17**), shown in Figure 3.19B) on two cell lines that are estrogen dependent (MCF-7 and T47D), as compared to an estrogen-independent cell line. As with PROTAC-A in the androgen-dependent cell lines, PROTAC-B acted specifically to inhibit the growth of the estrogen-dependent cell lines, while having no observable effect on the estrogen-independent cells (IC$_{50}$ at $72\,h$: MCF-7 $= 50\,\mu M$, T47D $= 16\,\mu M$).[44]

Interestingly, they found that although the polyarginine tag is not expressly required for PROTAC permeability, its addition significantly improves this property, resulting in improved PROTAC activity. PROTAC-AA (**18**, shown in Figure 3.19C) is otherwise identical to the AR-targeting PROTAC-A, but was found to be significantly more active than its polyarginine-free counterpart (IC$_{50}$ at $72\,h = 3.8\,\mu M$).[44]

A novel therapeutic application of the PROTAC technique was recently investigated by Kim and co-workers.[46,47] The aryl hydrocarbon receptor (AHR) is a ligand-activated transcription factor that facilitates many of the carcinogenic effects caused by environmental toxins, such as polycyclic and halogenated aromatic hydrocarbons. This process involves the binding of the hydrocarbon ligand to the AHR, which initiates a cascade of events including the induction of metabolizing enzymes and the formation of DNA adducts. In addition to mediating the carcinogenicity of these toxins, there is also evidence that the AHR facilitates tumor development and progression. Thus, a chemical probe that initiates AHR degradation would be useful both as a tool for understanding further the role of AHR, and potentially as an anti-cancer therapeutic.

In designing an AHR-targeting PROTAC, Kim and co-workers chose to exploit a known natural product ligand for the AHR known as apigenin. The apigenin–PROTAC (**19**) utilizes the HIF1α pentamer as the E3 ligase recognition motif, and is shown in Figure 3.20. The apigenin–PROTAC was found to promote significant degradation of the AHR by both western blot and immunofluorescence.[46] Further investigation into the effects on AHR activity showed that treatment with the apigenin–PROTAC also inhibited the agonist-induced transcription of AHR-controlled genes in both a time- and dose-dependent manner.[47] Additionally, the PROTAC was found to inhibit the formation of a protein–DNA complex crucial to AHR signaling.[47] The apigenin-PROTAC signifies an important advance in the study of the aryl hydrocarbon receptor, and will no doubt serve as an important tool in the further study of this protein. In addition, the development of yet another successful PROTAC–target protein pair emphasizes the fact that potentially

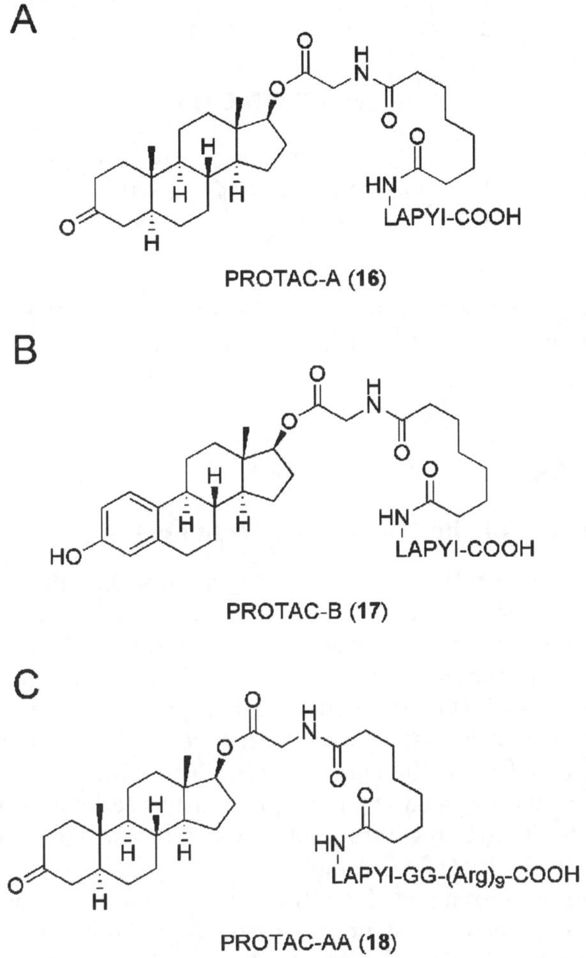

**Figure 3.19** (A) Structure of PROTAC-A (**16**). DHT binds the target protein, the androgen receptor (AR), while the HIF1α pentapeptide binds the E3 ligase VHL. (B) Structure of PROTAC-B (**17**). Estradiol binds the target protein, the estrogen receptor (ER), while the HIF1α pentapeptide binds the E3 ligase VHL. (C) Structure of PROTAC-AA (**18**). DHT binds the target protein, the androgen receptor (AR), while the HIF1α pentapeptide binds the E3 ligase VHL. A polyarginine tag provides added cell permeability.

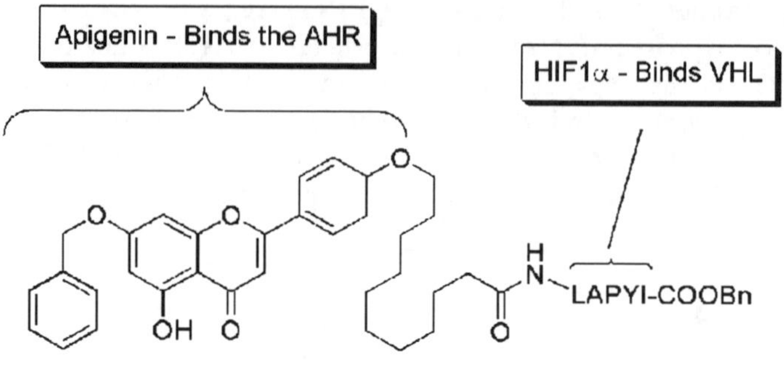

**Figure 3.20**  Structure of the apigenin–PROTAC (**19**). Apigenin binds the target protein, the aryl hydrocarbon receptor (AHR), while the HIF1α penta-peptide binds the E3 ligase VHL.

any protein for which we can identify a ligand of reasonable affinity can be successfully targeted by the PROTAC method.

## 3.3   Outlook

### 3.3.1   Strengths of the PROTAC Technique

A system for post-translational protein degradation has the potential to be invaluable in both the knowledge-driven study of the function and relevance of a particular protein in the cell, as well as in the treatment of a variety of disease states in which the presence or overexpression of a particular protein is a causative agent. The PROTAC method is the first technique involving post-translational protein degradation with the capacity to be useful in a laboratory environment as well as in the clinic. Alternative methods for targeted post-translational protein degradation have previously been developed, but these methods have limitations regarding both versatility and applicability that are not relevant with the PROTAC technique.[6–13]

One of the primary strengths of the PROTAC technique as compared to other known methods for post-translational protein degradation is the fact that no biochemical manipulations are required for activity. Other techniques require the creation of fusion proteins, which is not only time-consuming within the context of laboratory research, but also precludes their use in a clinical situation. While the proof-of-concept PROTAC experiments described in this chapter often utilized a GFP–target protein fusion, this was done simply for visualization purposes, and is not required for PROTAC activity. The fact that a PROTAC could potentially be exogenously administered to any cell type or organism and exert its full effect makes the technique a viable option for drug development.

Many of the specific characteristics of the PROTAC design make it well-suited to a variety of applications. For example, the PROTAC-mediated degradation of a target protein has been found to be dose-dependent. Thus, levels of the target protein in the cell can actually be tailored to meet the needs of the specific application. Whether the requirement is the complete removal of the protein or simply a slight decrease in cellular concentration, each PROTAC can be tested to determine the proper concentrations necessary to achieve the desired result. This is in contrast to all-or-nothing methods such as the genetic knockout, in which there can be no gradient of protein levels in the cell.

In addition to the dose-dependent nature of the PROTACs method, there is also the potential for significant kinetic versatility as well. Though it has not yet been fully explored, the choice of E3 ligase in the design of the PROTAC should have a strong influence on the kinetics of degradation. Even amongst the small selection of E3 ligases utilized in PROTACs thus far, there has been significant variation in the time required for degradation. Thus, as additional E3 ligases continue to be explored and targeted with this technique, a vast repertoire of degradation kinetics should become available, and could be utilized based on the requirement for each individual application.

Another significant strength of the PROTAC method, particularly with regard to drug development, is the fact that it can be used to target proteins without the identification of an active site inhibitor, or otherwise "functional" ligand. Current drug development focuses on a small portion of possible protein targets, such as enzymes, based on the availability of an active site that can be targeted. And while the PROTAC method can certainly utilize a ligand that also functions as an active site inhibitor to recruit the target protein, it is not necessary for the ligand to have such an activity. Any ligand that binds the target protein on any unique surface could potentially be exploited for use in a PROTAC. This greatly increases the utility of various types of compound libraries, as any compound that shows binding activity to the target protein could be optimized for use in a PROTAC.

Finally, one of the greatest strengths of the PROTAC technique is simply the diversity of potential applications for which it could be useful. As a tool for molecular biology, PROTACs could prove invaluable. The chemically induced down-regulation or elimination of a particular protein in the cell provides significant insight into the details and importance of the roles played by the protein, and aids in the quest to understand the interconnectivity of functions and components within the cell. Though protein knockdown can also be produced using the RNAi technique, there are many specific applications in which the post-translational nature of the PROTAC would make it a far superior option, some of which will be discussed in the next section. In addition to its utility as a tool for elucidating the many mysteries of the cell, the PROTAC method has obvious and significant implications in the field of drug discovery. The modular design of the PROTAC guarantees that given the identification of the proper ligands, potentially any protein in the cell can be targeted for degradation. And though ligand identification is rarely a straight-forward process, the likelihood of identifying a "binding" ligand versus identifying an "active" ligand suggests that it should be a more surmountable task than the current methods utilized in drug development.

## 3.3.2 Potential Applications of the PROTAC Technique

Due to the fact that many diseases, including cancers, are dependent on the presence or over-expression of a small number of proteins, PROTACs designed to target these proteins could be of potential use in the clinic. Some of the most exciting applications of the PROTAC technique involve the exploitation of its many unique features in the development of PROTAC-based therapeutics.

One exciting possibility is the development of therapeutic PROTAC molecules that would be active only in a particular tissue of interest. The identification of a ligand for an E3 ligase that is expressed in a tissue-specific manner would enable highly specific PROTAC-based therapeutics to be developed, minimizing potential off-target effects of the drug.

Another exciting application that is only possible with a post-translational protein-targeting technique is the development of a conformationally specific PROTAC. Many therapeutically relevant proteins exert their disease-causing effects as a result of an alteration to, or prolongation of, a particular conformational state. For example, the small GTPase Ras alternates between an active (GTP-bound) and inactive (GDP-bound) conformation.[48–50] In quiescent adult cells, Ras signaling would normally be turned off. However, in up to 30% of all cancers, a point mutation in Ras prevents it from returning to the inactive, GDP-bound state.[51] If a ligand could be identified that would bind selectively to the active, GTP-bound conformation of Ras, it could be incorporated into a PROTAC that would target only "Ras active" cancer cells, and not the healthy quiescent cells in which Ras signaling is turned off. A conformationally specific PROTAC could also potentially find application in the targeting of infectious prion proteins, which differ from their normal counterparts in three-dimensional conformation only.[52] The ability to specifically target only the diseased form of an abundant cellular protein could have significant clinical advantages, limiting dangerous or unpleasant side-effects that might result from eliminating the normal form of the protein as well. The capacity to design a conformationally specific PROTAC sets this method apart from other techniques that act at the DNA or RNA level.

Another potential application of the PROTAC technique exploits its unique mechanism of action. As mentioned previously, one advantage of the PROTAC method is the fact that it is not necessary to identify an "active" ligand for a target protein. This characteristic of the technique will allow it to be utilized to target the "un-druggable" proteome. Many types of important and disease-relevant proteins are currently considered "un-druggable" due to the difficulty of designing screens that would allow for the identification of drug leads. Protein classes such as transcription factors and scaffolding proteins often play critical roles in the development of a disease, particularly cancer. However, because they do not function as enzymes that can be inhibited, it can be difficult to ascertain how best to target them with pharmaceutical compounds. Thus, these types of proteins would be ideal targets for the PROTAC method. Once a particular protein has been identified as being crucial to the progression of a disease state, it could be screened against large libraries of compounds by simple fluorescence polarization in order to identify lead compounds, since it does not matter where or how the compound might bind to the protein.

Finally, one could envision the creation of a large PROTAC library which could be used to identify new therapeutically relevant protein targets. This chemical genetic strategy would utilize a library of PROTAC molecules with identical E3 ligase ligands attached to a wide variety of chemically diverse target ligands. This PROTAC library could then be screened against cells for a loss-of-function phenotype resulting from selective protein degradation. Following PROTAC library incubation with cells and the detection of the desired cellular phenotype (*e.g.* inhibition of pro-inflammatory signaling), the next step would be the identification of the specific protein target of the active PROTAC. This could easily be achieved with a simple technique such as affinity chromatography. Additionally, the PROTAC library could also be used to screen for novel ligands for a specific protein target. This could be done by monitoring degradation of the target protein (*e.g.* loss of a GFP fusion protein), and then identifying the specific PROTAC binding domain that elicited the effect.

### 3.3.3    Continued Development of the PROTAC Technique

In addition to all of the exciting potential applications described above, there are some basic areas of PROTAC design that will require further exploration. For example, more work is needed to understand the role that linker length plays in PROTAC activity. It is reasonable to assume that the appropriate linker length will depend greatly on the specific E3 ligase, as well as the depth of ligand binding within the three-dimensional structure of both the E3 ligase and the target protein. Thus far this issue has not really been explored, and a variety of different linker lengths have been employed with success. However, a detailed understanding of the relationship between linker length and PROTAC activity will be necessary in order to fully optimize the capabilities of the technique.

The biggest issue currently impeding the broad use of the PROTAC method is the limited availability of small molecule E3 ligase ligands. Now that E3 ligases have been identified as valid targets for drug development, it is likely that more small molecule ligands for specific E3s (such as the Nutlin ligands for MDM2) will be regularly identified. But as with the target protein-recruiting ligand of the PROTAC, it is not necessary for the E3 ligase ligand to have a particular "activity," such as blocking the binding of the E3's natural substrate. Thus, those who wish to pursue the application of the PROTAC technique would be well-served to pursue independent screening of compound libraries in the search for E3 ligase ligands to add to the PROTAC repertoire.

# Acknowledgement

CMC acknowledges financial support from the NIH (R33CA118631).

# References

1. K. M. Sakamoto, K. B. Kim, A. Kumagai, F. Mercurio, C. M. Crews and R. J. Deshaies, *Proc. Natl. Acad. Sci. U. S. A.*, 2001, **98**, 8554.
2. S. Vijay-Kumar, C. E. Bugg, K. D. Wilkinson, R. D. Vierstra, P. M. Hatfield and W. J. Cook, *J. Biol. Chem.*, 1987, **262**, 6396.
3. H. C. Ardley and P. A. Robinson, *Essays Biochem.*, 2005, **41**, 15.
4. J. Myung, K. B. Kim and C. M. Crews, *Med. Res. Rev.*, 2001, **21**, 245.
5. B. Dahlmann, *Essays Biochem.*, 2005, **41**, 31.
6. P. Zhou, R. Bogacki, L. McReynolds and P. M. Howley, *Mol. Cell*, 2000, **6**, 751.
7. M. Scheffner, K. Munger, J. M. Huibregtse and P. M. Howley, *EMBO J.*, 1992, **11**, 2425.
8. J. S. Grimley, D. A. Chen, L. A. Banaszynski and T. J. Wandless, *J. Bioorg. Med. Chem. Lett.*, 2008, **18**, 759.
9. B. W. Chu, L. A. Banaszynski, L. C. Chen and T. J. Wandless, *J. Bioorg. Med. Chem. Lett.*, 2008, **18**, 5941.
10. L. A. Maynard-Smith, L. C. Chen, L. A. Banaszynski, A. G. Ooi and T. J. Wandless, *J. Biol. Chem.*, 2007, **282**, 24866.
11. K. Stankunas, J. H. Bayle, J. J. Havranek, T. J. Wandless, D. Baker, G. Crabtree and J. E. Gestwicki, *ChemBioChem*, 2007, **8**, 1162.
12. L. A. Banaszynski, L. C. Chen, L. A. Maynard-Smith, A. G. Ooi and T. J. Wandless, *J. Cell*, 2006, **126**, 995.
13. R. J. Dohmen, P. Wu and A. Varshavsky, *Science*, 1994, **263**, 1273.
14. D. M. Spencer, T. J. Wandless, S. L. Schreiber and G. R. Crabtree, *Science*, 1993, **262**, 1019.
15. P. J. Belshaw, S. N. Ho, G. R. Crabtree and S. L. Schreiber, *Proc. Natl. Acad. Sci. U. S. A.*, 1996, **93**, 4604.
16. E. T. Kipreos and M. Pagano, *Genome Biol.*, 2000, **1**, Reviews 3002.
17. A. Yaron, H. Gonen, I. Alkalay, A. Hatzubai, S. Jung, S. Beyth, F. Mercurio, A. M. Manning, A. Ciechanover and Y. Ben-Neriah, *EMBO J.*, 1997, **16**, 6486.
18. E. C. Griffith, Z. Su, B. E. Turk, S. Chen, Y. H. Chang, Z. Wu, K. Biemann and J. O. Liu, *Chem. Biol.*, 1997, **4**, 461.
19. K. M. Sakamoto, K. B. Kim, R. Verma, A. Ransick, B. Stein, C. M. Crews and R. J. Deshaies, *Mol. Cell Proteomics*, 2003, **2**, 1350.
20. A. Howell, S. J. Howell and D. G. Evans, *Cancer Chemother. Pharmacol.*, 2003, **52**(Suppl 1), S39.
21. J. D. Debes, L. J. Schmidt, H. Huang and D. J. Tindall, *J. Cancer Res.*, 2002, **62**, 5632.
22. C. P. Cardozo, C. Michaud, M. C. Ost, A. E. Fliss, E. Yang, C. Patterson, S. J. Hall and A. J. Caplan, *Arch. Biochem. Biophys.*, 2003, **410**, 134.

23. Z. Nawaz, D. M. Lonard, A. P. Dennis, C. L. Smith and B. W. O'Malley, *Proc. Natl. Acad. Sci. U. S. A.*, 1999, **96**, 1858.
24. A. C. Epstein, J. M. Gleadle, L. A. McNeill, K. S. Hewitson, J. O'Rourke, D. R. Mole, M. Mukherji, E. Metzen, M. I. Wilson, A. Dhanda, Y. M. Tian, N. Masson, D. L. Hamilton, P. Jaakkola, R. Barstead, J. Hodgkin, P. H. Maxwell, C. W. Pugh, C. J. Schofield and P. J. Ratcliffe, *J. Cell*, 2001, **107**, 43.
25. W. C. Hon, M. I. Wilson, K. Harlos, T. D. Claridge, C. J. Schofield, C. W. Pugh, P. H. Maxwell, P. J. Ratcliffe, D. I. Stuart and E. Y. Jones, *Nature*, 2002, **417**, 975.
26. S. Fawell, J. Seery, Y. Daikh, C. Moore, L. L. Chen, B. Pepinsky and J. Barsoum, *Proc. Natl. Acad. Sci. U. S. A.*, 1994, **91**, 664.
27. A. Breiman and I. Camus, *Transgenic Res.*, 2002, **11**, 321.
28. W. Yang, L. W. Rozamus, S. Narula, C. T. Rollins, R. Yuan, L. J. Andrade, M. K. Ram, T. B. Phillips, M. R. van Schravendijk, D. Dalgarno, T. Clackson and D. A. Holt, *J. Med. Chem.*, 2000, **43**, 1135.
29. J. S. Schneekloth Jr., F. N. Fonseca, M. Koldobskiy, A. Mandal, R. Deshaies, K. Sakamoto and C. M. Crews, *J. Am. Chem. Soc.*, 2004, **126**, 3748.
30. C. T. Rollins, V. M. Rivera, D. N. Woolfson, T. Keenan, M. Hatada, S. E. Adams, L. J. Andrade, D. Yaeger, M. R. van Schravendijk, D. A. Holt, M. Gilman and T. Clackson, *Proc. Natl. Acad. Sci. U. S. A.*, 2000, **97**, 7096.
31. Y. Yang, R. L. Ludwig, J. P. Jensen, S. A. Pierre, M. V. Medaglia, I. V. Davydov, Y. J. Safiran, P. Oberoi, J. H. Kenten, A. C. Phillips, A. M. Weissman and K. H. Vousden, *Cancer Cell*, 2005, **7**, 547.
32. J. Kitagaki, K. K. Agama, Y. Pommier, Y. Yang and A. M. Weissman, *Mol. Cancer Ther.*, 2008, **7**, 2445.
33. J. Huang, J. Sheung, G. Dong, C. Coquilla, S. Daniel-Issakani and D. G. Payan, *Methods Enzymol.*, 2005, **399**, 740.
34. M. J. Lee, K. Pal, T. Tasaki, S. Roy, Y. Jiang, J. Y. An, R. Banerjee and Y. T. Kwon, *Proc. Natl. Acad. Sci. U. S. A.*, 2008, **105**, 100.
35. Y. Sun, *Cancer Biol. Ther.*, 2003, **2**, 623.
36. A. L. Talis, J. M. Huibregtse and P. M. Howley, *J. Biol. Chem.*, 1998, **273**, 6439.
37. F. Bernassola, M. Karin, A. Ciechanover and G. Melino, *Cancer Cell*, 2008, **14**, 10.
38. P. Guedat and F. Colland, *BMC Biochem.*, 2007, **8**(Suppl 1), S14.
39. J. Piette, H. Neel and V. Marechal, *Oncogene*, 1997, **15**, 1001.
40. L. T. Vassilev, B. T. Vu, B. Graves, D. Carvajal, F. Podlaski, Z. Filipovic, N. Kong, U. Kammlott, C. Lukacs, C. Klein, N. Fotouhi and E. A. Liu, *Science*, 2004, **303**, 844.
41. C. A. Marhefka, W. Gao, K. Chung, J. Kim, Y. He, D. Yin, C. Bohl, J. T. Dalton and D. D. Miller, *J. Med. Chem.*, 2004, **47**, 993.
42. A. R. Schneekloth, M. Pucheault, H. S. Tae and C. M. Crews, *Bioorg. Med. Chem. Lett.*, 2008, **18**, 5904.
43. P. Bargagna-Mohan, S. H. Baek, H. Lee, K. Kim and R. Mohan, *Bioorg. Med. Chem. Lett.*, 2005, **15**, 2724.

44. A. Rodriguez-Gonzalez, K. Cyrus, M. Salcius, K. Kim, C. M. Crews, R. J. Deshaies and K. M. Sakamoto, *Oncogene*, 2008, **27**, 7201.
45. Y. Q. Tang, B. M. Han, X. Q. Yao, Y. Hong, Y. Wang, F. J. Zhao, S. Q. Yu, X. W. Sun and S. J. Xia, *Asian J. Androl.*, 2009, **11**, 119.
46. H. Lee, D. Puppala, E. Y. Choi, H. Swanson and K. B. Kim, *ChemBioChem*, 2007, **8**, 2058.
47. D. Puppala, H. Lee, K. B. Kim and H. I. Swanson, *Mol. Pharmacol.*, 2008, **73**, 1064.
48. S. L. Campbell, R. Khosravi-Far, K. L. Rossman, G. J. Clark and C. J. Der, *Oncogene*, 1998, **17**, 1395.
49. P. M. Campbell and C. J. Der, *Semin. Cancer Biol.*, 2004, **14**, 105.
50. J. F. Hancock, *Nat. Rev. Mol. Cell Biol.*, 2003, **4**, 373.
51. K. Giehl, *Biol. Chem.*, 2005, **386**, 193.
52. D. Peretz, R. A. Williamson, Y. Matsunaga, H. Serban, C. Pinilla, R. B. Bastidas, R. Rozenshteyn, T. L. James, R. A. Houghten, F. E. Cohen, S. B. Prusiner and D. R. Burton, *J. Mol. Biol.*, 1997, **273**, 614.

# Chemical Biology of Stem Cell Modulation

STEPHEN G. DAVIES[a] AND ANGELA J. RUSSELL[a, b,]*

[a] Department of Chemistry, Chemistry Research Laboratory, University of Oxford, Mansfield Road, Oxford OX1 3TA, UK; [b] Department of Pharmacology, University of Oxford, Mansfield Road, Oxford OX1 3QT, UK

## 4.1  Introduction

Harnessing the potential of stem cells is one of the next major challenges in medicine and healthcare. Stem cells may be derived from an embryo, fetus, umbilical cord blood, or adult tissue, and have recently been accessed from reprogramming of adult somatic or lineage-committed cells.[1,2] They exhibit a number of remarkable qualities: they may be proliferated under appropriate conditions, whilst retaining the ability to differentiate *via* a number of steps to adult tissues.[3] These unique properties represent a significant potential to revolutionise medicine by offering treatment options applicable to a wide range of degenerative diseases such as neuromuscular diseases, bone marrow disorders, diabetes, heart disease, and vision and hearing loss.[4] Options for regenerative therapies range from cell-replacement therapy and tissue-engineering to the *in vivo* activation of endogenous stem cell populations (Figure 4.1). Chemistry can impact on all of these treatment strategies.

The ability to isolate, expand and manipulate stem cells also represents a major enabling tool for drug discovery. They provide model systems to study disease pathology, dissect and validate new drug targets, and a rapid means for

RSC Drug Discovery Series No. 5
New Frontiers in Chemical Biology: Enabling Drug Discovery
Edited by Mark E. Bunnage

Published by the Royal Society of Chemistry, www.rsc.org

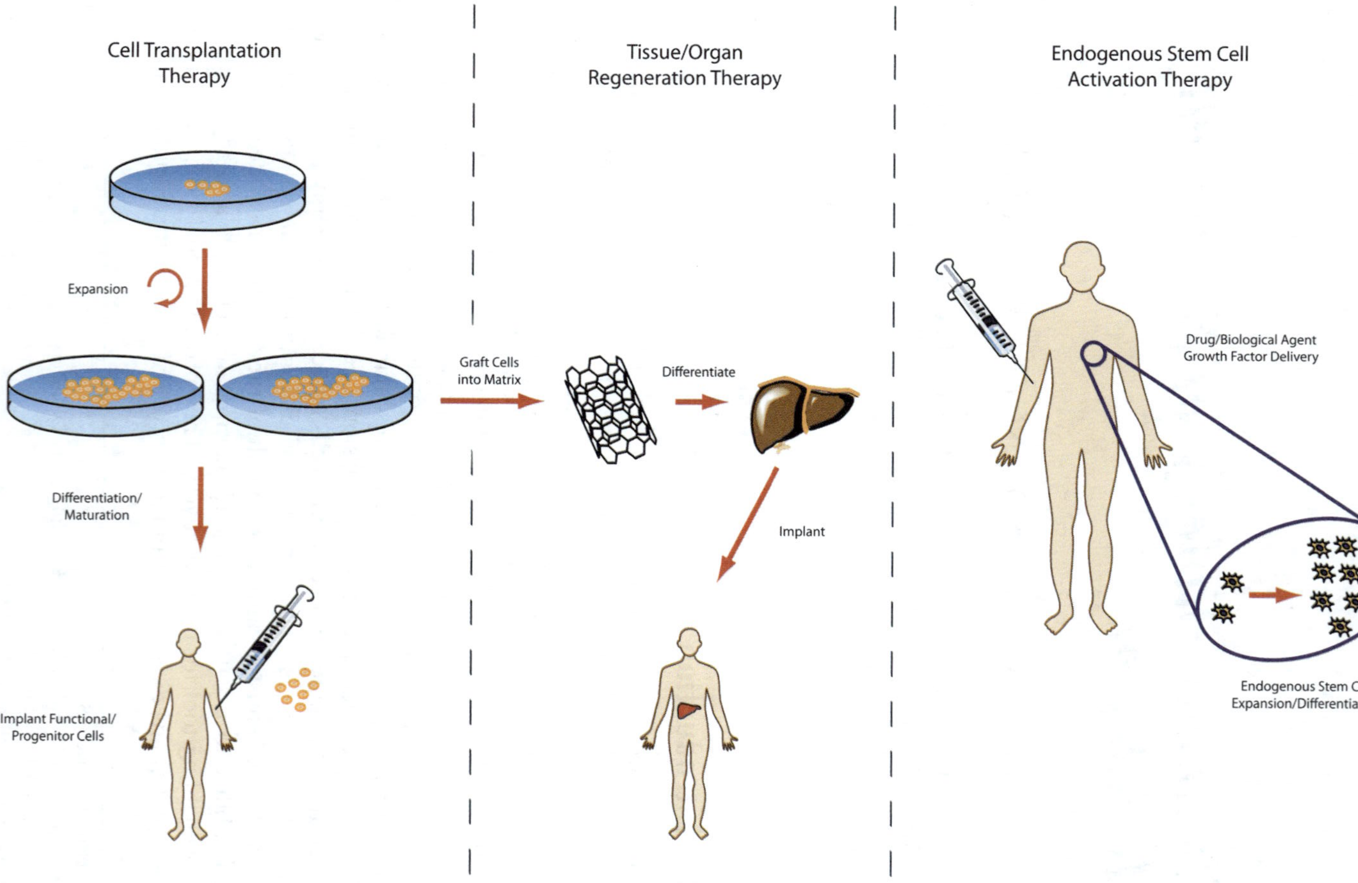


**Figure 4.1**  The role of chemistry in stem cell manipulation and regenerative therapies.

drug efficacy and toxicology testing, which may be established as an alternative to animal testing.[5,6] With the means to generate patient-specific stem cells through cellular reprogramming technology, these applications together have the potential to provide the ultimate personalised medicine platform.

The transformation of a stem cell to a fully differentiated cell within adult tissue occurs *via* a number of steps. The ability to control each step in the differentiation process and to manipulate exquisitely a cell's lineage commitment *in vitro* and *in vivo* would allow the selective production of different cell and tissue types and would represent a major advance in the treatment of disease. The application of chemistry to control the destiny of stem cells is a rapidly emerging discipline.[7] The effects of small molecules to modulate biological function are rapid, and often reversible, providing an unparalleled means to control dose and timing of delivery, both critical parameters in the determination of cell fate.

This chapter will review the use of small molecules to elucidate and control stem cell development and also their impact on drug discovery and regenerative medicine. The use of biological agents will be referred to, especially as they relate to the design of small molecule modulators of stem cell fate, although a comprehensive survey is beyond the scope of this chapter. Similarly, the use of polymers and scaffolds, while an important area of stem cell and tissue engineering will not be reviewed.[8] An overview of stem cells will be provided, followed by strategies and representative examples of small molecules to enable stem cell expansion, directed differentiation, dedifferentiation or transdifferentiation. Applications of stem cell manipulation *in vitro* will be discussed first, then the identification of small molecules acting *in vivo* to activate endogenous stem cells and promote tissue repair mechanisms.

## 4.2 Stem Cells: Ontogeny, Characterisation and Clinical Utility

The term stem cell refers to a broad collection of diverse cell types with a number of common characteristics: they may be defined as any cell capable of both self-renewal (producing at least one copy of themselves upon cell division) and differentiating into one or more fully differentiated adult cell types. Stem cells may be derived from a number of sources including embryos or fetal tissue, from umbilical cord or placenta, or from adult somatic tissue or gonadal tissue.[3] The range of cell types or lineages any given stem cell may differentiate into is termed its potency. Pluripotent stem cells, which include those derived from embryos, are capable of giving rise to cells representative of all embryonic and adult tissue, including the germline. Pluripotent cells differentiate through more specialised, lineage-committed cell types (termed multipotent) which, while still retaining the ability to self-renew, differentiate to a restricted set of cell types. These multipotent stem cells in turn can give rise to progressively more lineage-restricted cell types to ultimately produce fully differentiated, mature somatic cell types (Figure 4.2). Dedifferentiation refers to a process in

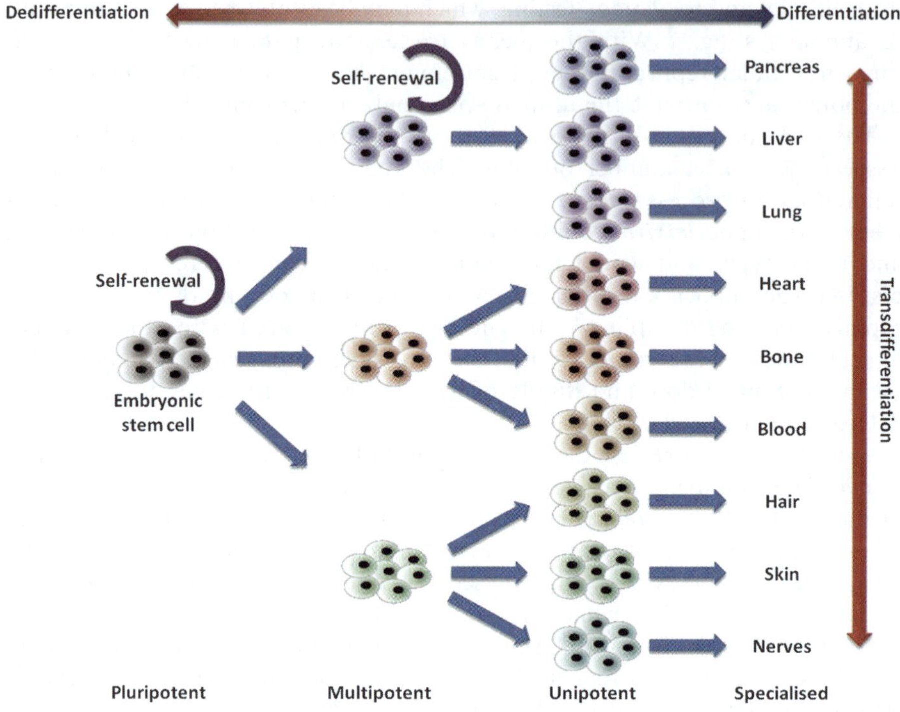


**Figure 4.2**  Stem cell ontogeny.

which a cell reverts to a less mature state with greater potency, capable of differentiating to a number of other cell types, and transdifferentiation to a process whereby a cell transforms directly to another lineage without first reverting to a less mature state.

Adults retain populations of multipotent stem cells within many tissues, for example haematopoietic stem cells (HSCs) within bone marrow or neural stem cells (NSCs) within the brain, which still possess the capacity to self-renew and the ability to generate multiple cell types, although often apparently restricted to lineages associated with the organ or system in which they reside; for example, HSCs give rise to blood lineages, and NSCs to neuronal cell types, but not usually to other lineages. The mechanisms determining stem cell fate have been investigated for many years, and major advances have been made in elucidating many of the intrinsic and extrinsic factors that influence cell maintenance and lineage commitment both *in vitro* and *in vivo*, although significant gaps still remain in our understanding.

The detection and characterisation of stem cells represents a challenge, and is often the cause for dispute within the literature, yet it is a central question to be addressed for the generation of reliable, reproducible and widely applicable methods to efficiently manipulate stem cell fate. Cellular morphology may be used as an indicator in some instances; for example, in neurogenesis observing the formation of characteristic neurites, or cardiomyogenesis, observing the formation of spontaneously beating cardiomyocytes. However, morphology alone is not indicative of cell type, nor does it necessarily indicate a homogenous cell population; for example, a colony of beating cells needs only contain a small fraction of cardiomyocytes.[9] More stringent assessments are therefore necessary to characterise cell populations and to assess the yield of desired cells, or the efficiency of conversion. Currently methods to detect and characterise stem cell populations and their differentiated progeny include immunocytochemistry or RT-PCR to detect diagnostic biomarkers. For example, pluripotent embryonic stem cells are characterised by expression levels of transcription factors such as Oct4 and Nanog; and Oct4 in particular has been shown to be a critical factor in induced pluripotency. Differentiation may be monitored by the loss of these pluripotency markers and the gain of markers indicative of a given lineage, for example neuron-specific β-tubulin for neuronal differentiation, or MyoD for muscle differentiation. The ultimate means to determine cell fate comes from functional analyses. For example, a rigorous assessment of pluripotency demonstrates that cells contribute to all tissue types *in vitro* and *in vivo*. For mouse embryonic stem cells and induced pluripotent cells this includes the ability of the cells to form chimeras and contribute to the germline upon implantation into mouse embryos,[2,10] and for human embryonic stem cells, where the analogous *in vivo* experiment is forbidden on ethical grounds, this includes the ability to form teratomas upon implantation. Gene array analysis is increasingly used to characterise and compare cell populations; however, epigenetic variations within cells also influence their behaviour and a routine, rigorous characterisation of cells at this level would be of great benefit to the field.

Specific types of stem cells have already been applied clinically for a variety of purposes:[11] HSCs have been used in the clinic for transplantation purposes for several years,[12,13] derived either from adult bone marrow or more recently from umbilical cord blood, reducing the risk of graft versus host disease.[14] The use of multipotent stromal cells (MSCs) has also been explored: MSCs are a multipotent cell type, found in mesenchymal (mesoderm-derived) tissue in development, but have been found to persist in numerous tissues in adults where they play key contributions to tissue repair and homeostasis.[15] In recent years,[16] embryonic stem cells have been widely investigated and touted as sources for allogeneic cell therapy and drug discovery as a result of their ability to self-renew indefinitely under appropriate culture conditions and differentiate to all tissue types in the adult.[17] While it is unclear whether these cells will represent a safe therapy in the clinic due to the risk of teratoma formation, nonetheless they represent readily accessible common sources of other more mature cell types such as cardiomyocytes or insulin-secreting β-cells for application in clinical therapies[18] or drug discovery.[9] The advent of cellular reprogramming and induced pluripotency also provides an opportunity for personalised cell therapy, exploiting the derivation and manipulation of cells derived from the patient.[19]

The isolation, and *in vitro* expansion and manipulation of all these classes of stem cell could have enormous benefits for cell replacement therapy, tissue engineering and drug discovery applications. Moreover, endogenous multi-potent tissue stem cells themselves represent a potential target for *in vivo* manipulation for adult tissue and organ repair and regeneration.


## 4.3  Stem Cell Manipulation *In Vitro*

Although recent years have seen an explosion of interest in the identification and development of small molecule modulators of stem cell fate, early observations in this area were made over 30 years ago describing the effects of naturally derived molecules on cell fate. In 1978, two independent reports demonstrated embryonal carcinoma (EC) cells could be differentiated towards parietal endoderm-like cells with either *all-trans*-retinoic acid (ATRA)[20] or hexamethylenebis-acetamide.[21] ATRA has since been shown to exhibit pleiotropic effects and to promote differentiation to multiple cell types,[22] and has been used in the clinic for the differentiation therapy of acute promyelogenous leukaemia.[23]

Since these early observations, there have been ever increasing numbers of reports of small molecule manipulation of stem cell fate.[24-27] This increased interest has arisen from both the enhancement of our understanding of the molecular pathways controlling stem cell development[28] and advances in screening and bioanalytical methodologies, particularly high throughput, high content phenotypic screens.[29] In line with these developments, the main approaches to the identification of stem cell modulators have been the application of small molecules to inhibit or activate signalling cascades or to promote epigenetic remodelling, or to perform high-throughput screening of compound libraries in phenotypic assays. Both approaches and their roles in enabling or enhancing our understanding of stem cell biology will be reviewed.


### 4.3.1  Small Molecules to Promote Stem Cell Expansion

For allogeneic or autologous stem cell therapies, as many as $10^8$–$10^9$ cells may be required depending on the therapeutic application and the efficiency of transplantation. To overcome limitations in the application of embryonic, induced pluripotent or adult multipotent stem cells in clinical therapies and drug discovery, efficient and cost effective methods for expanding these cell populations are needed. Human ESCs, for example, have stringent requirements for their culture, requiring either the presence of a feeder cell layer or the presence of an artificial matrix substratum; both in combination with a cocktail of growth factors and careful control of cell density.[30] Human ESCs are thus exquisitely sensitive to their microenvironment, with multiple factors combining to maintain pluripotency and block apoptosis, or to initiate differentiation, including the culture medium, growth factors/cytokines, oxygen tension, salt concentration, pH, and surface properties, as well as temperature, pressure, shear stress and mechanical stimuli.[31]

Much of the early work on the development of small molecules to allow ESC expansion has used murine cells, building on our knowledge of self-renewal pathways in mouse ESCs. Mouse ESCs have been shown to undergo self-renewal upon treatment with just two morphogens, leukaemia inhibitory factor (LIF)[32] and bone morphogenetic protein (BMP).[33] LIF was shown to act *via* activation of STAT3 signalling which promotes self-renewal and inhibition of ERK signalling blocking differentiation towards mesoderm and endoderm lineages. LIF alone, however, was not sufficient to block neurectodermal differentiation, and BMP-mediated inhibition of mitogen activated protein kinase (MAPK) activity was found to be required to maintain pluripotency of mouse ESCs. Later, the use of 6-bromoindirubin-3′-oxime (BIO) was reported to enhance mouse ESC self-renewal; reportedly through inhibition of glycogen synthase kinase-3β (GSK-3β) and activation of the Wnt signalling pathway.[34] The role of GSK-3β inhibition and Wnt signalling in enhancing ESC self-renewal was supported by the use of the more specific GSK-3β inhibitor, CHIR99021,[10] and through the observation that **IQ-1** also promotes ESC self-renewal, through activation of Wnt signalling.[35] The application of PD98059,[36] an inhibitor of MEK signalling, and either SB203580 or PD169316,[37] both p38 MAPK inhibitors, were also shown to enhance self-renewal, and provided a mechanistic rationale for LIF-mediated anti-apoptotic effects. These observations together have led to the definition of culture conditions for mouse ESCs using small molecules in the absence of exogenous growth factors:[10] the combination employs a GSK-3β inhibitor (CHIR99021) with a MEK inhibitor (PD0325901) with or without the presence of a fibroblast growth factor (FGF) receptor inhibitor (PD173074) (termed "3i" or "2i" media, respectively). This work elegantly demonstrated that mouse ESCs have an innate programme for self-renewal, and illustrated the feasibility of ESC expansion under well defined conditions using small molecules (Table 4.1).

An understanding of the underlying pathways influencing ESC self-renewal has thus allowed the application of specific small molecule modulators of their constituent molecular targets. However, this approach is inherently limited to cases where the pathway is known, and specific small molecule modulators of the pathway may be sourced. Furthermore, it makes the assumption that modulation of a single molecular target, or defined set of targets alone will have the desired biological effect.

An alternative approach draws on phenotypic screening, where no assumption is made regarding the molecular target(s). A small molecule, **SC-1** (pluripotin), capable of expanding mouse ESCs in the absence of LIF was identified through high-throughput screening using an Oct4-GFP reporter assay. **SC-1** was reported to act *via* inhibition of two pathways, ERK and RasGAP, negating the requirement for LIF supplementation.[38] Phenotypic screens have successfully identified other small molecules to promote self-renewal of ES cells, *e.g.* **ID-8**;[39] however, elucidation of their mechanism of action has proved more challenging.

Human ESCs have been shown to display significant differences to mouse ESCs;[40] they do not respond to LIF, instead relying on FGF and activin/nodal signalling.[41] Small molecules identified to allow the expansion of mouse ESCs

**Table 4.1**  Selected small molecules to promote stem cell expansion/proliferation.

| Compound | Structure | Role | Possible target | Notes | Reference |
|---|---|---|---|---|---|
| 1 | 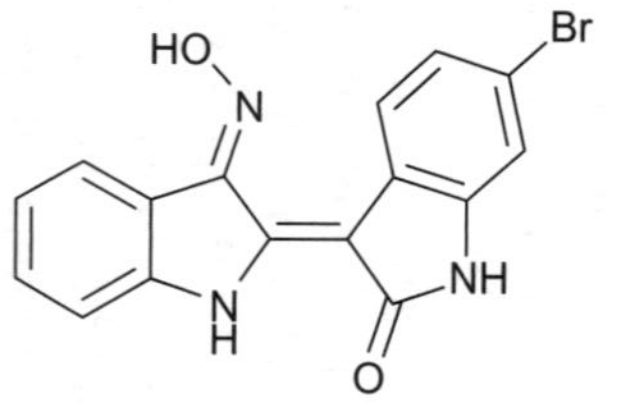 **BIO** | Embryonic stem cell self-renewal | GSK3-β inhibitor | — | Sato et al.[34] |
| 2 | 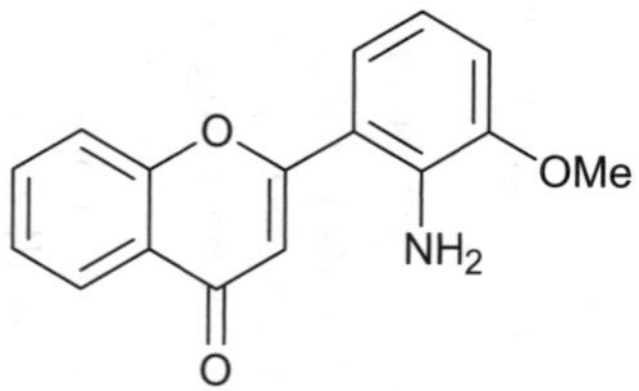 **PD98059** | Embryonic stem cell self renewal | MEK inhibitor | — | Qi et al.[36] |
| 3 | 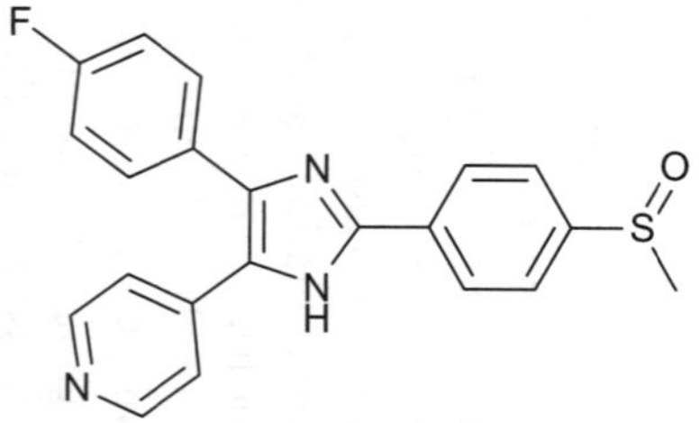 **SB203580** | Embryonic stem cell self-renewal | P38 MAPK inhibitor | Acts down-stream of BMP4 | Qi et al.[36]; Duval et al.[37] |

| # | Structure | | | | |
|---|---|---|---|---|---|
| 4 | 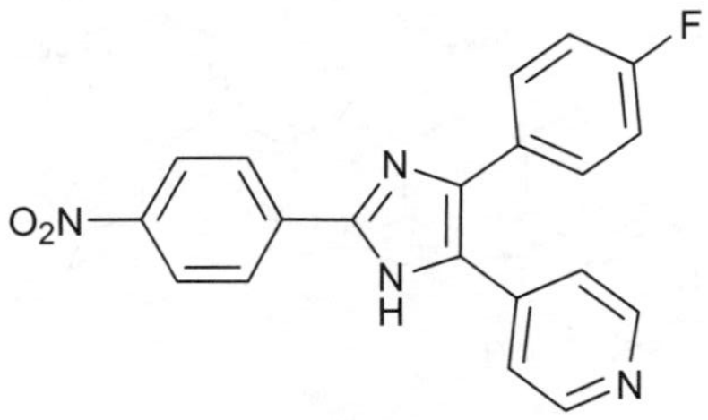 **PD169316** | Embryonic stem cell self-renewal | P38 MAPK inhibitor | Acts down-stream of BMP4 | Duval *et al.*[37] |
| 5 | 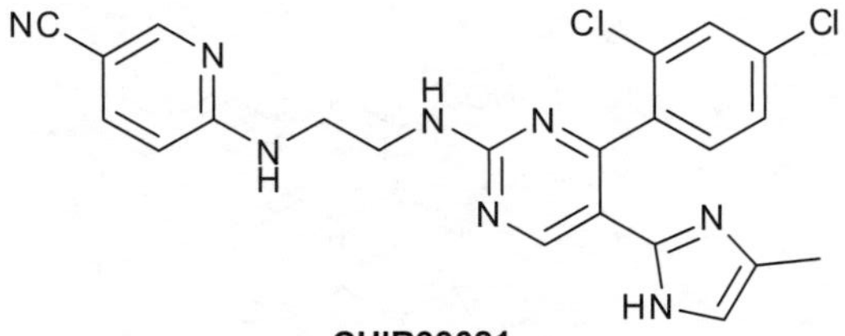 **CHIR99021** | Embryonic stem cell self-renewal | GSK3-β inhibitor | "3i" medium | Ying *et al.*[10] |
| 6 | 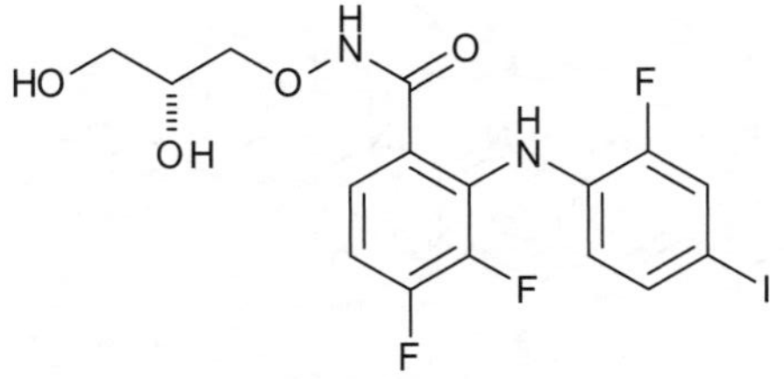 **PD0325901** | Embryonic stem cell self-renewal | MEK inhibitor | "3i" medium | Ying *et al.*[10] |

**Table 4.1**  *Continued*

| Compound | Structure | Role | Possible target | Notes | Reference |
|---|---|---|---|---|---|
| 7 | 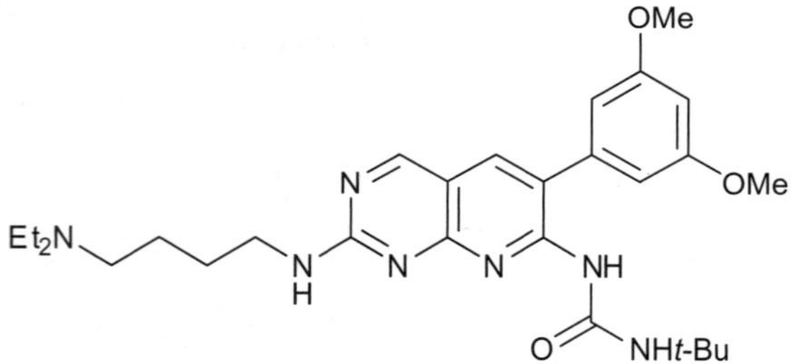 **PD173074** | Embryonic stem cell self-renewal | FGFR inhibitor | "3i" medium | Ying *et al.*[10] |
| 8 | 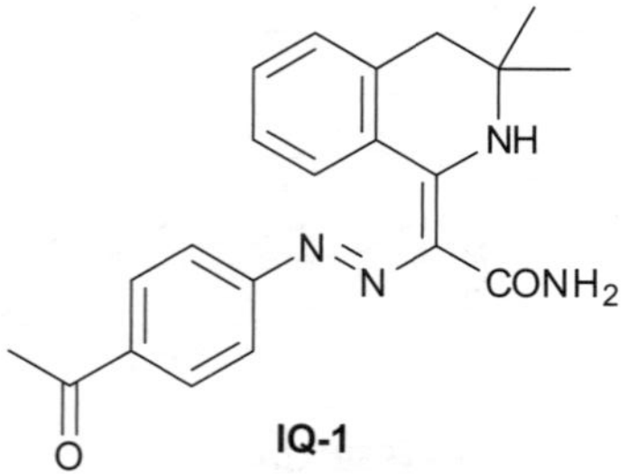 **IQ-1** | Embryonic stem cell self-renewal | PP2A inhibitor | — | Miyabayashi *et al.*[35] |
| 9 | 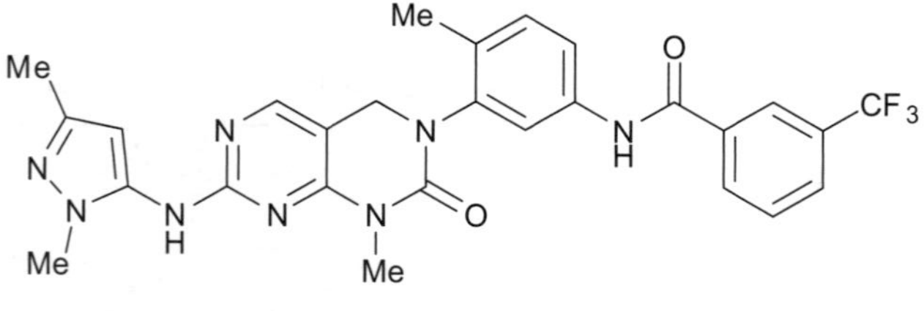 **SC1 (Pluripotin)** | Expansion of embryonic stem cells | Dual Ras-GAP and ERK1 inhibition | — | Chen *et al.*[38] |

| | | | | | |
|---|---|---|---|---|---|
| **10** | 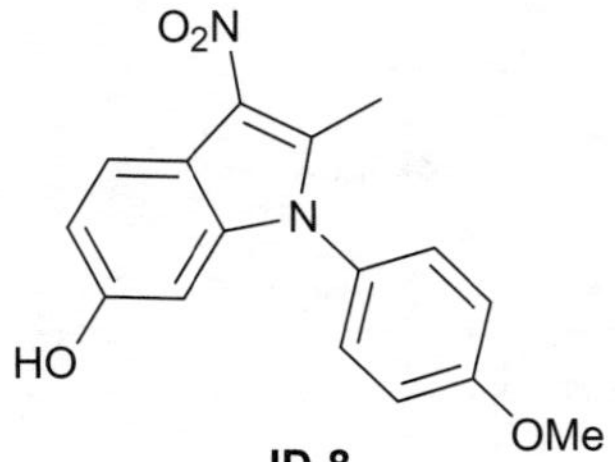 ID-8 | Embryonic stem cell self-renewal | Unknown | — | Miyabayashi *et al.*[39] |
| **11** | 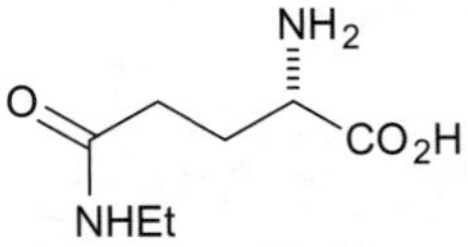 Theanine | Embryonic stem cell self-renewal | Unknown | — | Desbordes *et al.*[43] |
| **12** | 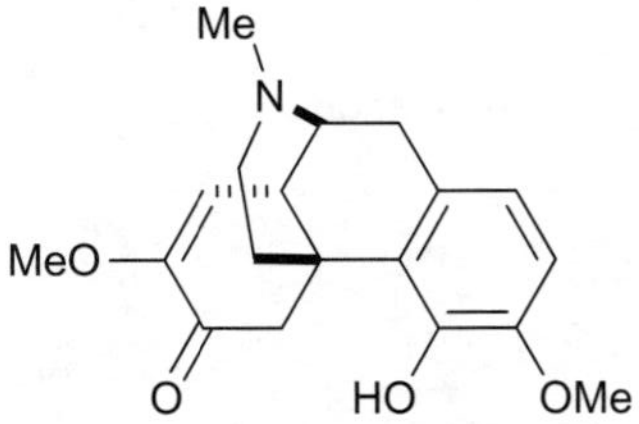 Sinomenine | Embryonic stem cell self renewal | Unknown | — | Desbordes *et al.*[43] |

**Table 4.1** *Continued*

| Compound | Structure | Role | Possible target | Notes | Reference |
|---|---|---|---|---|---|
| **13** | 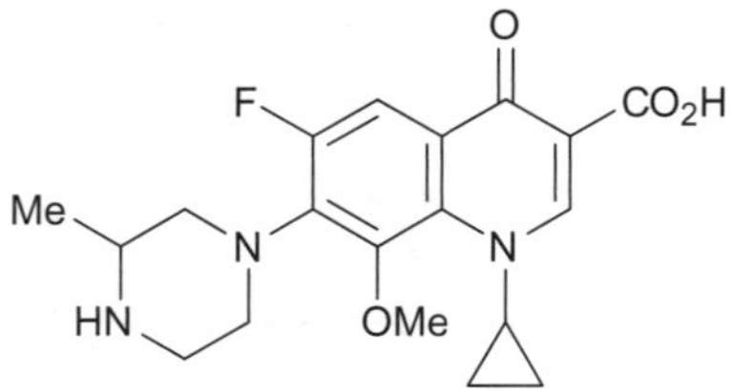 **Gatifloxac in** | Embryonic stem cell self-renewal | Unknown | — | Desbordes et al.[43] |
| **14** | 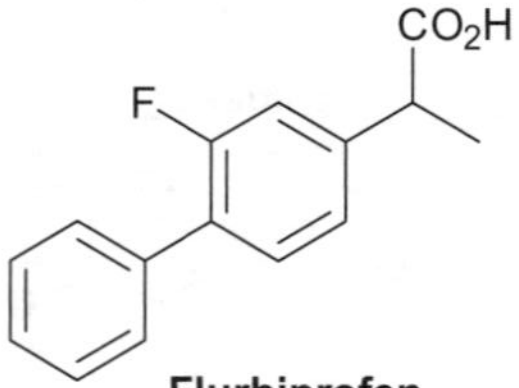 **Flurbiprofen** | Embryonic stem cell self renewal | Unknown | — | Desbordes et al.[43] |
| **15** | 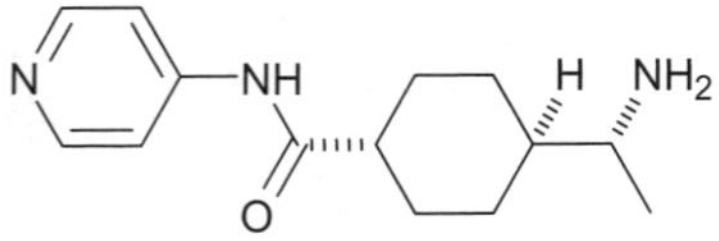 **Y27632** | Expansion of embryonic stem cells | Rho kinase inhibitor | Promotes survival of dissociated ES cells | Watanabe et al.[44] |

| | | | | | |
|---|---|---|---|---|---|
| **16** | 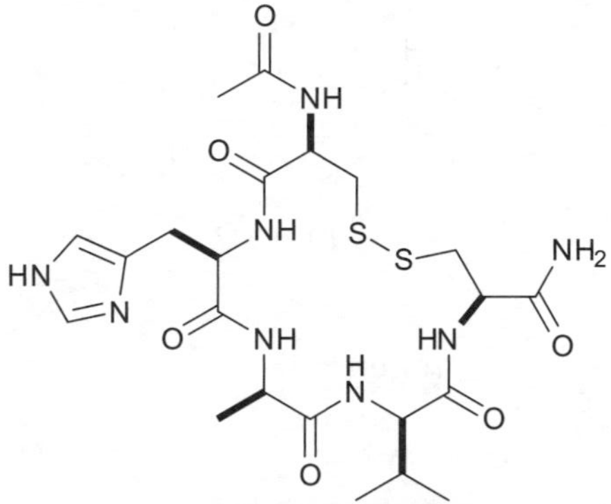 cyclic **N-Ac-CHAVC-NH₂** | Embryonic stem cell self renewal | E-cadherin inhibition | — | Soncin *et al.*[45] |
| **17** | 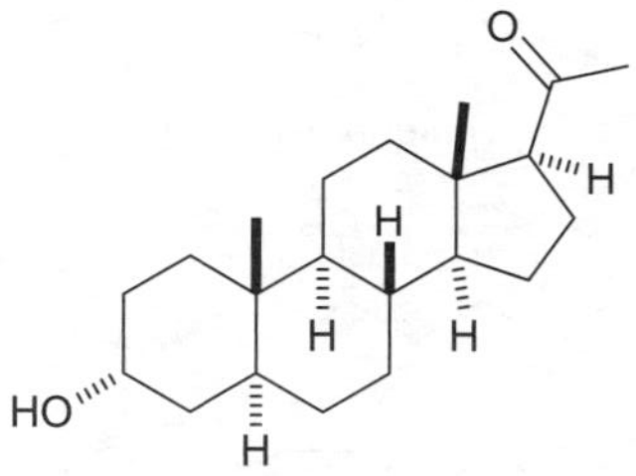 **Allopregnanolone** | Expansion of neural stem cells | Unknown | — | Wang *et al.*[47] |
| **18** | 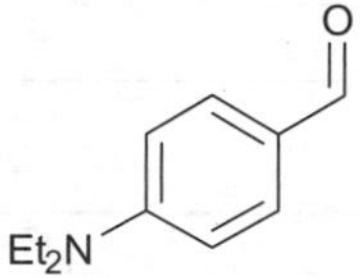 **Diethylaminobenzaldehyde** | Expansion of HSCs | Aldehyde dehydrogenase inhibitor | — | Chute *et al.*[50] |

**Table 4.1** *Continued*

| Compound | Structure | Role | Possible target | Notes | Reference |
|---|---|---|---|---|---|
| 19 | 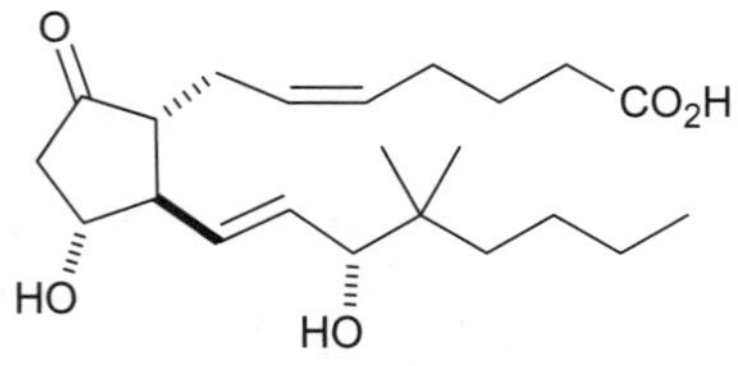 16,16-Dimethyl-PGE$_2$ | Expansion of HSCs | — | — | North et al.[52] |
| 20 | 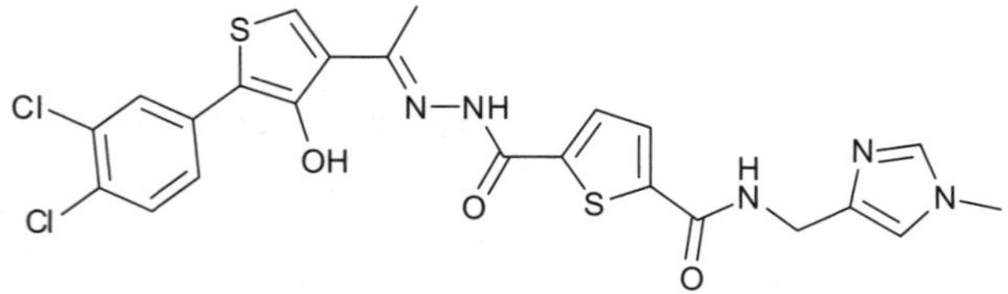 NR-101 | Expansion of HSCs | c-MPL agonist | — | Nishino et al.[51] |
| 21 | 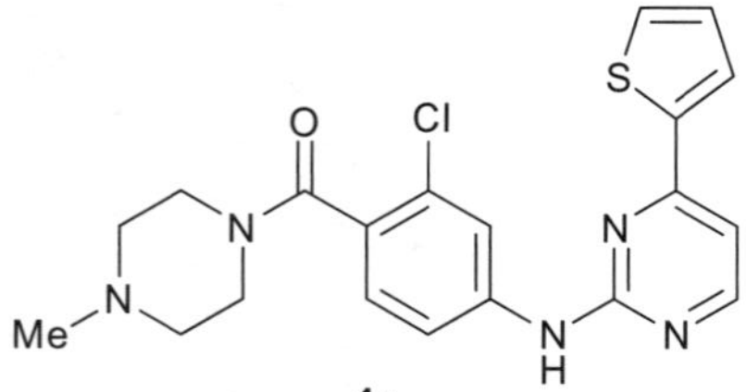 1a | Expansion of pancreatic beta-cells | Wnt activation | — | Wong et al.[48] |

| 22 | 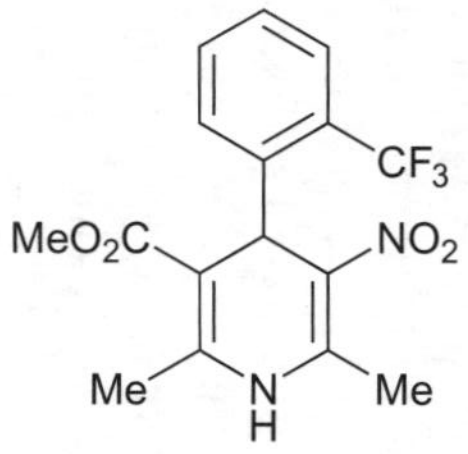 (structure 2a) | Expansion of pancreatic beta-cells | L-type calcium channel activation | Exerts an additive effect with GLP-1 receptor agonist, Exendin-4 | Wong *et al.*[48] |

have generally not proved directly transferable to human ESC culture,[42] although BIO is reported to also enhance human ESC self-renewal in the absence of feeder cells.[34] Phenotypic screening for small molecules to circumvent the requirement for FGF signalling in human ESC self-renewal identified four candidates: theanine, sinomenine, gatifloxacin and flurbiprofen, although in each case no mode of action has yet been elucidated.[43]

Currently, ESCs are routinely cultured in two dimensions,[30] which severely limits the availability of sufficiently large numbers of high quality stem cells. Three-dimensional (3D) culture techniques, employing either 3D matrices, or suspended colonies, have recently been reported, but at present human ESCs cannot be grown in 3D culture conditions as single cell suspensions as they either do not survive or rapidly differentiate.[31] However, it has recently been reported that treatment of human ESCs with a Rho kinase inhibitor, such as Y27632, enhances survival of dissociated stem cell populations, through suppression of apoptotic pathways.[44] Furthermore, blocking E-cadherin-mediated cell–cell contact, using a small molecule cyclic peptide CHAVC, was recently shown to maintain pluripotency and promote self-renewal in mouse ESCs upon withdrawal of LIF.[45] Blocking cell–cell contact using a small molecule while promoting self-renewal and maintaining pluripotency can potentially provide a means for the single cell suspension culture of ESCs, enabling their expansion on a clinically relevant scale.

The expansion of adult, tissue-derived stem cells has proved harder to achieve. Studies on the stem cell niche (*in vivo* microenvironment) revealed that tissue homeostasis *in vivo* was often achieved through successive asymmetric cell divisions of the endogenous stem cell populations, *i.e.* each cell division produced an identical daughter cell and a differentiated cell. Factors to promote symmetrical cell division, akin to self-renewal in embryonic stem cells, were generally unknown. However, in 2005, Conti *et al.* demonstrated that neural stem cells could undergo symmetrical cell division, and expansion *in vitro*, using defined extrinsic growth factors.[46] Small molecules to enhance the proliferation of neural stem cells *in vitro* such as allopregnanolone have also since been reported.[47]

Expansion of other adult cell populations have also been investigated for potential clinical application, including the expansion of pancreatic β-cells with small molecules such as the Wnt agonist **1a**, indicating a further role for Wnt signalling in pancreatic cell proliferation, and the L-type calcium channel activator **2a**, which was observed to exert an additive effect when used in conjunction with a GLP-1 receptor agonist, Exendin-4.[48]

Efficient methods to allow the *ex vivo* expansion of HSCs would be of enormous clinical utility; an expansion of even 3- to 4-fold would provide clinical benefit. Although various cytokines, such as SCF and thrombopoietin have been shown to expand HSCs and demonstrate efficacy in murine models, these have so far not translated into efficacy in a clinical setting.[49] Diethyl-aminobenzaldehyde has been used to expand HSCs *in vitro*: its mode of action was reported to be, at least in part, through inhibition of aldehyde dehydrogenase.[50] In combination with inhibition of retinoic acid signalling, this intervention was proposed to lead to blockade of HSC differentiation. The *ex vivo* expansion of HSCs has also been recently reported using a small molecule agonist of c-MPL.[51] Through an *in vivo* screen for HSC expansion in zebrafish embryos, North *et al.* demonstrated a role for prostaglandin signalling and the inflammatory response in HSC homeostasis, a role which they found to be conserved across multiple vertebrates. This observation led to the identification of 16,16-dimethyl-$PGE_2$ as a candidate for the *ex vivo* expansion of HSCs which may also be applied to HSC activation *in vivo* (see Section 4.4.1).[52]

## 4.3.2 Small Molecules to Direct Stem Cell Differentiation

The directed differentiation of stem cells to somatic cells has typically involved the use of growth factors or cytokines and subsequent purification/extraction from the remaining undifferentiated cells or unwanted somatic cells.[53] Processes are generally extremely low yielding, and the proportion of the desired somatic cells is often very low, even after partial purification.[53] For such cells to be applicable in the clinic or to provide reliable model systems for drug discovery, robust, efficient and high yielding methods to routinely direct their production from stem cell precursors are needed. Improved protocols employing small molecules to direct differentiation are emerging, as outlined below. Several approaches have been employed for the identification of small molecules to control differentiation, using *in vitro* phenotypic screens or targeted approaches drawing on inspiration from the study of endogenous pathways for stem cell specification during development.[54] Directed differentiation using small molecules to a variety of multipotent and somatic cell types has been reported including adipocytes,[55] keratinocytes,[56] cardiomyocytes, osteoblasts, neurons, retinal cells and pancreatic cells (Table 4.2).[24–27]

A number of small molecules exerting multiple effects have been identified, including retinoic acid,[20] epigenetic modifiers such as 5-azacytidine, valproic acid and trichostatin A, and antioxidants such as ascorbic acid. However, such molecules have proved to be useful tools for stem cell fate manipulation, with directed differentiation to a variety of cell types reported, dependent on the stem cell precursor employed and the timing of administration; for example 5-azacytidine, a DNA methyltransferase inhibitor, has been used to direct differentiation towards muscle cell types, including cardiac muscle,[57] and can even induce transdifferentiation to muscle lineages from other somatic cell types (see Section 4.2).[58] Modification of histone deacetylase (HDAC) activity has also been shown to have multiple effects on directing cell fate; valproic acid, suberoylanilide hydroxamic acid (SAHA), and trichostatin A have all been exploited in a variety of contexts, including cellular reprogramming (see Section 4.2.3), for example trichostatin A has been shown to induce differentiation towards cardiac lineages, putatively through accumulation of acetylated GATA4, a cardiac-specific factor.[59] Altering the redox potential of stem cells through the use of antioxidants such as ascorbic acid has also been shown to enhance differentiation to several cell types, and particularly towards bone and cardiac muscle cells.[60,61]

**Table 4.2**  Selected small molecules to direct stem cell differentiation.

| Compound | Structure | Role | Possible target | Notes | Reference |
| --- | --- | --- | --- | --- | --- |
| 23 | 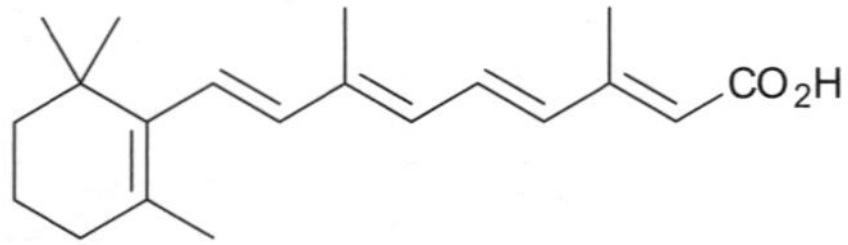 All *trans* Retinoic acid (ATRA) | Trophoblast differentiation; neural differentiation | Retinoic acid receptor | Shown to exert pleiotropic effects | Desbordes et al.[43] |
| 24 | 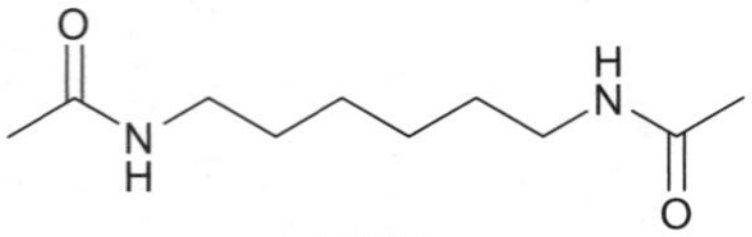 HMBA | Parietal endoderm differentiation of EC cells | Unknown | — | Jakob et al.[21] |
| 25 | 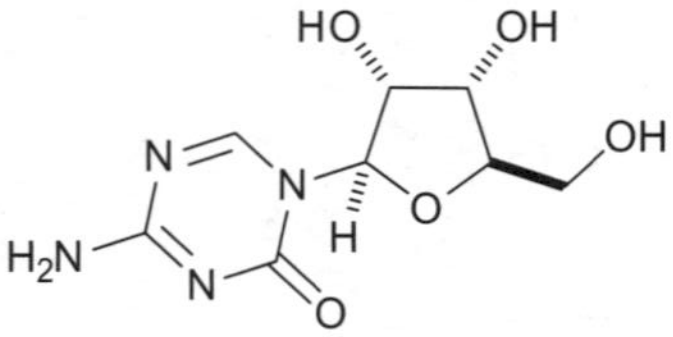 5-Azacytidine | Muscle differentiation (including cardiac); induces dedifferentiation | DNA methyltransferase | Shown to exert pleiotropic effects | Xu et al.[57] |
| 26 | 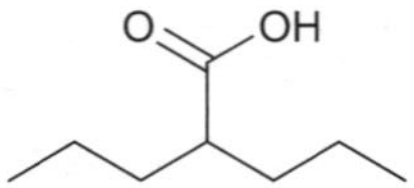 Valproic acid | Differentiation to a range of cell types; enhanced reprogramming | HDAC inhibitor | Shown to exert pleiotropic effects | Huangfu et al.[106] |

| | Structure | Target/Mechanism | Application | Notes | Reference |
|---|---|---|---|---|---|
| 27 | SAHA | HDAC inhibitor | Differentiation to a range of cell types; enhanced reprogramming | Shown to exert pleiotropic effects | Huangfu et al.[106] |
| 28 | 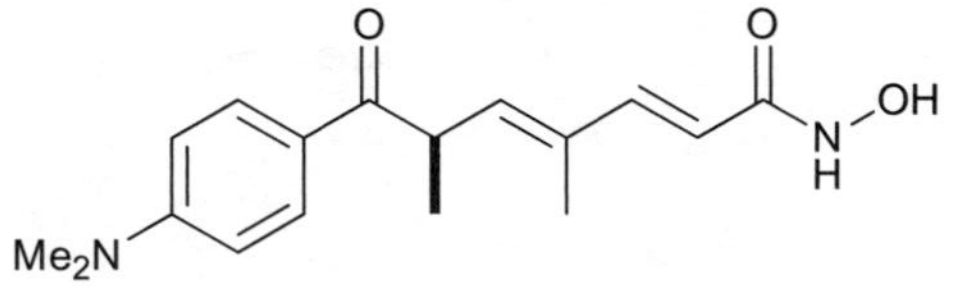 Trichostatin A | HDAC inhibitor | Cardiac differentiation; enhanced reprogramming | Results in accumulation of acetylated GATA4 | Kawamura et al.[59] |
| 29 | 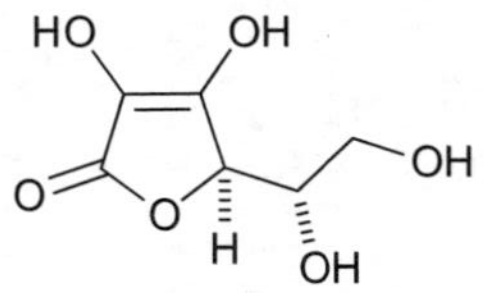 Ascorbic acid | Unknown | Muscle differentiation (including cardiac) | Shown to exert pleiotropic effects | Takahashi et al.[60]; Passier et al.[61] |
| 30 | 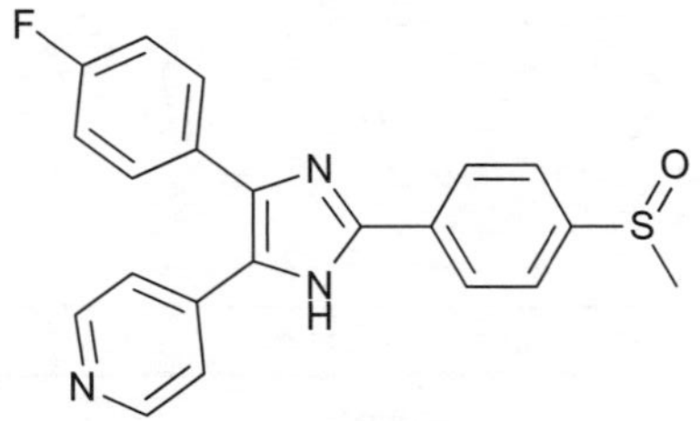 SB203580 | P39-MAPK inhibitor | Cardiac differentiation | — | Graichen et al.[62] |

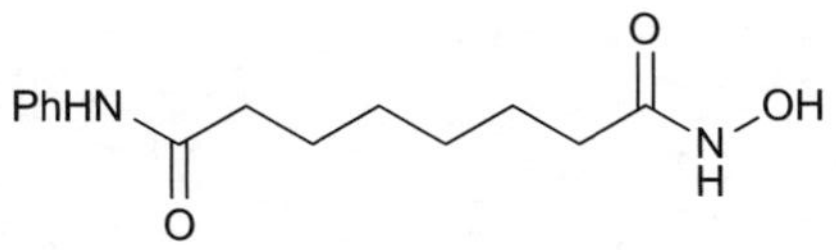

**Table 4.2** *Continued*

| Compound | Structure | Role | Possible target | Notes | Reference |
|---|---|---|---|---|---|
| **31** | 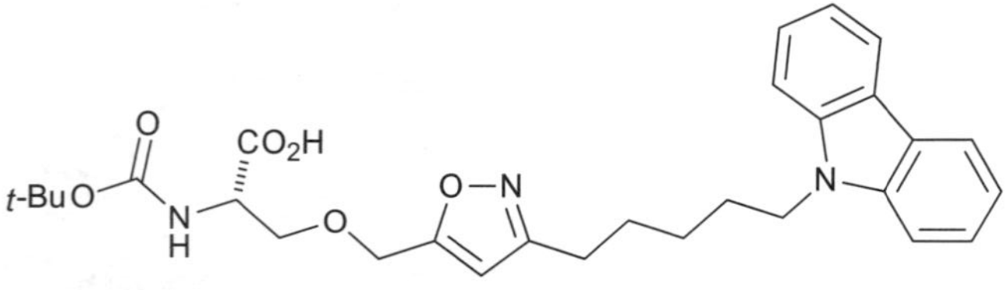 **Isoxazole-serine** | Cardiac differentiation | PPARα agonist | | Wei *et al.*[64] |
| **32** | 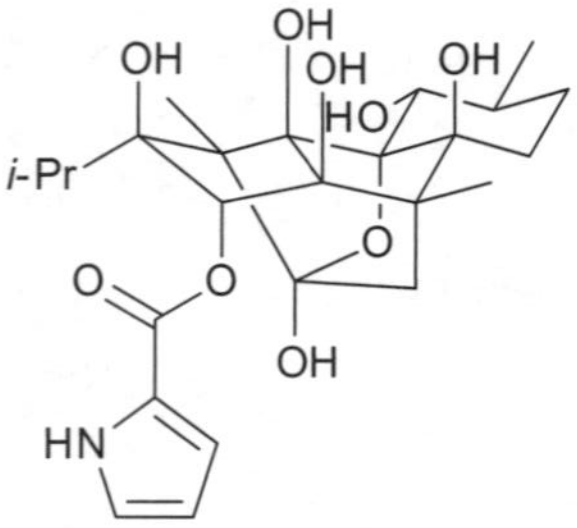 **Ryanodine** | Cardiac differentiation | Ryanodine receptor inhibitor | Induces differentiation from mouse ES cells | Sachinidis *et al.*[65] |
| **33** | 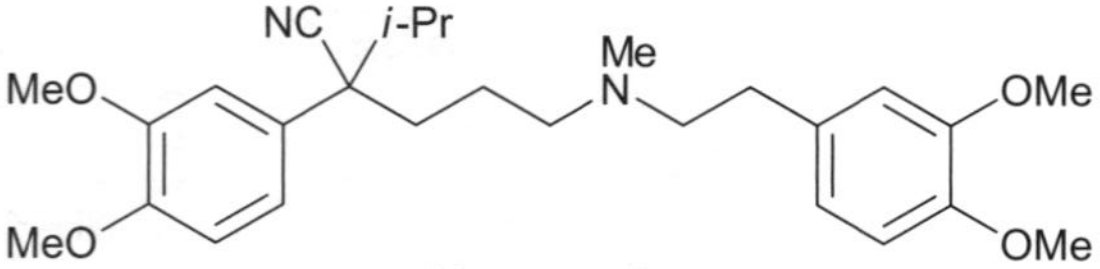 **Verapramil** | Cardiac differentiation | L-type calcium channel blocker | Induces differentiation from mouse ES cells | Sachinidis *et al.*[65] |

| | | | | | |
|---|---|---|---|---|---|
| **34** | 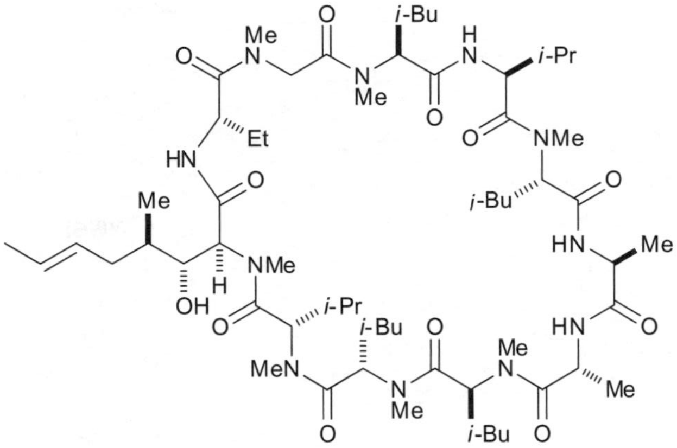 Cyclosporin A | Cardiac differentiation | Calcineurin/ PP2A/PLC/ PKC inhibitor | Induces differentiation from mouse ES cells | Sachinidis et al.[65] |
| **35** | 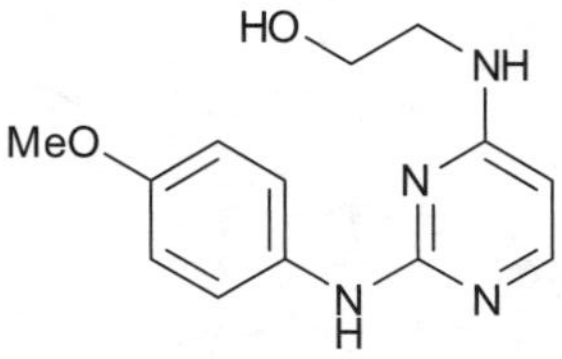 Cardiogenol | Cardiac differentiation | — | — | Wu et al.[66] |

**Table 4.2** *Continued*

| Compound | Structure | Role | Possible target | Notes | Reference |
|---|---|---|---|---|---|
| 36 | **Dexamethasone** | Osteogenic differentiation | Glucocorticoid receptor | Induces osteogenic differentiation from MSCs; shown to exert pleiotropic effects | Jaiswal *et al.*[67] |
| 37 | **IBMX** | Osteogenic differentiation | Non-selective PDE inhibitor | From MSCs | Jaiswal *et al.*[67] |
| 38 | **Phenamil** | Osteogenic differentiation | Stimulates BMP signalling; Trb3 | Additive effect with BMP | Park *et al.*[69] |

| | | | | | |
|---|---|---|---|---|---|
| 39 | 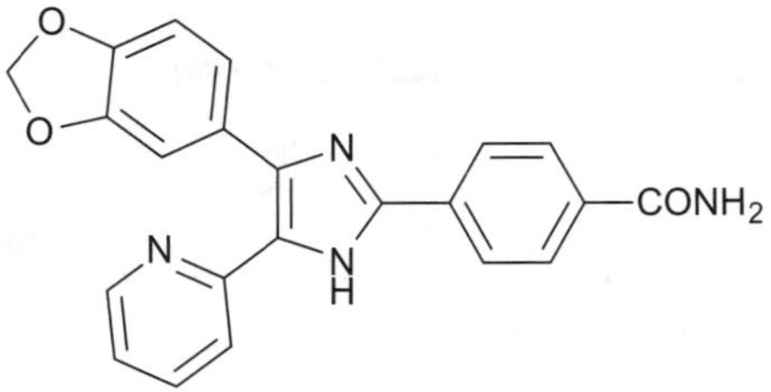 SB431542 | Neural differentiation | ALK4/5/7 inhibitor | Acts synergistically with noggin | Chambers *et al.*[74] |
| 40 | 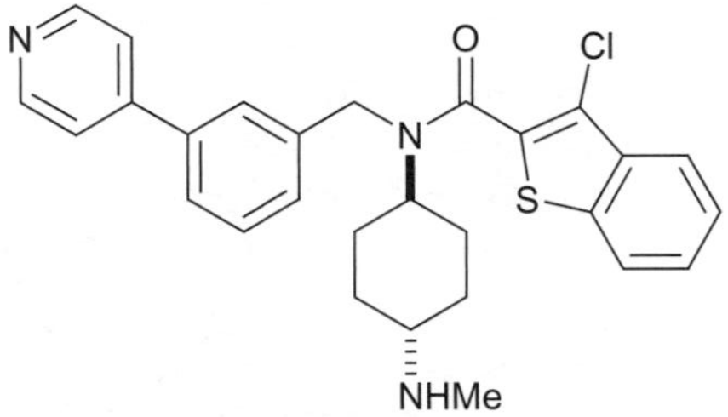 SAG 1.3 | Neural differentiation | Sonic hedgehog signalling agonist | Used in conjunction with ATRA | Wichterle *et al.*[82] |
| 41 | 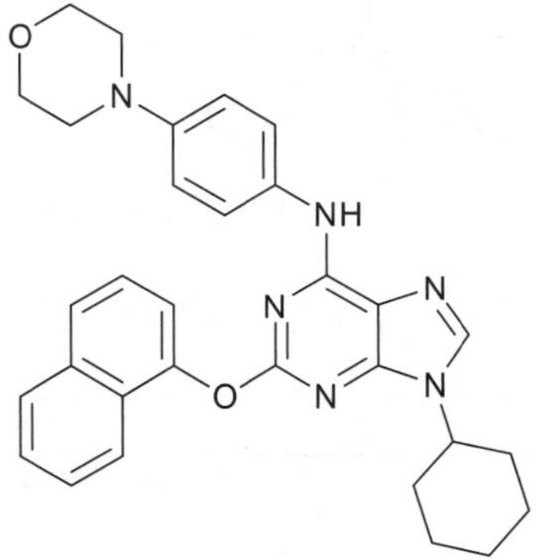 Purmorphamine | Neural differentiation | Smoothened agonist | — | Li *et al.*[72] |

**Table 4.2**  *Continued*

| Compound | Structure | Role | Possible target | Notes | Reference |
| --- | --- | --- | --- | --- | --- |
| **42** | **TWS119** | Neural differentiation | GSK3β | — | Ding *et al.*[75] |
| **43** | **Kenpaullone** | Neural differentiation | Wnt activation | — | Castelo-Branco *et al.*[70] |
| **44** | **Neuropathiazol** | Neural differentiation | Unknown | — | Warashina *et al.*[76] |

| 45 | 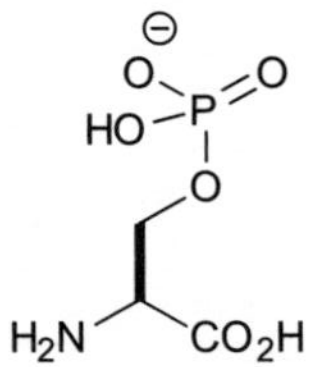 Phosphoserine | Neural differentiation | mGluR4 | — | Saxe *et al.*[77] |
| 46 | 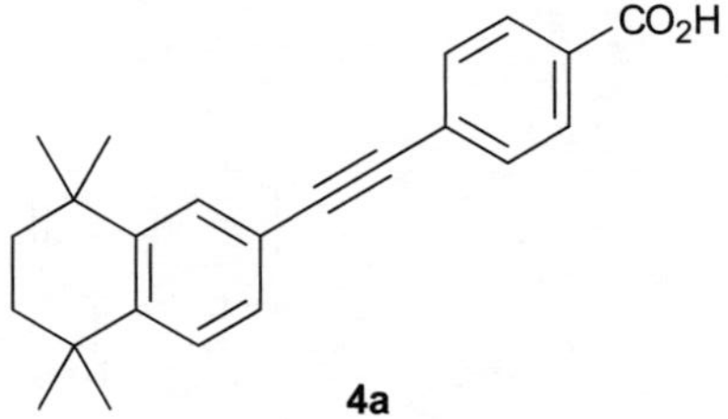 **4a** | Neural differentiation | RAR agonist | ATRA mimic | Christie *et al.*[79] |
| 47 | 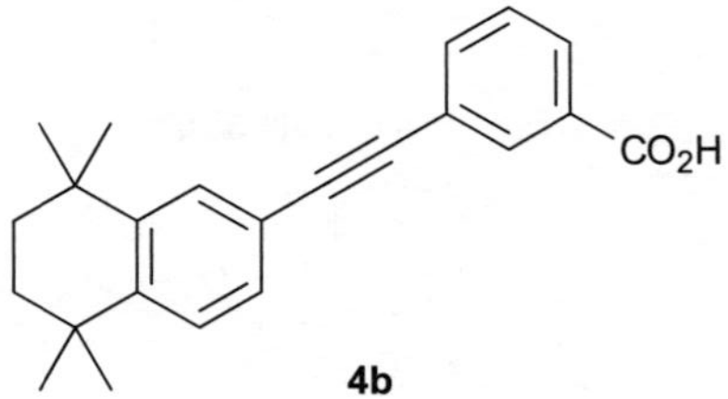 **4b** | Epithelial differentiation | Unknown | — | Christie *et al.*[79] |

**Table 4.2**  *Continued*

| Compound | Structure | Role | Possible target | Notes | Reference |
|---|---|---|---|---|---|
| **48** | 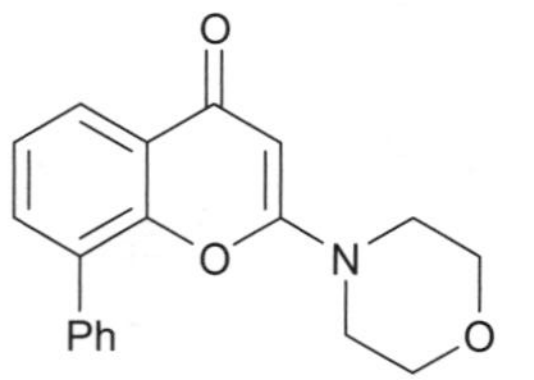 **LY294002** | Insulin secreting β-cell differentiation | PI3K | — | Hori *et al.*[84] |
| **49** | 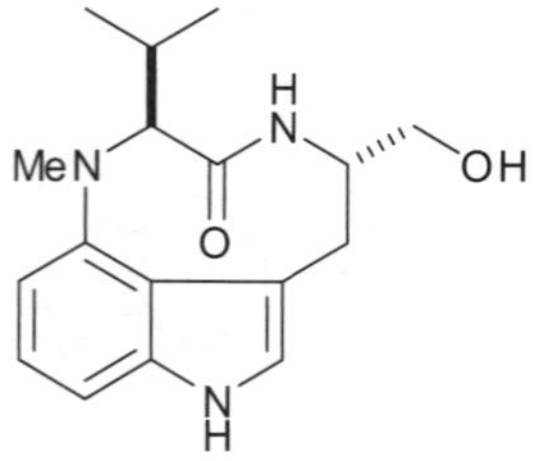 **Indolactam V** | Pancreatic differentiation | Unknown | — | Chen *et al.*[87] |
| **50** | 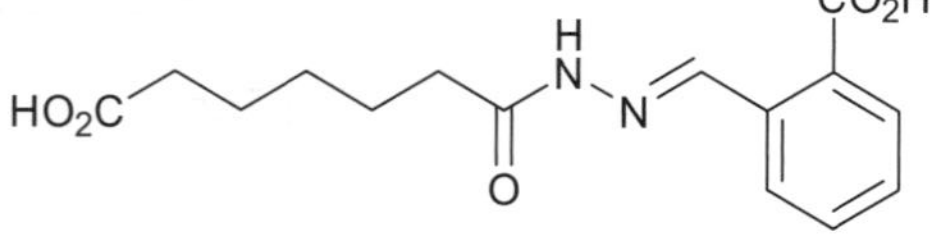 **IDE1** | Endoderm differentiation | Unknown | — | Borowiak *et al.*[86] |

| 51 | 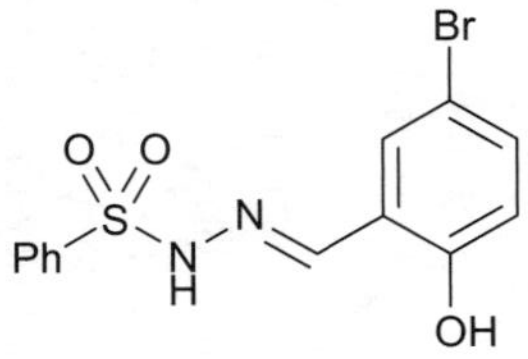 Shz-1 | Cardiac differentiation | Unknown | Activates *Nkx2.5* expression | Sadek *et al.*[121] |

The directed differentiation of ESCs and multipotent stem cells towards cardiomyocytes has been the focus of many reports; the controlled production of such cells in high purity would have a huge impact in the treatment of cardiac dysfunction such as acute myocardial infarction, and also as a source of cells to enable cardiotoxicity screening. Several small molecules have been trialled and shown to enhance or promote cardiomyogenesis, including SB203580, potentially through inhibition of p38 MAPK signalling[62] and dorsomorphin, putatively through inhibition of bone morphogenetic protein signalling, although other pathways may also be involved.[63] A series of isoxazolylserines, originally designed and screened as peroxisome proliferator-activated receptor (PPAR) agonists were found to promote cardiomyogenesis. However, their action may be independent of PPAR agonism, as they are relatively weak PPAR agonists, and the use of other PPAR ligands was not reported.[64] A number of unbiased screens of compound libraries to seek other small molecules to promote cardiomyogenesis have also been reported. A screen for effects of natural products and drugs identified a number of small molecules capable of promoting cardiomyogenesis in mouse embryonic stem cells including ryanodine, an antagonist of ryanodine receptors on sarcoplasmic reticulum, verapramil, an L-type calcium channel antagonist and cyclosporin A, which exerts pleiotropic effects including inhibition of calcineurin, phospholipase and protein kinase C.[65] However, the mechanisms through which these molecules act to promote cardiomyogenesis was unclear. A screen of a combinatorial compound library of purines and their isosteres also identified a series of small molecules, termed the cardiogenols, which induced cardiomyocyte formation from ESCs, although again the mechanism of action is not yet clear.[66]

Small molecules that induce osteogenesis have also been investigated as such molecules would show great potential in bone remodelling and reconstruction. A number of small molecules have been investigated for their effects on bone mineralisation acting through a variety of mechanisms, for example dexamethasone, a synthetic glucocorticoid, PPAR agonists such as rosiglitazone and isobutylmethylxanthine (IBMX), a non-specific phosphodiesterase inhibitor, have all been shown to promote osteogenesis.[67] Purmorphamine was originally identified as inducing osteogenesis in mesenchymal progenitor cells, and was later shown to be an agonist of hedgehog signalling, implicating a role for this signalling pathway in bone development.[68] More recently, a small molecule shown to activate BMP signalling has also been shown to exert osteogenic activity, consistent with the established role of BMP signalling in bone formation and remodelling.[69]

Differentiation towards a range of neuronal cell types including dopaminergic neurons,[70] cholinergic neurons,[71] motor neurons[72] and astrocytes amongst others has also been investigated using either pluripotent stem cells,[72] MSCs[73] or NSCs.[70] Several neurological diseases are characterised by progressive loss of neurons and the ability to regenerate functional neurons would represent a hugely important opportunity for therapy. Small molecules such as SB431542, which shows activin-like kinase inhibitory activity has been used in conjunction with another biological inhibitor of Smad signalling, noggin, to induce neurogenesis with high efficiency.[74] Several examples of small molecule agonists of hedgehog signalling have also been shown to induce neurogenesis, such as purmorphamine, consistent with the central role of hedgehog signalling in neurogenesis *in vivo* in embryo development.[72] Wnt signalling has also been shown to play a role in neurogenesis, for example with small molecules such as TWS119, originally identified through a high-throughput screen, found to act *via* activation of Wnt signalling through inhibition of GSK-3β activity.[75] Kenpaullone was found to produce dopaminergic neurons from neural precursor cells, which was proposed to occur through activation of Wnt signalling.[70] Small molecule screening also identified neuropathiazol which was shown to be capable of directing differentiation towards mature neurons from neural precursors whilst inhibiting differentiation towards the astroglial lineage.[76]

Endogenous small molecules such as phosphoserine[77] and ATRA[78] were also found to promote neural differentiation. Although ATRA has been shown to promote neurogenesis *in vivo* it shows relatively low levels of efficiency *in vitro*, and is also unstable upon irradiation. In an attempt to circumvent this problem, Christie *et al.* prepared stable ATRA mimics based on biaryl acetylenes.[79] Intriguingly, *para*-substituted **4a** was observed to give rise to a neural phenotype *in vitro*, but the corresponding *meta*-isomer **4b** was observed to give primarily epithelial phenotype upon treatment of the same precursor cells.

Directed differentiation towards retinal lineages has also been investigated for application in cell-based therapies to treat retinopathies and degenerative disorders such as age-related macular dystrophy. The production of retinal cells has recently been reported from ESCs and induced pluripotent stem cells with a cocktail of small molecules in serum-free conditions: an ALK inhibitor, a Rho kinase inhibitor and a casein kinase inhibitor.[80]

A huge advantage of the use of small molecules to direct the differentiation of stem cells is the ability to control both the dose and the timing of administration. This is particularly important in multistep differentiation protocols, where the intention is to recapitulate endogenous development pathways *in vitro* with small molecules.[28] Inspired by the roles of retinoic acid and hedgehog signalling in the production of motor neurons *in vivo*, a protocol has been devised and optimised for the selective production of motor neurons from ESCs *in vitro* involving sequential treatment of embryoid bodies derived from ESCs with ATRA and SAG 1.3,[81] an agonist of sonic hedgehog signalling.[82]

Following the report of a multistep protocol to programme pancreatic islets from mouse ESCs,[83] it was reported that the protocol could be modified through incorporating a small molecule, LY294002 which acts, amongst other mechanisms, as an inhibitor of PI3K.[84] Similarly, a two-step protocol for the production of pancreatic β-cells using small molecules has been developed, through screening for molecules to recapitulate a stepwise differentiation protocol using growth factors.[85] A screen for small molecules to direct differentiation to definitive endoderm from mouse or human ES cells was first performed, resulting in the identification of **IDE-1**.[86] Further exploration identified indolactam V, which promoted the production of immature pancreatic endocrine cells.[87] These molecules were then employed sequentially to convert ESCs directly into pancreatic lineages.[87]

An alternative to the development of protocols which simply recapitulate endogenous mechanisms or painstakingly unravel stepwise programming events would be to screen for molecules or methods to directly convert one cell type to another. This approach has proved successful both in the identification of new molecules and biological pathways to direct differentiation towards particular lineages and particularly in the identification of new methods to reprogramme cells through dedifferentiation and transdifferentiation processes, where the underlying biology is much less well understood.

### 4.3.3 Small Molecules to Promote Dedifferentiation and Transdifferentiation

Developmental processes and the transformation of pluripotent stem cells through progressively more specialised, lineage-committed cells to fully ("terminally") differentiated somatic cells has been traditionally represented as irreversible.[88] However, many examples have emerged of cellular reprogramming events, illustrating that even fully differentiated cells can exhibit plasticity when exposed to particular stimuli.[19] Technologies that allow full or partial dedifferentiation of somatic cells, or even direct transdifferentiation to other lineages, will have a huge impact on their use in regenerative medicine and drug discovery. Such technologies negate the ethical issues surrounding the use of human embryonic stem cells, and the technological barriers associated with accessing sufficient quantities of adult stem cells. Furthermore, as the approach involves the reprogramming of adult somatic cells it also represents an emerging opportunity for personalised medicine, for example through autologous cell therapies.[89]

One of the earliest observations of small molecule-mediated cellular reprogramming was that upon treatment of mouse fibroblasts with 5-azacytidine, a DNA methyltransferase inhibitor, stable myoblasts were produced. This was later shown to be due to demethylation at a particular locus in the genomic DNA.[58] This DNA modification was determined to lead to increased expression levels of *MyoD*, which is now known to be a master regulator of muscle development. Thus, the application of this small molecule in cellular reprogramming enabled the elucidation of the fundamental role of *MyoD* in myogenesis. Other early examples of cellular reprogramming included the use of dexamethasone, a synthetic glucocorticoid, which was observed to transform pancreatic cells into hepatocytes,[90] and also oligodendrocytes into multipotent stem cells, capable of subsequently differentiating into astrocytes or neurons or re-differentiating into oligodendrocytes.[91]

The most fruitful approach to the identification of small molecules to reprogramme cell fate has been through the screening of combinatorial libraries. In 2002, Schultz and co-workers, identified the purine myoseverin in a screen designed to identify small molecules capable of dedifferentiating myotubes.[92] Although myoseverin is now postulated to act through cytoskeletal remodelling to generate a myoblast-like cell type, and does not promote a reprogramming step to a multipotent cell type, it nonetheless established a precedent for screening for small molecules to reprogramme somatic cells. Their initial report was followed by a second screen which identified reversine, another purine, which they demonstrated was able to dedifferentiate murine myoblasts to precursor cells which were capable of differentiation to adipocytes and osteoblasts when subsequently treated with the appropriate corresponding differentiation media.[93] A further study has demonstrated that reversine-treated myoblasts may be directed towards neuroectodermal cell types.[94] It has also since been shown that reversine is capable of dedifferentiating fibroblasts and that the reprogrammed cells may be differentiation to myogenic cell types both *in vitro* and *in vivo* in a mouse model.[95] Although the mechanism of action of reversine is still not known, it has been shown to date to inhibit non-muscle myosin heavy chain (MHC) and protein kinases including Aurora kinases and MEK1, consistent with some of its effects on cytoskeletal remodelling and altering ATP levels (Table 4.3).[94,96,97]

The seminal report from Yamanaka and co-workers in 2006 that transfection of mouse embryonic fibroblasts with a cocktail of just four factors, *Sox2*, *Oct4*, *Klf4* and *c-Myc*, allowed their transformation into pluripotent cells,[11] demonstrated for the first time that it was possible to fully reprogramme differentiated cells to regain pluripotency. This report, and the later demonstration that these induced pluripotent stem cells (iPSCs) were germline competent,[2] was followed by several others reporting the generation of human iPSCs using the same four factors, or combinations thereof with other factors such as *Lin 28* and *Nanog*.[98,99] However, there are several drawbacks with this method employing viral transduction, most notably efficiency and the time required for reprogramming (2 weeks of treatment produces <0.01% iPSC forming colonies), with many of the remaining cells displaying partially differentiated phenotypes.[100] There is also a concern that introduction of known oncogenes such as *c-Myc* could have serious safety consequences; for example, virally reprogrammed cells show a higher incidence of tumour formation when reintroduced *in vivo* due to reactivation of the *c-Myc* transgene.[1] However, emerging technologies are addressing these issues,[101] including improved methods for gene delivery using non-integrating methods[102] and genetic modification-free methods such as the direct introduction of the corresponding tagged recombinant proteins.[103]

Several approaches employing small molecules to enhance reprogramming or to substitute for the introduction of one or more transgenes in the reprogramming of somatic cells to pluripotent cells have been trialed in an attempt to improve the overall speed and efficiency of the reprogramming process. In an effort to elucidate the requirements for reprogramming, a detailed study was undertaken to compare the characteristics of fully and partially reprogrammed iPSCs with ESCs.[104] It was observed that upon viral transfection early transcriptional changes are observed producing a partially reprogrammed state, but that full reprogramming to cells virtually indistinguishable, genetically and

**Table 4.3** Selected small molecules to reprogramme stem cell fate.

| Compound | Structure | Role | Possible target | Notes | Reference |
|---|---|---|---|---|---|
| **52** | 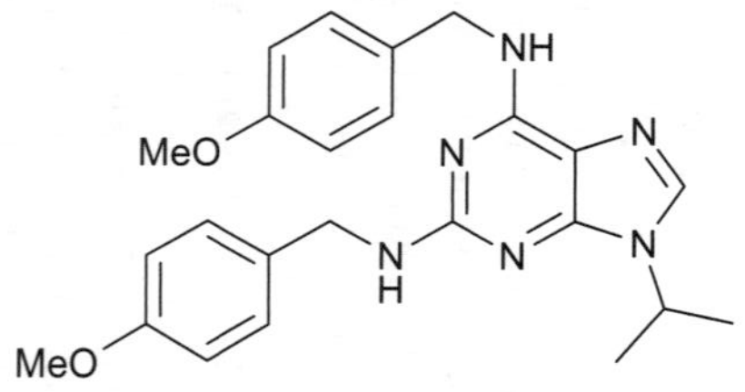 Myoseverin | Myotube remodelling | | Thought to alter myotube morphology rather than true dedifferentiation | Rosania et al.[92] |
| **53** | 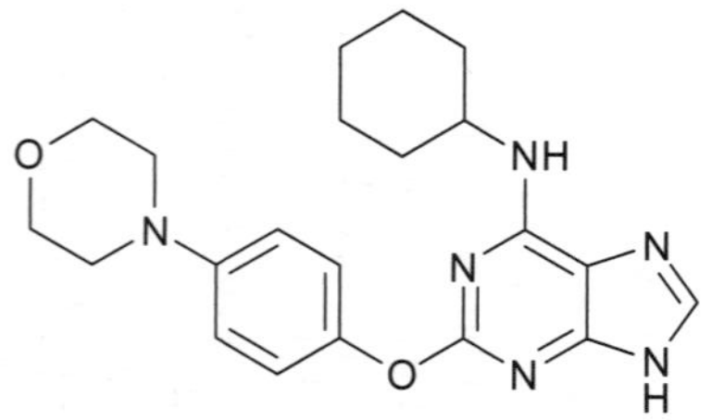 Reversine | Myoblast dedifferentiation | Non-muscle MHC; MEK1; aurora kinase | — | Chen et al.[93] |
| **54** | 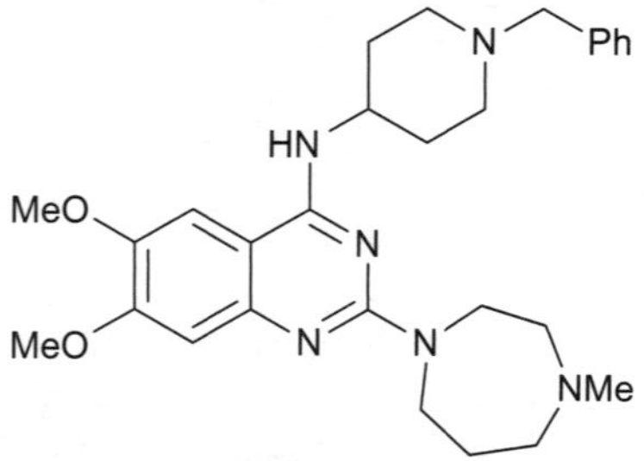 BIX-01294 | iPS cell generation | G9a histone methyltransferase | Enhances iPS cell generation from NPCs transduced with Oct 3/4 and Klf4 | Shi et al.[107] |

| | Structure | | | | |
|---|---|---|---|---|---|
| 55 | 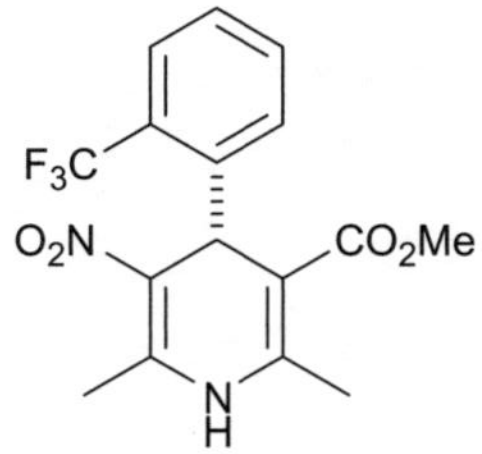 (+)-Bay K8644 | iPS cell generation | L-Type calcium channel blocker | With BIX-01294 iPS cells from fibroblasts transduced with *Oct 3/4* & *Klf4* | Shi *et al.*[108] |
| 56 | 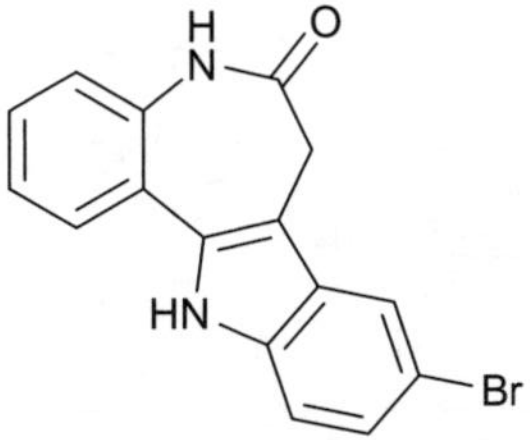 Kenpaullone | iPS cell differentiation | Unknown | Acts as a replacement for transduction of *Klf4* | Lyssiotis *et al.*[109] |
| 57 | 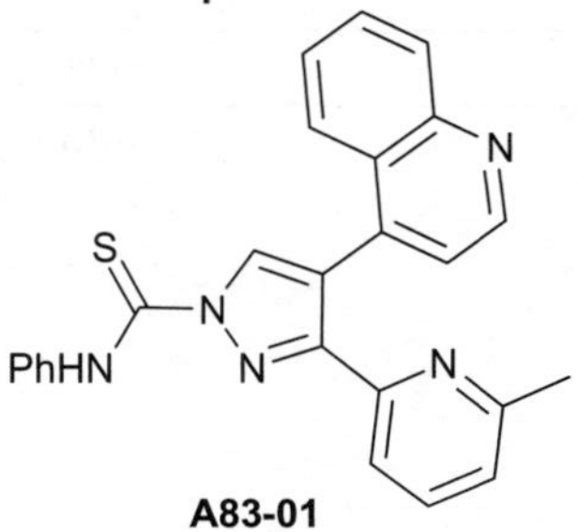 A83-01 | iPS cell self renewal | ALK5/7 inhibitor | Used in conjunction with PD0325901 and CHIR99021 | Li *et al.*[110] |

**Table 4.3**  *Continued*

| Compound | Structure | Role | Possible target | Notes | Reference |
|---|---|---|---|---|---|
| **58** | 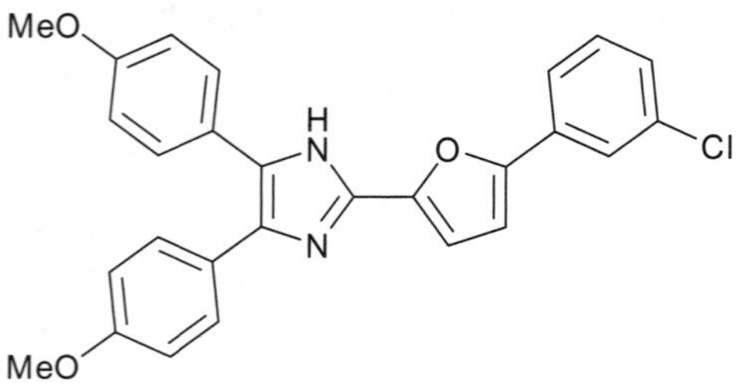 **Neurodazine** | Neuronal trans-differentiation from skeletal muscle cells | — | — | Williams et al.[115] |

epigenetically, from ESCs was extremely slow. The authors further demonstrated that suppression of DNA methylation using either shRNA knockdown of DNA methyltransferase 1 (Dnmt1) or the DNA methyltransferase inhibitor, 5-azacytidine, resulted in improved reprogramming efficiency (Table 4.3).[104] Consistent with the observation that epigenetic modifications can enhance reprogramming, histone deacetylase (HDAC) inhibitors such as valproic acid, trichostatin A and suberoylanilide hydroxamic acid (SAHA) have also been observed to dramatically enhance the efficiency of reprogramming; valproic acid enhancing reprogramming efficiency by more than 100-fold compared to viral transfection alone.[105] Further studies showed that valproic acid also serves to reduce the requirement for the transfection of all four reprogramming factors; *Oct4* and *Sox2* transfection alone allowed the reprogramming of human fibroblasts when used in conjunction with valproic acid.[106]

In an alternative approach, other cell types have been sought as starting points for reprogramming which already contain high endogenous levels of one or more of the desired transcription factors, for example neural stem cells (NSCs) expressing high levels of *Sox2* may be reprogrammed with high efficiencies to iPSCs through transduction with *Oct 3/4* and *Klf4* with a small molecule histone methyltransferase inhibitor, BIX-01294.[107] This approach is limited, however, by the availability of NSCs; while they may be accessed by biopsy in small quantities, they do not represent a practical starting point compared to fibroblasts. Further small molecules are therefore necessary to negate the requirement for the forced introduction of other transcription factors in fibroblasts, in particular molecules to suppress the induction of differentiation and to promote expression of pluripotency genes are needed.[104] iPSCs have recently been generated from fibroblasts using just *Oct4* and *Klf4* transduction and small molecules by combination with BIX-01294 and an L-type calcium channel antagonist, Bay K8644.[108] While Bay K8644 does not act to induce *Sox2* expression, it appears to compensate for *Sox2* transduction in this system. Similarly, kenpaullone appears to negate the requirement for *Klf4* transduction in fibroblast reprogramming.[109] The mechanism of action of kenpaullone is intriguing as it was previously identified as a GSK-3β inhibitor, and is also known to inhibit a number of cyclin-dependent kinases; however, other methods to inhibit the function of these kinases do not recapitulate the effects of kenpaullone. This suggests either that the molecule is exerting its effects through an as yet unknown mechanism or that the pleiotropic effects may be required for kenpaullone's activity, highlighting the value of phenotypic screens where no assumption about the molecular target is made.

Finally, small molecules have been employed to selectively maintain iPSC colonies in the presence of non-reprogrammed and/or partially reprogrammed somatic cells. It has been found that the MEK inhibitors PD0325901, ALK5,7 inhibitor A83-01, and GSK-3β inhibitor CHIR99021 together can promote the growth of iPSC colonies, applying a selection pressure to the heterogenous population.[110]

As yet there have been no reports of a single small molecule or combination of small molecules capable of dedifferentiating fibroblasts to a pluripotent state, although methods to enhance and accelerate reprogramming are known.[111] *Oct4* transfection has so far proved to be essential for reprogramming to pluripotency, and it remains to be seen when small molecules will emerge to compensate for this factor, and whether such compounds can be applied either alone or in combination with other small molecules to reprogramme somatic cells to pluripotency.

The direct reprogramming of one somatic or unipotent cell type to another – transdifferentiation – would circumvent the need to dedifferentiate to multi-potent or pluripotent states at all, and as such represents a highly attractive and potentially more efficient paradigm for drug discovery and regenerative medi-cine.[112,113] While there are examples of transdifferentiation occurring *in vivo* (see Section 4.3.2), there is little precedent *in vitro*. An early example involved the introduction of the master muscle regulator gene *MyoD* into a range of cultured cells: fibroblasts, chondroblasts and retinal epithelial cells were observed to convert into contracting muscle cells *in vitro* upon introduction of MyoD.[114] A small molecule, neurodazine, capable of promoting transdiffer-entiation of myoblasts into neurons has been reported, although the mechan-ism of action is unknown.[115] The authors also observed that myotubes could be reprogrammed to neurons with similar levels of efficiency by sequential treat-ment with myoseverin (to induce myoblast formation) and neurodazine. In a very recent report, Vierbuchen *et al.* demonstrated that fibroblasts may be converted directly into functional neurons by reprogramming intro-ducing just three factors identified from combinatorial analysis of a set of neuron-specific transcription factors, *Ascl1*, *Brn2* and *Myt1l*.[116] It remains to be seen whether this observation may be recapitulated with small mole-cules, or whether the strategy proves to be applicable to a wider range of cell types.


## 4.4   Stem Cell Manipulation *In Vivo*

One of the potentially most powerful applications of using small molecules to direct stem cell fate would be the ability to manipulate these cells *in vivo*. Such techniques could be effective in principle both for the activation of exogenous administered stem cells in cell-based therapy, and for the direct manipulation of endogenous stem cells. There would be enormous advantages in having the capability to safely activate, proliferate and specify the fate of a patient's own endogenous stem cells in order to repair damaged or diseased tissue *solely through administering a drug* (*or drugs*), thereby precluding the need for cell or tissue transplantation therapies.

The presence of adult multipotent stem cells has been established in many tissue types in humans including haematopoietic stem cells (HSCs), pancreatic progenitor cells, retinal progenitor cells, neural stem cells (NSCs), epidermal stem cells, cardiomyocytes, oval cells (liver) and satellite cells (muscle), although their regenerative capacity and even their existence have been con-troversial in many cases (Figure 4.3).[25] In most cases in humans these tissue

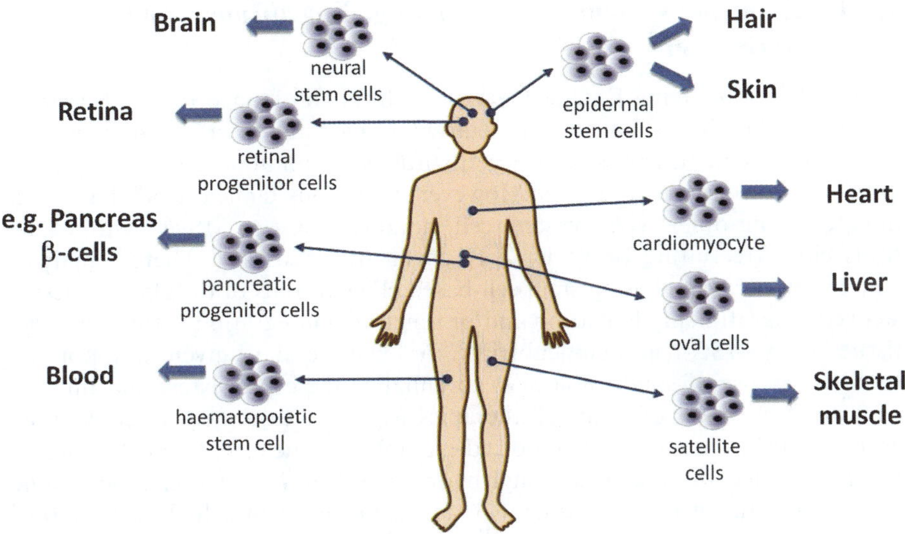


**Figure 4.3**   Examples of human endogenous stem cell populations potentially capable of regeneration through activation *in vivo*.

stem cells are believed to contribute to tissue repair and homeostasis, although many appear to have a limited regenerative capacity, as evidenced in such disorders as muscular dystrophy, where muscle satellite cells' regenerative capacity is exhausted as the disease progresses.[25] There are many examples in nature, however, where organisms such as fish and amphibians exhibit more widespread tissue regeneration. A notable example is the salamander, which is capable of regrowing its own limbs in response to injury, through reprogramming of endogenous cells.[117] To be able to recapitulate or activate such repair mechanisms, or enhance existing tissue repair mechanisms in humans would have far-reaching consequences on treatment strategies for a wide range of diseases. Stem cells in adult tissue reside within distinct microenvironments (niches) which are believed to regulate stem cell fate *in vivo*.[118] This provides several potential therapeutic options for *in vivo* stem cell activation and manipulation, either directly through targeting the endogenous stem cells, or indirectly through manipulation of their microenvironment.

Although the concept of endogenous stem cell activation for regenerative therapy is only recently gathering momentum, there are in fact already several examples of agents acting on endogenous stem cells in clinical use. For example, erythropoietin (EPO) and granulocyte colony-stimulating factor (G-CSF) have been employed to stimulate the production of red and white blood cells, respectively, through the activation of HSCs. HSCs for transplantation are now most commonly obtained from either umbilical cord blood, or from treatment of a patient with G-CSF to stimulate HSC mobilisation from the bone marrow.[119]

### 4.4.1 Exogenous Stem Cells: Homing, Engraftment and Activation

Clinical trials involving ESC-derived cells, although planned for the treatment of spinal cord injuries, have not yet started, in part due to safety concerns over the possible consequences of trace quantities of undifferentiated ESC contaminants in administered cells. However, numerous clinical trials have been initiated using other types of stem cell, notably HSCs or MSCs, with >2500 trials either recruiting or under way as of April 2010.[120] There remains a concern, even with ongoing stem cell-based clinical trials of the effectiveness of stem cell engraftment, activation and/or functional integration within the target tissue; many MSC-based clinical trials, for example, have given disappointing results. The use of small molecules to enhance these processes, and improve clinical effectiveness of stem cell therapies would be extremely valuable; however, as yet there are very few reported examples.[52] One recent report, however, has highlighted the use of sulfonyl hydrazone, **Shz-1**, which enhanced functional engraftment upon treatment of cardiac progenitor cells in a rat model, albeit by an unknown mechanism.[121]

### 4.4.2 Endogenous Stem Cell Manipulation

The effects of small molecule modulators on a number of endogenous stem cell populations have been investigated to date, including HSCs, NSCs, hair follicle cells, retinal precursor cells and cardiomyocytes. Although G-CSF is widely exploited for the mobilisation of HSCs in the clinic, it is not effective across all patients, and small molecules capable of mobilising HSCs from bone marrow, either alone or in combination with G-CSF, have also been identified. Examples include AMD3100 (Plerixafor), a bismacrocycle which has been shown to be a partial agonist of the chemokine receptor CXCR4, which has recently completed Phase III clinical trials (Table 4.4).[122] G-CSF has also since been reported to potentially play a number of roles in regeneration, and it is likely small molecules recapitulating its effects will follow.

Studies to investigate key players and pathways involved in HSC emergence have also been performed *in vivo* in the zebrafish, a combination of small molecule modulators and genetic manipulation was able to elucidate the roles of hedgehog and BMP signalling in HSC formation.[123] The zebrafish was also used as a model system to perform an *in vivo* screen for small molecules capable of expanding HSCs *in vivo*, revealing the role of $PGE_2$ signalling and the role of the inflammatory response.[52] 16,16-Dimethyl-$PGE_2$ was found to expand HSCs *in vivo* and *in vitro*, and to enhance kidney marrow recovery after irradiation injury in adult zebrafish, while indomethacin, a cyclooxygenase-2 inhibitor, suppressed HSC numbers. 16,16-Dimethyl-$PGE_2$ and compounds acting through similar mechanisms show great promise for patients undergoing HSC transplantation, or potentially to allow endogenous HSC activation and expansion *in vivo* in patients. Furthermore, clinical defects in haematopoiesis result from over-activation of TGF-β signalling in haematopoietic precursor

**Table 4.4**  Small molecules to manipulate stem cell fate *in vivo*.

| Compound | Structure | Role | Possible target | Notes | Reference |
|---|---|---|---|---|---|
| 59 | 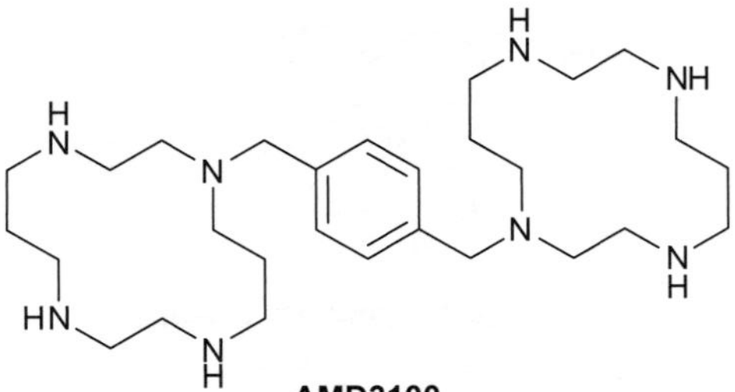 **AMD3100** | HSC mobilisation | CXCR4 antagonist | Works *in vivo* with or without G-CSF | Flomenberg *et al.*[122] |
| 60 | 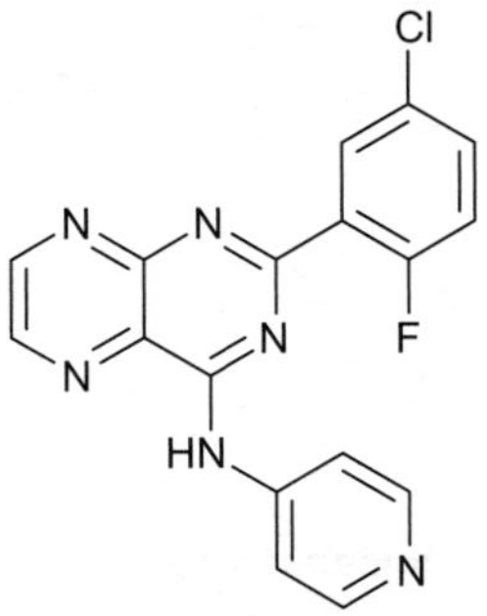 **SD-208** | Haematopoiesis | TGFβ receptor kinase I | Rescues defect in haematopoiesis *in vivo* | Zhou *et al.*[124] |
| 61 | 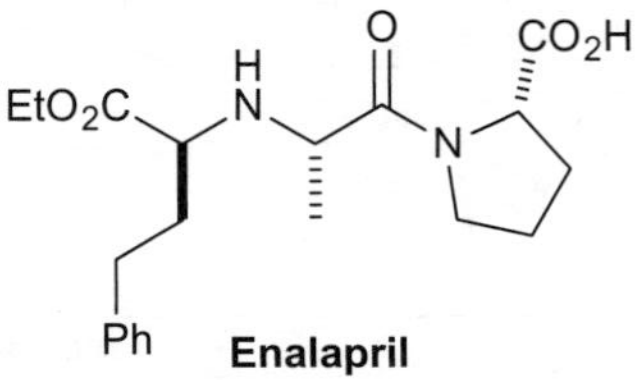 **Enalapril** | Endothelial pro-genitor cell mobilisation | ACE inhibitor | — | Wang *et al.*[126] |

**Table 4.4** *Continued*

| Compound | Structure | Role | Possible target | Notes | Reference |
|---|---|---|---|---|---|
| **62** | 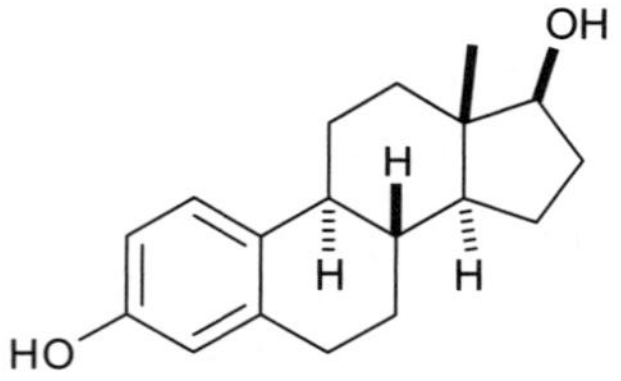 **17β-estradiol** | Endothelial progenitor cell mobilisation | — | — | Strehlow et al.[127] |
| **63** | 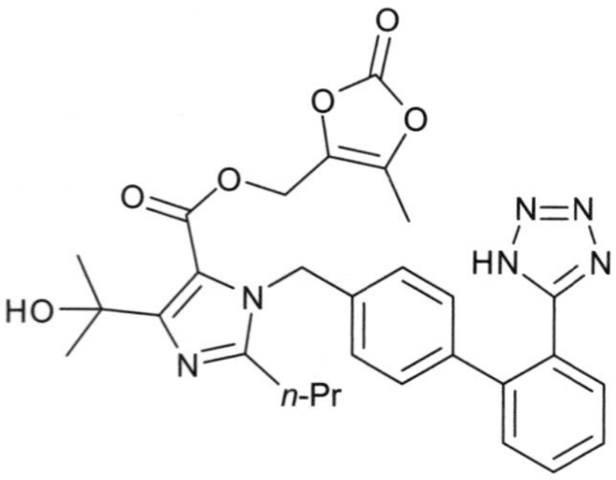 **Olmesartan medoxomil** | Increase in endothelial progenitor cells | Angiotensin Receptor 1 antagonist | — | Bahlmann et al.[128] |
| **64** | 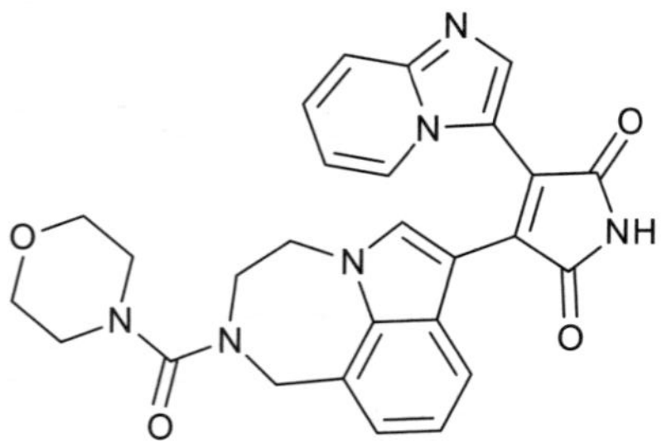 **603281-31-8** | Osteogenic differentiation | GSK-3β | Wnt activation, works *in vivo* | Kulkami et al[129] |

| | Structure | | | | |
|---|---|---|---|---|---|
| 65 | 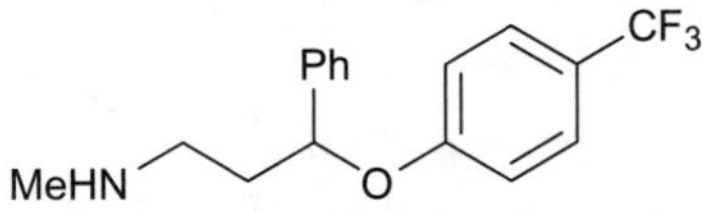 **Fluoxetine** | Neural differentiation | Selective serotonin reuptake inhibitor | — | Malberg *et al.*[132] |
| 66 |  **Tranylcypromine** | Neural differentiation | Monoamine oxidase inhibitor/demethylase inhibitor | — | Malberg *et al.*[133] |
| 67 | 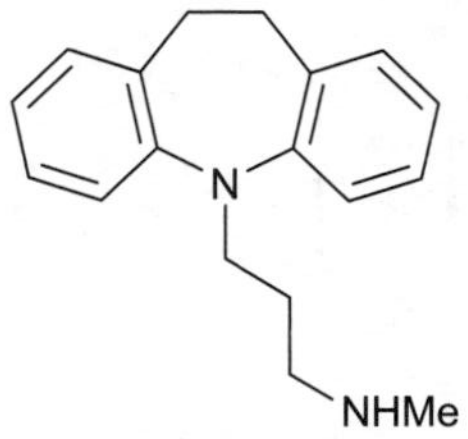 **Desipramine** | Neural differentiation | Norepinephrine reuptake inhibitor | — | Malberg *et al.*[133] |
| 68 | 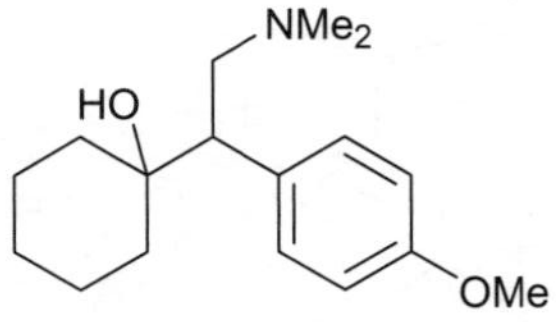 **Venlafaxine** | Neural differentiation | Selective nor-epinephrine reuptake inhibitor | — | Malberg *et al.*[133] |

**Table 4.4**  *Continued*

| Compound | Structure | Role | Possible target | Notes | Reference |
|---|---|---|---|---|---|
| **69** | 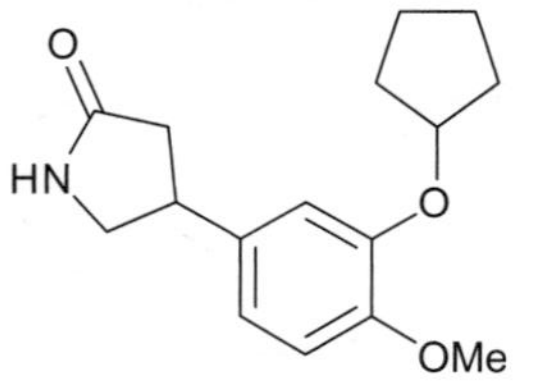 Rolipram | Neural differentiation | PDE4 inhibitor | — | Malberg et al.[133] |
| **70** | 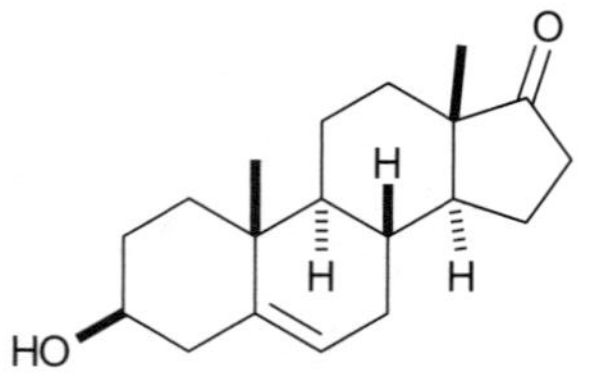 Dehydroepiandrosterone(DHEA) | Neurogenesis | Unknown | — | Karishma and Herbert[135] |
| **71** | 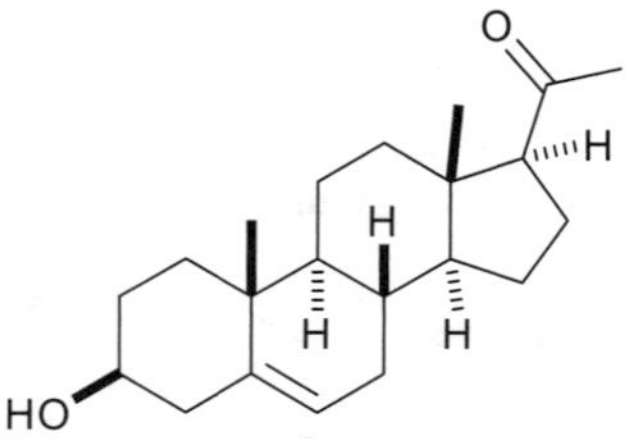 Pregnenolone | Neurogenesis | Unknown | — | Mayo et al.[136] |

| | | | | | |
|---|---|---|---|---|---|
| **72** | 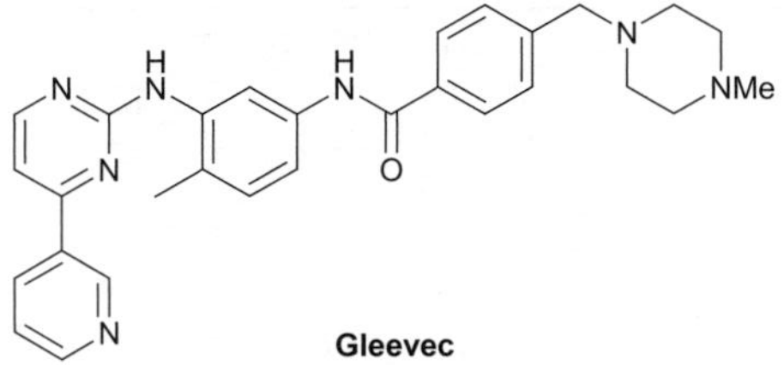 **Gleevec** | Tumour proliferation | BCR-Abl fusion; c-Kit; PDGFR | CML differentiation therapy | Druker *et al.*[147] |
| **73** | 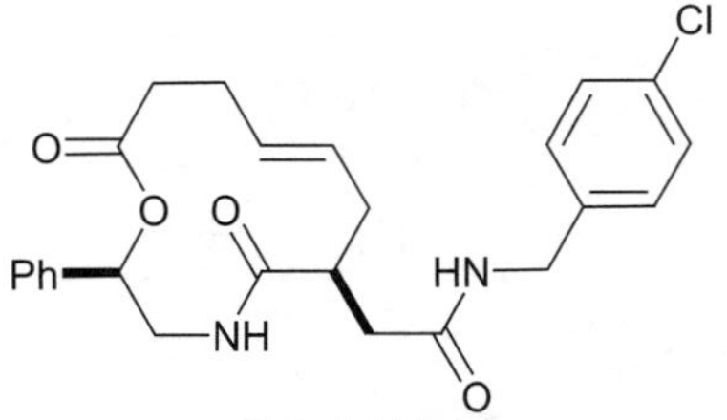 **Robotnikinin** | Tumour proliferation | Hh signalling inhibitor | — | Stanton *et al.*[148] |
| **74** | 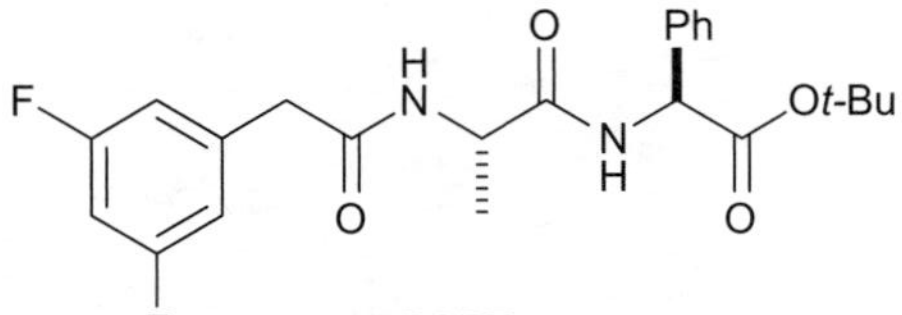 **DAPT** | Promotes hair follicle growth; blocks retinal regeneration | Notch inhibitor | — | Yamamoto *et al.*[140]; Hayes *et al.*[141] |
| **75** | 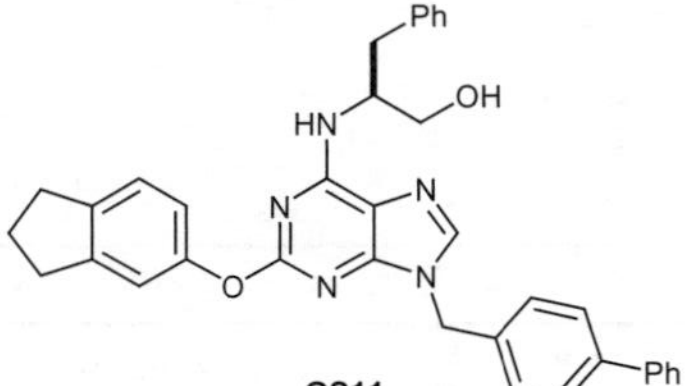 **QS11** | Blocks Gastrointestinal regeneration | Tankyrase 1/2 inhibitor | Axin stabiliser/ Wnt inhibitor | Chen *et al.* |

**Table 4.4** *Continued*

| Compound | Structure | Role | Possible target | Notes | Reference |
|---|---|---|---|---|---|
| **76** | 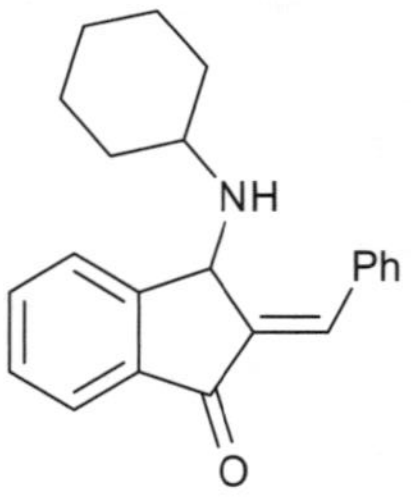 **BCl** | Expansion of cardiac cell lineages | Dusp6 inhibitor | FGF agonist identified *in vivo* | Molina *et al.*[143] |
| **77** | 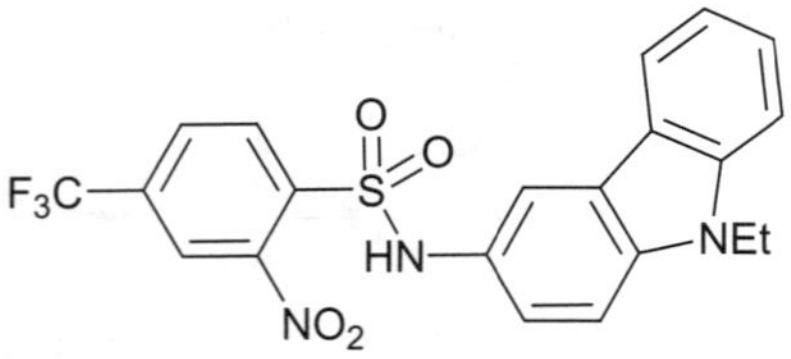 **Cardiosulfa** | Impairs cardiac development | — | *In vivo* | Ko *et al.*[149] |
| **78** | 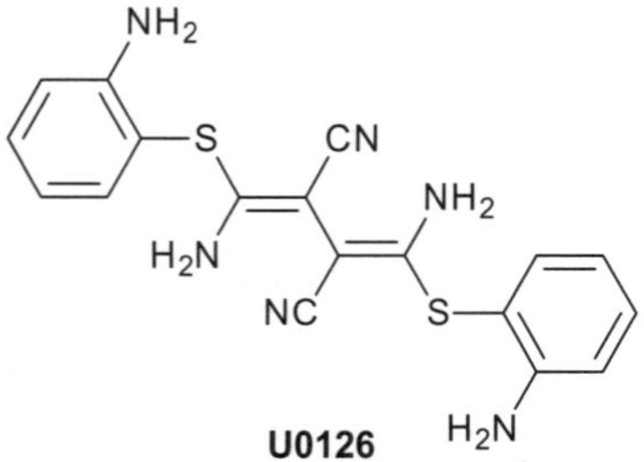 **U0126** | Germ cell reprogramming | MAPK | *In vivo* in *C. elegans* | Morgan *et al.*[145] |

cells. It has been demonstrated that this defect may be overcome by treatment *in vivo* with **SD-208**, a TGF-β inhibitor.[124]

Besides HSCs, mobilisation of other types of stem cell has also been achieved *in vivo* using small molecules including endogenous endothelial progenitor cells (also an action of EPO[125]) using enalapril,[126] an angiotensin II converting enzyme inhibitor, or oestrogens such as 17β-estradiol.[127] In patients with type II diabetes mellitus, treatment with olmesartan, an angiotensin receptor 1 antagonist, were also reported to give an increase in endothelial progenitor cells.[128]

Methods to manipulate bone remodelling *in vivo* have also attracted much interest due to their potential application in injuries or diseases such as osteoporosis. Several pathways and opportunities for therapeutic intervention have been identified, including hedgehog, bone morphogenetic protein and Wnt signalling. Wnt activators such as 603281-31-8 have been shown to promote mineralisation and increase in bone mass *in vitro* and *in vivo*.[129] Oestrogens such as 17β-estradiol have been shown to promote osteogenesis of bone marrow stromal cells while suppressing adipogenesis,[130] and selective estrogen receptor modulators have been used in the clinic for the treatment of osteoporosis.[131]

Investigations into many anti-depressants currently on the market have also recently revealed a previously unknown effect on neurogenesis *in vivo*,[132,133] an observation hypothesised to translate into clinical activity.[134] A range of anti-depressants with diverse mechanisms were all observed to enhance neural proliferation *in vitro* and promote neurogenesis *in vivo* in rats including fluoxetine (selective serotonin reuptake inhibitor, SSRI), tranylcypromine (monoamine oxidase inhibitor, demethylase inhibitor), desipramine (nor-epinephrine/serotonin reuptake inhibitor), venlafaxine (selective nor-epinephrine reuptake inhibitor) and rolipram (phosphodiesterase 4 inhibitor). Hippocampal neurogenesis was also observed to increase *in vivo* in the rat with dehydroepiandrosterone (DHEA) treatment[135] or pregnenolone treatment.[136] The role of hedgehog signalling in stimulating neurogenesis has been exploited by the use of small molecule hedgehog agonists to produce medial motor column neurons *in vitro*[137] and *in vivo* in a rat model of spinal cord injury.[138]

While there are many examples of small molecules capable of manipulating endogenous stem cell fate which have been shown to be safe, a concern remains over the potential dangers of stimulating stem cell proliferation or differentiation *in vivo*. Fears have been raised over potential undesired effects of these molecules on other cell types leading to proliferative diseases, and it is likely that the dosing of any such drugs will need to be carefully controlled. A possible means to side-step this issue is to apply this strategy to locations where drugs would be administered either topically, or into self-contained systems such as intraocularly. For example, sonic hedgehog agonists[139] and notch inhibitors such as DAPT[140] have been applied for the treatment of hair loss, and have been shown to promote hair follicle regeneration. While there are no small molecules as yet capable of regenerating retina within the eye, compounds have been used as tools to probe discrepancies in animals capable of regenerating retina, such as fish and chick, and those that cannot in order to elucidate pathways involved in retinal regeneration. Notch signalling has been shown to play an important role in this context.[141]

A number of other *in vivo* screens have been performed to reveal small molecules to manipulate stem cell fate, mostly using model organisms such as transparent, externally developing zebrafish or xenopus embyros. The small molecule synergist of Wnt signalling, **QS11**, was observed to act *in vivo* in *Xenopus* to promote axis duplication.[142] An *in vivo* screen in the zebrafish revealed small molecules capable of expanding cardiac lineages *in vivo*, these function at least in part through activation of FGF signalling through the inhibition of the dual specificity phosphatase Dusp6.[143]

Cellular reprogramming has also been recently achieved directly *in vivo*. In a pioneering study by Zhou *et al.* mature pancreatic exocrine (acinar) cells were reprogrammed *in vivo* in mice to functional insulin-secreting endocrine β-cells using viral transduction of three factors together from nine candidate genes believed to be required for β-cell specification.[144] Furthermore, a small molecule approach enabled the identification of MAPK inhibitors such as U0126 to promote the reprogramming of germ cells *in vivo* in masculinised worms.[145] These studies demonstrate the feasibility of manipulating stem cell fate, or reprogramming cell fate *in vivo*, the next years and decades will show whether these strategies may translate into new clinical therapies.

### 4.4.3   Stem Cells in Dysplasia: Cancer and Teratogenicity

It is now clear that there are many striking parallels between stem cell development and oncogenesis. Similarly, cancer stem cells (CSCs) share many common features with adult stem cells: they are capable of self-renewal whilst retaining the capability to differentiate into all cell types required to propagate the tumour. CSCs were originally detected in leukaemia, but they have since been reported in a range of other haematological and solid tumours.[146] A significant problem exists as the majority of current chemotherapeutic agents target the bulk of the tumour cells, but do not target CSCs. Therapeutic strategies to address this issue have been evaluated, particularly in leukaemias where the use of differentiation therapy aims to promote the specification of CSCs towards more mature lineages which will be susceptible to apoptosis.[23] Amongst others, the use of Gleevec[147] for chronic myeloid leukaemia and retinoic acid for acute promyelocytic leukaemia have both been evaluated as differentiation therapies in the clinic.[23] It is anticipated that there will be a role for numerous other antiproliferative agents in stem cell manipulation *in vitro* and *in vivo* such as signalling pathway modulators hedgehog (*e.g.* robotnikinin[148]) and notch.

Zebrafish have been used to screen for small molecules giving rise to teratogenic effects; for example, cardiosulfa was found to give rise to severe heart defects in early zebrafish development.[149] Such molecules will provide valuable tools in elucidating key events in development and may prove to be useful tools themselves in manipulating stem cell fate. Similarly there has been a resurgence of interest in the effect and mode of action of teratogenic molecules such as thalidomide, including its immunomodulatory effects. There are a variety of other small molecules and drug candidates that have been abandoned due to teratogenic effects observed *in vivo* and it is likely that many of these will emerge as useful candidates for the manipulation of stem cells in the future.

## 4.5   Conclusions and Future Prospects

It is clear that stem cells represent a huge opportunity to revolutionise medicine in the future, providing a range of treatment options to regenerate or repair tissue due to injury or degenerative diseases. Stem cells also potentially embody an immensely valuable tool to enable drug discovery through disease pathway modelling, target validation and drug efficacy and toxicology assessment. With the discovery of induced pluripotent stem cell technology and methods to promote cellular reprogramming there is a genuine prospect to realise personalised healthcare. Harnessing stem cells for application in clinical therapies and drug discovery will be a major challenge. The use of small molecules to control the destiny of stem cells *in vitro* and *in vivo* is an emerging discipline which offers unprecedented advantages over other techniques in terms of speed, cost, reproducibility and the ability to influence stem cell fate reversibly. Since the earliest observations of the impact of naturally occurring small molecules on manipulating stem cell fate, there have been numerous examples reported of small molecules efficiently and rapidly controlling the destiny of stem cells through a variety of mechanisms both *in vitro* and *in vivo*. In the coming years and decades these technologies will undoubtably have a huge impact, increasing our understanding of the fundamental biology of stem cell regulation, and on the exploitation of stem cells for regenerative medicine and drug discovery.

## Acknowledgements

The authors wish to acknowledge Research Councils UK for the provision of a fellowship (A.J.R.), Dr David Wilson for assistance in preparing many of the figures, and Dr Carole Bataille, Professor Dame Kay Davies, Professor Leonard Seymour, Dr Robert Westwood and Dr Graham Wynne for helpful advice and comments, and for proof reading this manuscript.

## References

1. K. Takahashi and S. Yamanaka, *Cell*, 2006, **126**, 663.
2. K. Okita, T. Ichisaka and S. Yamanaka, *Nature*, 2007, **448**, 313.
3. R. Lanza, J. Gearhart, B. Hogan, D. Melton, R. Pederson, J. Thomson and M. West (Ed.), *Handbook of Stem Cells*, Elsevier Academic Press, 2004.
4. C. W. Pouton and J. M. Haynes, *Adv. Drug Deliv. Rev.*, 2005, **57**, 1918.

5. J. D. McNeish, *Curr. Opin. Pharmacol.*, 2007, **7**, 515.

6. C. W. Pouton and J. M. Haynes, *Nat. Rev. Drug Discovery*, 2007, **6**, 605.

7. (a) N. Emre, R. Coleman and S. Ding, *Curr. Opin. Chem. Biol.*, 2007, **11**, 252; (b) T. E. Allsopp, M. E. Bunnage and P. V. Fish, *Med. Chem. Comm.*, 2010, **1**, 16.

8. N. S. Hwang, S. Varghese and J. Elisseeff, *Adv. Drug Deliv. Rev.*, 2008, **60**, 199.

9. S. E. Harding, N. N. Ali, M. Brito-Martins and J. Gorelik, *Pharmacol. Ther.*, 2007, **113**, 341.

10. Q. L. Ying, J. Wray, J. Nichols, L. Batlle-Morera, B. Doble, J. Woodgett, P. Cohen and A. Smith, *Nature*, 2008, **453**, 519.

11. R. E. Ploemacher, *Ballière's Clin. Haematol.*, 1997, **10**, 429.

12. J. E. Talmadge, *Int. Immunopharmacol.*, 2003, **3**, 1121.

13. A. S. Tsiftsoglou, I. D. Bonovolias and S. A. Tsiftsoglou, *Pharmacol. Ther.*, 2009, **122**, 264.

14. S. S. Grewel, 3J. N. Barker, S. M. Davies and J. E. Wagner, *Blood*, 2003, **101**, 4233.

15. F. Anjos-Afonso, E. K. Siapati and D. Bonnet, *J. Cell Sci.*, 2004, **117**, 5655.

16. A. G. Smith, *Semin. Cell Biol.*, 1992, **3**, 385.

17. C. Améen, R. Strehl, P. Björquist, A. Lindahl, J. Hyllner and P. Sartipy, *Crit. Rev. Oncol. Hematol.*, 2008, **65**, 54.

18. Y.-H. Liu, R. Karra and S. M. Wu, *Drug Discov. Today: Ther. Strat.*, 2008, **5**, 201.

19. L. Anastasia, G. Pelissero, B. Venerando and G. Tettamanti, *Cell Death Differ.*, 2010, **17**, 1230.

20. S. Strickland and V. Mahdavi, *Cell*, 1978, **15**, 393.

21. H. Jakob, P. Dubois, H. Eisen and F. Jacob, *C. R. Acad. Sci. Hebd. Seances Acad. Sci. D*, 1978, **286**, 109.

22. G. Bain, D. Kitchens, M. Yao, J. E. Huettner and D. I. Gottlieb, *Dev. Biol.*, 1995, **168**, 342.

23. S. Sell, *Stem Cell Rev.*, 2005, **1**, 197.

24. Y. Xu, Y. Shi and S. Ding, *Nature*, 2008, **453**, 338.

25. K. Sakurada, F. M. McDonald and F. Shimada, *Angew. Chem. Int. Ed.*, 2008, **47**, 5718.

26. A. J. Firestone and J. K. Chen, *ACS Chem. Biol.*, 2010, **5**, 15.

27. A. I. Lukaszewicz, M. K. MacMillan and M. Kahn, *J. Med. Chem.*, 2010, **53**, 3439.

28. C. E. Murry and G. Keller, *Cell*, 2008, **132**, 661.

29. G. H. Underhill and S. N. Bhatia, *Curr. Opin. Chem. Biol.*, 2007, **11**, 357.

30. S. I. Nishikawa, L. M. Jakt and T. Era, *Nature Rev: Mol. Cell. Biol.*, 2007, **8**, 502.

31. J. A. King and W. M. Miller, *Curr. Opin. Chem. Biol.*, 2007, **11**, 394.

32. H. Niwa, T. Burdon, I. Chambers and A. Smith, *Genes Dev.*, 1998, **12**, 2048.

33. Q. L. Ying, J. Nichols, I. Chambers and A. Smith, *Cell*, 2003, **115**, 281.

34. N. Sato, L. Meijer, L. Skaltsounis, P. Greengard and A. H. Brivanlou, *Nat. Med.*, 2004, **10**, 55.
35. T. Miyabayashi, J. L. Teo, M. Yamamoto, M. McMillan, C. Nguyen and M. Kahn, *Proc. Natl. Acad. Sci. U. S. A.*, 2007, **104**, 5668.
36. X. Qi, T. G. Li, J. Hao, J. Hu J. Wang, H. Simmons, S. Miura, Y. Mishina and G. Q. Zhao, *Proc. Natl. Acad. Sci. U. S. A.*, 2004, **101**, 6027.
37. D. Duval, M. Malaise, B. Reinhardt, C. Kedinger and H. A. Boeuf, *Cell Death Differ.*, 2004, **11**, 331.
38. S. Chen, J. T. Do, Q. Zhang, S. Yao, F. Yan, E. C. Peters, H. R. Schöler, P. G. Schultz and S. Ding, *Proc. Natl. Acad. Sci. U. S. A.*, 2006, **103**, 17266.
39. T. Miyabayashi, M. Yamamoto, A. Sato, S. Sakano and Y. Takahashi, *Biosci. Biotechnol. Biochem.*, 2008, **72**, 1242.
40. L. G. Chase and M. T. Firpo, *Curr. Opin. Chem. Biol.*, 2007, **11**, 367.
41. L. Vallier, M. Alexander and R. A. Pedersen, *J. Cell Sci.*, 2005, **118**, 4495.
42. P. Sartipy, R. Strehl, P. Björquist and J. Hyllner, *Pharmacol. Res.*, 2008, **58**, 152.
43. S. C. Desbordes, D. G. Placantonakis, A. Ciro, N. D. Socci, G. Lee, H. Djaballah and L. Studer, *Cell Stem Cell*, 2008, **2**, 602.
44. K. Watanabe, M. Ueno, D. Kamiya, A. Nishiyama, M. Matsumura, T. Wataya, J. B. Takahashi, S. Nishikawa, S. Nishikawa, K. Muguruma and Y. Sasai, *Nat. Biotechnol.*, 2007, **25**, 681.
45. F. Soncin, L. Mohamet, D. Eckardt, S. Ritson, A. M. Eastham, N. Bobola, A. Russell, S. Davies, R. Kemler, C. L. R. Merry and C. M. Ward, *Stem Cells*, 2009, **27**, 2069.
46. L. Conti, S. M. Pollard, T. Gorba, E. Reitano, M. Toselli, G. Biella, Y. Sun, S. Sanzone, Q.-L. Ying, E. Cattaneo and A. Smith, *PLoS Biol.*, 2005, **3**, 1594.
47. J. M. Wang, P. B. Johnstone, B. G. Ball and R. D. Brinton, *J. Neurosci.*, 2005, **25**, 4706.
48. W. Wang, J. R. Walker, X. Wang, M. S. Tremblay, J. W. Lee, X. Wu and P. G. Schultz, *Proc. Natl. Acad. Sci. U. S. A.*, 2009, **106**, 1427.
49. J. Jaroscak, K. Goltry, A. Smith, B. Waters-Pick, P. L. Martin, T. A. Driscoll, R. Howrey, N. Chao, J. Douville, S. Burhop, P. Fu and J. Kurtzberg, *Blood*, 2003, **101**, 5061.
50. J. P. Chute, G. G. Muramoto, J. Whitesides, M. Colvin, R. Safi, N. J. Chao and D. P. McDonnell, *Proc. Natl. Acad. Sci. U. S. A.*, 2006, **103**, 11707.
51. T. Nishino, K. Miyaji, N. Ishiwata, K. Arai, M. Yui, Y. Asai, H. Nakauchi and A. Iwama, *Exp. Hematol.*, 2009, **37**, 1364.
52. T. E. North, W. Goessling, C. R. Walkley, C. Lengerke, K. R. Kopani, A. M. Lord, G. J. Weber, T. V. Bowman, I. H. Jang, T. Grosser, G. A. Fitzgerald, G. Q. Daley, S. H. Orkin and L. I. Zon, *Nature*, 2007, **447**, 1007.
53. N. Korin and S. Levenberg, *Biotechnol. Genet. Eng. Rev.*, 2007, **24**, 243.
54. D. A. F. Loebel, C. M. Watson, R. A. De Young and P. P. L. Tam, *Dev. Biol*, 2003, **264**, 1.

55. C. Xiong, C. Q. Xie, L. Zhang, J. Zhang, K. Xu, M. Fu, W. E. Thompson, L.-J. Yang and Y. E. Chen, *Stem Cells Dev.*, 2005, **14**, 671.
56. J. Hong, J. Lee, K. H. Min, J. R. Walker, E. C. Peters, N. S. Gray, C. Y. Cho and P. G. Schultz, *ACS Chem. Biol.*, 2007, **2**, 171.
57. C. Xu, S. Police, N. Rao and M. K. Carpenter, *Circ. Res.*, 2002, **91**, 501.
58. A. B. Lassar, B. M. Paterson and H. Weintraub, *Cell*, 1986, **47**, 649.
59. T. Kawamura, K. Ono, T. Morimoto, H. Wade, M. Hirai, K. Hidaka, T. Morisaki, T. Heike, T. Nakahata, T. Kita and K. Hasegawa, *J. Biol. Chem.*, 2005, **280**, 19682.
60. T. Takahashi, B. Lord, P. C. Schulze, R. M. Fryer, S. S. Sarang, S. R. Gullans and R. T. Lee, *Circulation*, 2003, **107**, 1912.
61. R. Passier, D. W. Oostward, J. Snapper, J. Kloots, R. J. Hassink, E. Kuijk, B. Roelen, A. B. de la Riviere and C. Mummery, *Stem Cells*, 2005, **23**, 772.
62. R. Graichen, X. Xu, S. R. Braam, T. Balakrishnan, S. Norfiza, S. Sieh, S. Y. Soo, S. C. Tham, C. Mummery, A. Colman, R. Zweigerdt and B. P. Davidson, *Differentiation*, 2007, **76**, 357.
63. J. Hao, M. A. Daleo, C. K. Murphy, P. B. Yu, J. N. Ho, J. Hu, R. T. Peterson, A. K. Hatzolpoulos and C. C. Hong, *PLoS ONE*, 2008, **3**, e2904.
64. Z. L. Wei, P. A. Petukhov, F. Bizik, J. C. Teixeira, M. Mercola, E. A. Volpe, R. I. Glazer, T. M. Willson and A. P. Kozikowski, *J. Am. Chem. Soc.*, 2004, **126**, 16714.
65. A. Sachinidis, S. Schwengberg, R. Hippler-Altenburg, D. Mariappan, N. Kamisetti, B. Seelig, A. Berkessel and J. Hescheler, *Cell Physiol. Biochem.*, 2006, **18**, 303.
66. X. Wu, S. Ding, Q. Ding, N. S. Gray and P. G. Schultz, *J. Am. Chem. Soc.*, 2004, **126**, 1590.
67. N. Jaiswal, S. E. Haynesworth, A. I. Kaplan and S. P. Bruder, *J. Cell Biochem.*, 1997, **64**, 295.
68. X. Wu, S. Ding, Q. Ding, N. S. Gray and P. G. Schultz, *J. Am. Chem. Soc.*, 2002, **124**, 14520.
69. K. W. Park, H. Waki, W.-K. Kim, B. S. J. Davies, S. G. Young, F. Parhami and P. Tontonoz, *Mol. Cell. Biol.*, 2009, **29**, 3905.
70. G. Castelo-Branco, N. Rawal and E. Arenas, *J. Cell Sci.*, 2004, **117**, 5731.
71. N. R. Kim, S. K. Kang, H. H. Ahn, S. W. Kwon, W. S. Park, K. S. Kim, S. S. Kim, H. J. Jung, S. U. Choi, J. H. Ahn and K. R. Kim, *J. Med. Chem.*, 2009, **52**, 7931.
72. X. J. Li, B. Y. Hu, S. A. Jones, Y. S. Zhang, T. Lavaute, Z. W. Du and S. Zhang, *Stem Cells*, 2008, **26**, 886.
73. M. Dezawa, H. Kanno, M. Hoshino, H. Cho, N. Matsumoto, Y. Itokazu, N. Tajima, H. Yamada, H. Sawada, H. Ishikawa, T. Mimura, M. Kitada, Y. Suzuki and C. Ide, *J. Clin. Invest.*, 2004, **113**, 1701.
74. S. M. Chambers, C. A. Fasano, E. P. Papapetrou, M. Tomishima, M. Sadelain and L. Studer, *Nature Biotechnol.*, 2009, **27**, 275.

75. S. Ding, T. Y. Wu, A. Brinker, E. C. Peters, W. Hur, N. S. Gray and P.G. Schultz, *Proc. Natl. Acad. Sci. U. S. A.*, 2003, **100**, 7632.

76. M. Warashina, K. Hoon Min, T. Kuwabara, A. Huynh, F. H. Gage, P. G. Schultz and S. Ding, *Angew. Chem. Int. Ed.*, 2006, **45**, 591.

77. J. P. Saxe, H. Wu, T. K. Kelly, M. E. Phelps, Y. E. Sun, H. I. Kornblum and J. Huang, *Chem. Biol.*, 2007, **14**, 1019.

78. M. Kim, A. Habiba, J. M. Doherty, J. C. Mills, R. W. Mercer and J. E. Huettner, *Dev. Biol.*, 2009, **328**, 456.

79. V. B. Christie, J. H. Barnard, A. S. Batsanov, C. E. Bridgens, E. B. Cartmell, J. C. Collings, D. J. Maltman, C. P. F. Redfern, T. B. Marder, S. Przyborski and A. Whiting, *Org. Biomol. Chem.*, 2008, **6**, 3497.

80. F. Osakada, Z.-B. Jin, Y. Hirami, H. Ikeda, T. Danjyo, K. Watanabe, Y. Sasai and M. Takahashi, *J. Cell. Sci.*, 2009, **122**, 3169.

81. J. K. Chen, J. Taipale, K. E. Young, T. Maiti and P. A. Beachy, *Proc. Natl. Acad. Sci. U. S. A.*, 2002, **99**, 14071.

82. H. Wichterle, I. Lieberman, J. A. Porter and T. M. Jessell, *Cell*, 2002, **65**, 2537.

83. N. Lumelsky, O. Blondel, P. Laeng, I. Velasco, R. Ravin and R. McKay, *Science*, 2001, **292**, 1389.

84. Y. Hori, I. C. Rulifson, B. C. Tsai, J. J. Heit, J. D. Cahoy and S. K. Kim, *Proc. Natl. Acad. Sci. U. S. A.*, 2002, **99**, 16105.

85. K. A. D'Amour, A. G. Bang, S. Eliazer, O. G. Kelly, A. D. Agulnick, N. G. Smart, M. A. Moorman, E. Kroon, M. K. Carpenter and E. E. Baetge, *Nature Biotechnol.*, 2006, **24**, 1392.

86. M. Borowiak, R. Maehr, S. Chen, A. E. Chen, W. Tang, J. L. Fox, S. L. Schreiber and D. A. Melton, *Cell Stem Cell*, 2009, **4**, 348.

87. S. Chen, M. Borowiak, J. L. Fox, R. Maehr, K. Osafune, L. Davidow, K. Lam, L. F. Peng, S. L. Schreiber, L. L. Rubin and D. Melton, *Nature Chem. Biol.*, 2009, **5**, 258.

88. C. H. Waddington, *The Strategy of the Genes*, George, Allen & Unwin, London, 1957.

89. S. Cai, X. Fu and Z. Sheng, *BioScience*, 2007, **57**, 655.

90. C. N. Chen, J. M. Slack and D. Tosh, *Nat. Cell Biol.*, 2000, **2**, 879.

91. T. Kondo and M. Raff, *Science*, 2000, **289**, 1754.

92. G. R. Rosania, Y.-T. Chang, O. Perez, D. Sutherlin, H. Dong, D. J. Lockhart and P. G. Schultz, *Nat. Biotechnol.*, 2000, **18**, 304.

93. S. Chen, Q. Zhang, X. Wu, P. G. Schultz and S. Ding, *J. Am. Chem. Soc.*, 2004, **126**, 410.

94. E. K. Lee, G. U. Bae, J. S. You, J. C. Lee, Y. J. Jeon, J. W. Park, J. H. Park, S. H. Ahn, Y. K. Kim, W. S. Choi, J. S. Kang, G. Han and J. W. Han, *J. Biol. Chem.*, 2009, **284**, 2891.

95. L. Anastasia, M. Sampaolesi, N. Papini, D. Oleari, G. Lamorte, C. Tringali, E. Monti, D. Galli, G. Tettamanti, G. Cossu and B. Venerando, *Cell Death Differ.*, 2006, **13**, 2042.

96. S. W. Shan, M. K. Tang, P. H. Chow, M. Maroto, D. Q. Cai and K. K. Lee, *Proteomics*, 2007, **7**, 4303.

97. G. Amabile, A. M. D'Alise, M. Iovino, P. Jones, S. Santaguida, A. Musacchio, S. Taylor and R. Cortese, *Cell Death Differ.*, 2009, **16**, 321.

98. J. Yu, M. A. Vodyanik, K. Smuga-Otto, J. Antosiewicz-Bourget, J. L. Frane, S. Tian, J. Nie, G. A. Jonsdottir, V. Ruotti, R. Stewart, I. I. Slukvin and J. A. Thomson, *Science*, 2007, **318**, 1917.

99. V. Selvaraj, J. M. Plane, A. J. Williams and W. Deng, *Trends Biotechnol.*, 2010, **28**, 214, doi:10.1016/j.tibtech.2010.01.002.

100. B. Feng, J.-H. Ng, J.-C. D. Heng and H.-H. Ng, *Cell Stem Cell*, 2009, **4**, 301.

101. N. Maherali and K. Hochedlinger, *Cell Stem Cell*, 2008, **3**, 595.

102. N. M. Kane, S. McRae, C. Denning and A. H. Baker, *Drug Discov. Today: Technol.*, 2008, **5**, e107, doi: doi:10.1016/j.ddtec.2008.10.002.

103. H. Zhou, S. Wu, J. Y. Joo, S. Zhu, D. W. Han, T. Lin, S. Tranger, G. Bien, S. Yao, Y. Zhu, G. Siuzdak, H. R. Schöler, L. Duan and S. Ding, *Cell Stem Cell*, 2009, **4**, 381.

104. T. S. Mikkelsen, J. Hanna, X. Zhang, M. Ku, M. Wernig, P. Scorderet, B. E. Bernstein, R. Jaenisch, E. S. Lander and A. Meissner, *Nature*, 2008, **454**, 49.

105. D. Huangfu, R. Maehr, W. Guo, A. Eijkelenboom, M. Snitow, A. E. Chen and D. A. Melton, *Nature Biotechnol.*, 2008, **26**, 795.

106. D. Huangfu, K. Osafune, R. Maehrm W. Guo, A. Eijkelenboom, S. Chen, W. Muhlestein and D. A. Melton, *Nature Biotechnol.*, 2008, **26**, 1269.

107. Y. Shi, J. T. Do, C. Desponts, H. S. Hahm, H. R. Schöler and S. Ding, *Cell Stem Cell*, 2008, **2**, 525.

108. Y. Shi, C. Desponts, J. T. Do, H. S. Hahm, H. R. Schöler and S. Ding, *Cell Stem Cell*, 2008, **3**, 568.

109. C. A. Lyssiotis, R. K. Foreman, J. Staerk, M. Garcia, D. Mathur, S. Markoulaki, J. Hanna, L. L. Lairson, B. D. Charette, L. C. Bouchez, M. Bollong, C. Kunick, A. Brinker, C. Y. Cho, P. G. Schultz and R. Jaenisch, *Proc. Natl. Acad. Sci. U. S. A.*, 2009, **106**, 8912.

110. W. Li, W. Wei, S. Zhu, J. Zhu, Y. Shi, T. Lin, E. Hao, A. Hayek, H. Deng and S. Ding, *Cell Stem Cell*, 2009, **4**, 16.

111. T. Lin, R. Ambasudhan, X. Yuan, W. Li, S. Hilcove, R. Abujarour, X. Lin, H. S. Hahm, E. Hao, A. Hayek and S. Ding, *Nat. Methods*, 2009, **6**, 805.

112. J. M. W. Slack, *Nat. Rev. Mol. Cell Biol.*, 2007, **8**, 369.

113. Q. Zhou and D. A. Melton, *Cell Stem Cell*, 2008, **3**, 382.

114. J. Choi, M. L. Costa, C. S. Mermelstein, C. Chagas, S. Holtzer and H. Holtzer, *Proc. Natl. Acad. Sci. U. S. A.*, 1990, **87**, 7988.

115. D. R. Williams, M. R. Lee, Y. A. Song, S. K. Ko, G. H. Kim and I. Shin, *J. Am. Chem. Soc.*, 2007, **129**, 9258.

116. T. Vierbuchen, A. Ostermeier, Z. P. Pang, Y. Kokubu, T. C. Südhof and M. Wernig, *Nature*, 2010, **463**, 1035.

117. J. P. Brockes, *Science*, 1997, **276**, 81.

118. B. Ohlstein, T. Kai, E. Decotto and A. Spradling, *Curr. Opin. Cell Biol.*, 2004, **16**, 693.

119. C. Cutler and J. H. Antin, *Clin. Chest Med.*, 2005, **26**, 517.

120. http://www.clinicaltrial.gov.

121. H. Sadek, B. Hannack, E. Choe, J. Wang, S. Latif, M. G. Garry, D. J. Garry, J. Longgood, D. E. Frantz, E. N. Olson, J. Hsieh and J. W. Schneider, *Proc. Natl. Acad. Sci. U. S. A.*, 2008, **105**, 6063.

122. N. Flomenberg, S. M. Devine, J. F. DiPersio, J. L. Liesveld, J. M. McCarty, S. D. Rowley, D. H. Vesole, K. Badel and G. Calandra, *Blood*, 2005, **106**, 1867.

123. R. N. Wilkinson, C. Pouget, M. Gering, A. J. Russell, S. G. Davies, D. Kimelman and R. Patient, *Dev. Cell*, 2009, **16**, 909.

124. L. Zhou, A. N. Nguyen, D. Sohal, J. Y. Ma, P. Pahanish, K. Gundabolu, J. Hayman, A. Chubak, Y. Mo, T. D. Bhagat, B. Das, A. M. Kapoun, T. A. Navas, S. Parmar, S. Kambhampati, A. Pellagatti, I. Braunchweig, Y. Zhang, A. Wickrema, S. Medicherla, J. Boultwood, L. C. Platanias, L. S. Higgins, A. F. List, M. Bitzer and A. Verma, *Blood*, 2008, **112**, 3434.

125. F. H. Bahlmann, K. De Groot, J. M. Spandau, A. L. Landry, B. Hertel, T. Duckert, S. M. Boehm, J. Menne, H. Haller and D. Fliser, *Blood*, 2004, **103**, 921.

126. C. H. Wang, S. Verma, I. C. Hsieh, Y. J. Chen, L. T. Kuo, N. I. Yang, S. Y. Yang, M. Y. Wu, C. M. Hsu, C. W. Cheng and W. J. Cherng, *J. Mol. Cell Cardiol.*, 2006, **41**, 34.

127. K. Strehlow, N. Werner, J. Berweiler, A. Link, U. Dirnagl, J. Priller, K. Laufs, L. Ghaeni, M. Milosevic, M. Böhm and G. Nickenig, *Circulation*, 2003, **107**, 3059.

128. F. H. Bahlmann, K. De Groot, O. Mueller, B. Hertel, H. Haller and D. Fliser, *Hypertension*, 2005, **45**, 526.

129. N. H. Kulkami, J. E. Onyia, Q. Zeng, X. Tian, M. Liu, D. L. Halladay, C. A. Frolik, T. Engler, T. Wei, A. Kriauciunas, T. J. Martin, M. Sato, H. U. Bryant and Y. L. Ma, *J. Bone Miner. Res.*, 2006, **21**, 910.

130. M. Heim, O. Frank, G. Kampmann, N. Sochocky, T. Pennimpede, P. Fuchs, W. Hunziker, P. Weber, I. Martin and I. Bendik, *Endocrinology*, 2004, **145**, 848.

131. M. W. Draper, *Ann. N. Y. Acad. Sci.*, 2003, **997**, 373.

132. J. E. Malberg, A. J. Eisch, E. J. Nestler and R. S. Duman, *J. Neurosci.*, 2000, **20**, 9104.

133. J. E. Malberg and J. A. Blendy, *Trends Pharmacol. Sci.*, 2005, **26**, 631.

134. L. Santarelli, M. Saxe, C. Gross, A. Surget, F. Battaglia, S. Dulawa, N. Weisstaub, J. Lee, R. Duman, O. Arancio, C. Belzung and R. Hen, *Science*, 2003, **301**, 805.

135. K. K. Karishma and J. Herbert, *Eur. J. Neurosci.*, 2002, **16**, 445.

136. W. Mayo, V. Lemaire, J. Malaterre, J. J. Rodriguez, M. Cayre, M. G. Stewert, M. Kharouby, G. Rougon, M. Le Moal, P. V. Piazza and D. N. Abrous, *Neurobiol. Aging*, 2005, **26**, 103.

137. P. Soundararajan, G. B. Miles, L. L. Rubin, R. M. Brownstone and V. F. Rafuse, *J. Neurosci.*, 2006, **26**, 3256.

138. N. C. Bambakidis, E. M. Horn, P. Nakaji, N. Theodore, E. Bless, T. Dellovade, C. Ma, X. Wang, M. C. Preul, S. W. Coons, R. F. Spetzler and V. K. Sonntag, *J. Neurosurg. Spine*, 2009, **10**, 171.

139. R. D. Paladini, J. Saleh, C. Qian, G. X. Xu and L. L. Rubin, *J. Invest. Dermatol.*, 2005, **125**, 638.

140. N. Yamamoto, K. Tanigaki, M. Tsuji, D. Yabe, J. Ito and T. Honjo, *J. Mol. Med.*, 2006, **84**, 37.

141. S. Hayes, B. R. Nelson, B. Buckingham and T. A. Reh, *Dev. Biol.*, 2007, **312**, 300.

142. Q. Zhang, M. B. Major, S. Takanashi, N. D. Camp, N. Nishiya, E. C. Peters, M. H. Ginsberg, X. Jian, R. A. Randazzo, P. G. Schultz, R. T. Moon and S. Ding, *Proc. Natl. Acad. Sci. U. S. A.*, 2007, **104**, 7444.

143. G. Molina, A. Vogt, A. Bakan, W. Dai, P. Queiroz de Oliveira, W. Znosko, T. E. Smithgall, I. Bahar, J. S. Lazo, B. W. Day and M. Tsang, *Nat. Chem. Biol.*, 2009, **5**, 680.

144. Q. Zhou, J. Brown, A. Kanarek, J. Rajagopal and D. A. Melton, *Nature*, 2008, **455**, 627.

145. C. T. Morgan, M. Lee and J. Kimble, *Nat. Chem. Biol.*, 2010, **6**, 102.

146. M. Zhang and J. M. Rosen, *Curr. Opin. Genet. Dev.*, 2006, **16**, 60.

147. B. J. Druker, *Cancer Cell*, 2002, **1**, 31.

148. B. Z. Stanton, L. F. Peng, N. Maloof, K. Nakai, X. Wang, J. L. Duffner, K. M. Taveras, J. M. Hyman, S. W. Lee, A. N. Koehler, J. K. Chen, J. L. Fox, A. Mandinova and S. L Schreiber, *Nat. Chem. Biol.*, 2009, **5**, 154.

149. S. Ko, H. J. Jin, D. Jung, X. Tian and I. Shin, *Angew. Chem. Int. Ed.*, 2009, **48**, 7809.

# Chemical Biology of Histone Modifications

NATHAN R. ROSE,[a] CHRISTOPHER J. SCHOFIELD[a],* AND TOM D. HEIGHTMAN[b],*

[a] The Department of Chemistry and the Oxford Centre for Integrative Systems Biology, Chemistry Research Laboratory, University of Oxford, 12 Mansfield Road, Oxford, UK, OX1 3TA; [b] Structural Genomics Consortium, University of Oxford, Old Road Campus, Roosevelt Drive, Headington, Oxford, UK, OX3 7DQ

## 5.1  Histones, Epigenetics and Chromatin

In the nucleus DNA is packaged into nucleosomes, then into higher order chromatin structures, and finally into chromosomes (Figure 5.1). The nucleosomes are comprised of the DNA itself and the histone proteins around which the DNA is wrapped. In order for the DNA of individual genes to be transcribed into mRNA, it must be temporarily released from the packaging apparatus. Each nucleosome (Figure 5.2) consists of 146 base pairs of double-stranded DNA that wrap around a core histone octamer, which consists of pairs of the histones H2A, H2B, H3 and H4. The core histones themselves are small basic globular proteins (100–150 amino acids) with $N$-terminal tails that protrude from the nucleosome. These tails, in the absence of binding partners, are generally thought to be disordered, in contrast to the folded core regions of the histone octamer around which the DNA is wrapped. In addition to the eight core histones, the linker histone H1 binds across the DNA strands entering and exiting the nucleosome, and this linker histone contributes to the

RSC Drug Discovery Series No. 5
New Frontiers in Chemical Biology: Enabling Drug Discovery
Edited by Mark E. Bunnage
© Royal Society of Chemistry 2011
Published by the Royal Society of Chemistry, www.rsc.org

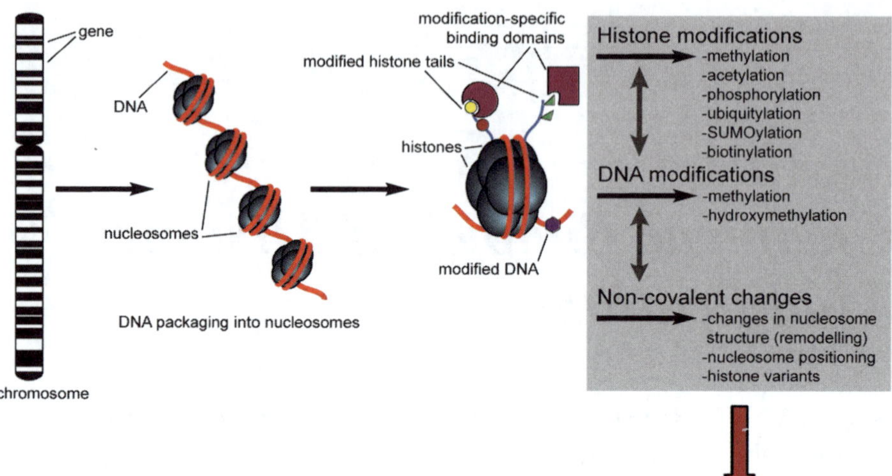



**Figure 5.1** The components of chromatin, *i.e.* DNA, and the histone proteins that organise it into nucleosomes, which enable the compaction of DNA into higher-order chromatin structures.

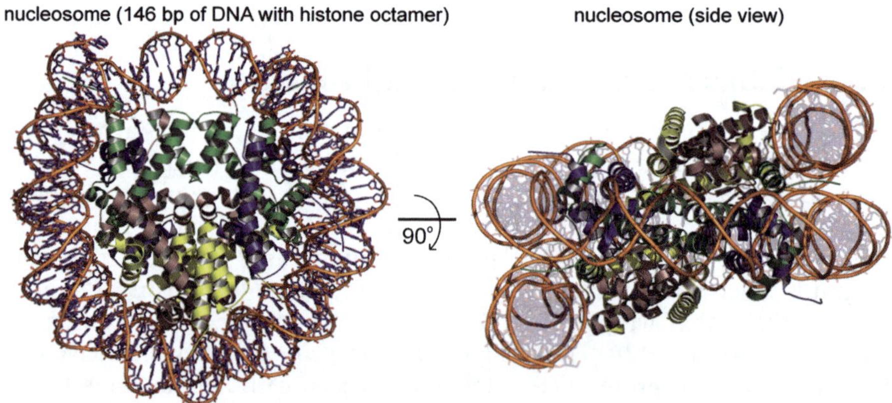



**Figure 5.2** Views from an X-ray crystal structure of a nucleosome (PDB ID 1KX3), consisting of 146 bp of DNA (orange/dark blue) wrapped around the histone octamer, which is made up of pairs of histone proteins H2A (pink), H2B (yellow), H3 (green) and H4 (blue).

formation of higher-order chromatin structures.[1] Modifications of the core histones, particularly in the N-terminal tail regions, are crucial regulators of whether or not genes are transcriptionally active, and are thus regulators of gene expression.

Many post-translational modifications to the *N*- and *C*-terminal tails of the core histones have been identified, which include acetylation (of lysines), methylation (of lysines and arginines), phosphorylation (of serines, threonines and tyrosines), and biotinylation, ubiquitylation and SUMOylation (of lysines). Some modifications to the folded core regions of the histones have also been identified, including methylation, acetylation and ubiquitylation. These post-translational modifications are specific to particular residues on particular histones, and are placed and removed by sequence-specific histone acetyl-transferases (HATs), histone deacetylases (HDACs), histone methyl-transferases (HMTs), histone demethylases (HDMs), kinases, and other enzymes. A summary of the patterns of modifications found in human histones is outlined in Figure 5.3. Although not a focus of this review, it is worth noting that in addition to covalent modifications to histones and DNA, non-covalent mechanisms, including conformational changes within nucleosomes, together with regulation of nucleosome assembly and positioning on DNA, contribute to histone-mediated regulation of gene expression.

In some cases, general roles for these covalent post-translational modifications have been identified, *e.g.* hyperacetylation of histones correlates with

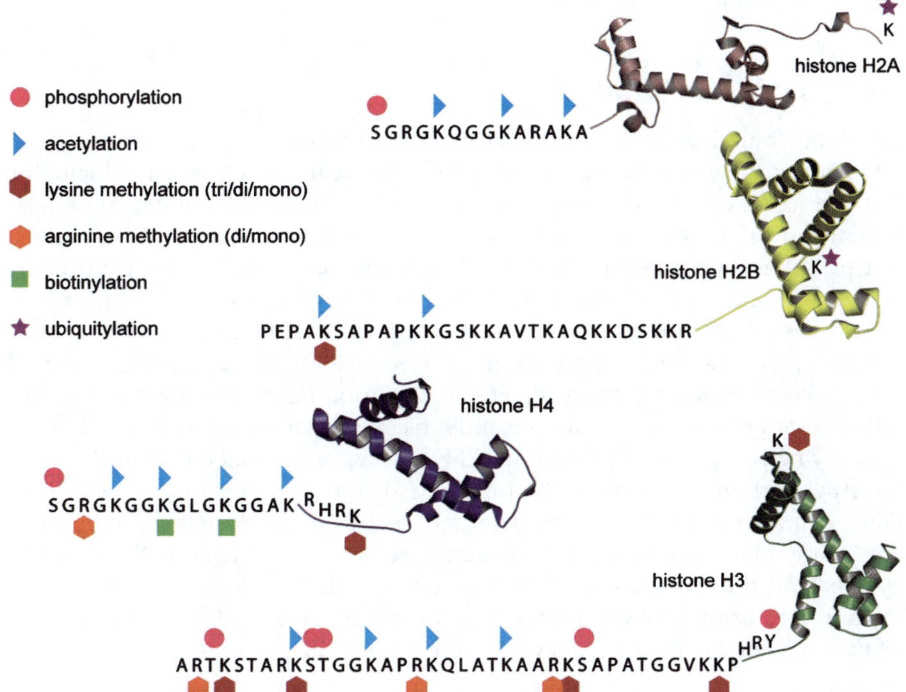


**Figure 5.3**   Selected post-translational modifications of human histones. Methylation, acetylation, phosphorylation, biotinylation and ubiquitylation sites have been identified on all four core histones. The modifications shown here should not be regarded as a complete or final list.

active expression of the associated genes. However, in many cases their roles are only beginning to emerge. Studies on how individual modifications regulate the expression of specific genes in a temporal manner are at a relatively early stage. The challenge of understanding how multiple histone post-translational modifications regulate expression is of fundamental interest from both biochemical and biological perspectives. The possibility of applying discoveries in the field to the development of new medicines has also attracted much interest, and histone deacetylase inhibitors have been recently introduced into clinical use for the treatment of some cancers.

Here we review current knowledge on the enzymes and related binding proteins that are involved in covalent modifications to histones. We begin with a brief overview of the field aimed at the non-expert, then focus on the covalent modifications themselves as catalysed by specific enzymes. We also focus on the consequences of these modifications for binding interactions with other proteins that modulate gene expression. The field is one that is rapidly evolving and we hope to convey some of the excitement that we feel about recent discoveries in histone science, particularly from a molecular perspective. Two inhibitors of histone modifying enzymes are in clinical use, with more likely to be introduced in the near future. Thus, the development of modifiers of histone biochemistry is of medicinal as well as basic interest.

### 5.1.1 DNA Modifications

The four canonical DNA bases themselves (adenine, cytosine, guanine and thymine) are subject to a number of chemical modifications, including methylation and hydroxymethylation (Figure 5.4). The most important of these is methylation of cytosine at the C-5 position, which is common in the context of CpG islands (regions enriched in CG dinucleotides), and which is normally associated with gene silencing. C-5 methylation of cytosine is catalysed by DNA methyltransferases, which utilise *S*-adenosylmethionine as a methyl source (as do histone methyltransferases; see below). As yet, no demethylases capable of reversing 5-methylcytosine methylation have been identified. Interestingly, 5-methylcytosine has recently been shown to be converted to 5-hydroxymethylcytosine by the human TET1 oxygenase, which hydroxylates the 5-methyl carbon;[2,3] however, the functions of this modification are not clear. DNA can also be methylated at heteroatoms by chemical methylating agents, and these methyl groups may be removed by DNA demethylases, which hydroxylate the methyl groups in a manner analogous to the histone demethylases (see below); as yet, no biological role for these modifications has been identified, though there are enzymes that reverse them.

### 5.1.2 Euchromatin and Heterochromatin

Chromatin may be crudely divided into two types, "euchromatin" and "heterochromatin". In general, euchromatin comprises transcriptionally active

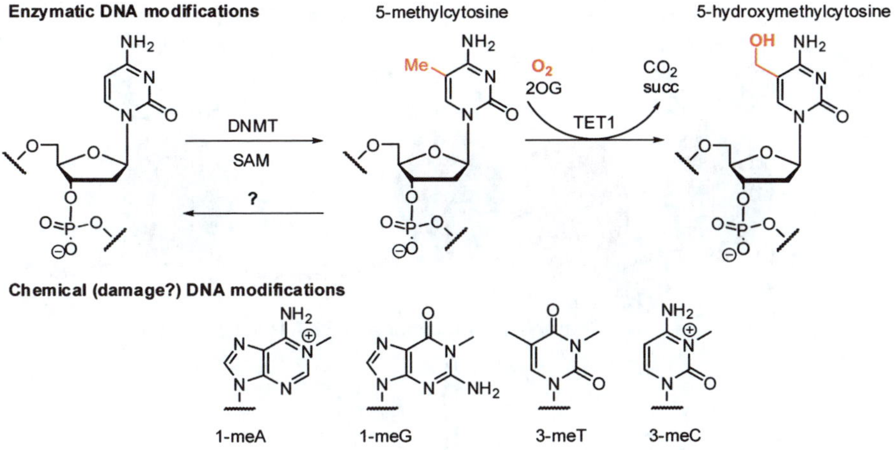

**Figure 5.4** Selected DNA modifications caused by enzymatic (5-methyl and 5-hydroxymethylcytosine) and chemical (*i.e.* damaging) methylation (1-meA, 1-meG, 3-meT, 3-meC). Enzymes capable of catalysing demethylation of 5-meC have not yet been discovered. DNMT = DNA methyltransferase; SAM = *S*-adenosylmethionine.

chromatin, which is accessible to transcription factors, RNA polymerases, *etc.* In contrast, heterochromatin can be defined as transcriptionally silent chromatin (Figure 5.5). "Constitutive" heterochromatin is characteristic of centromeres and telomeres, where it helps to ensure long-term chromatin stability and the necessary mechanical characteristics of chromosomes during mitosis and meiosis; further, constitutive heterochromatin serves to prevent transcription of the non-coding and repetitive sequences which form a large part of the genome. "Facultative" heterochromatin, in contrast, consists of genes which are silenced at certain times in the cell cycle, or in certain cell types as a result of differentiation.[4] Certain histone and DNA modifications serve to establish and propagate these different chromatin states.

## 5.1.3 Correlation of Histone Modifications with Transcriptional States

Genome-wide mapping of histone modifications has led to the correlation of certain histone modifications with particular chromosomal regions, and with particular transcriptional states (Figure 5.6). On a genome-wide scale, for example, lysine methylation at H3K4 and at H3K36 is associated with actively transcribed genes. In contrast, lysine methylation at H3K9 and H4K20 is associated with silenced genes, and with centromeric and pericentromeric heterochromatic regions. However, more detailed analysis of histone modification across silenced or actively transcribed genes has revealed that histone modifications vary also within the context of a single gene, with certain modifications associated with promoter regions, others with the body of actively

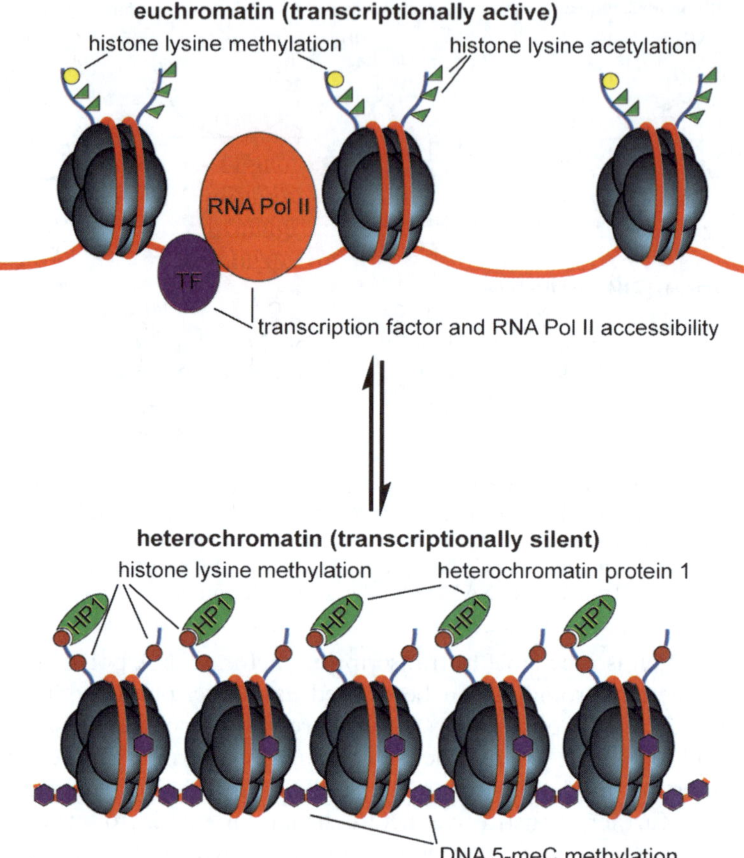



**Figure 5.5**  Schematic showing euchromatin and heterochromatin, illustrating the accessibility to transcription machinery (transcription factors and RNA polymerases) as a result of different histone and DNA modifications. H3K4 methylation and lysine (hyper)acetylation are generally associated with euchromatin formation, while H3K9, H3K27 and H4K20 methylation, together with lysine hypoacetylation and DNA 5-meC methylation, are characteristic of heterochromatin.

transcribed genes, and so on (reviewed in detail in Lee and Mahadevan[5]). The local effects of histone modifications on individual nucleosome structure and corresponding transcriptional availability, as well as the global effects on chromosome structure and stability, are thus of importance.

## 5.1.4  Mechanisms by Which Histone Modifications Achieve Transcriptional Regulation

The means by which individual histone modifications, or combinations thereof, achieve silencing or activation of the associated genes are not elucidated fully.

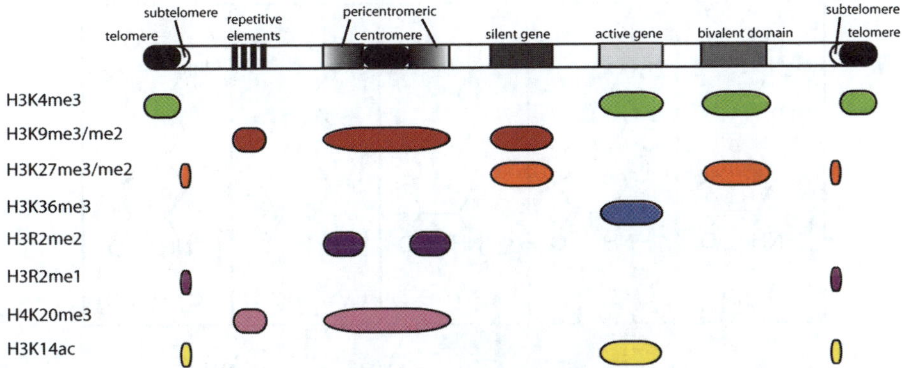


**Figure 5.6** The association of selected histone modifications with particular chromosomal regions (*e.g.* centromeres, telomeres, repetitive elements) or transcriptional states (active, silent and "bivalent" gene domains). (After Lee and Mahadevan.[5])

Histones are very basic, positively charged proteins (on average, 23 % of their total residues are lysine or arginine), which facilitates their interaction with the acidic DNA backbone. In the simplest case, modifications such as acetylation and phosphorylation cause reductions in the overall electrostatic charge of the positively charged histone proteins (Figure 5.7), thus changing the nature of the interactions between histones and DNA, leading to changes in nucleosome structure. A consequence of histone lysine deacetylation (which increases the positive charge on histones) is an increase in the electrostatic attraction between DNA (negatively charged) and histones (positively charged), resulting in tighter binding and thereby repression of transcription. However, the same does not apply to methylation of histones, because there is no associated change in overall electrostatic charge. Methylation can effect silencing or activation either through steric factors, or by facilitating or blocking recruitment of various other proteins which then further modify the histones and the overall chromatin structure. For example, methylation of certain lysine residues can recruit methyllysine-binding domains, *e.g.* HP1 (heterochromatin protein 1), which is recruited to H3K9me3. Once HP1 has bound to H3K9me3, it facilitates heterochromatin formation and spreading, by recruiting histone deacetylases, DNA methyltransferases and histone methyltransferases.[6] Likewise, multiprotein complexes such as the polycomb repressive complexes (PRC1/PRC2) bind to methylation marks at H3K27 in the course of transcriptional silencing.[7] In contrast, H3K4me3 methylation prevents binding of DNA methyltransferases whose methylation of CpG islands would lead to transcriptional silencing.[8] Acetylated and phosphorylated histones recruit other proteins in a manner analogous to methylated histones; the effects of acetylation and phosphorylation are not limited to changes in electrostatic charge. Further, histone and DNA modifications may affect the positioning of nucleosomes on the genomic DNA strands, by a variety of steric and/or mechanical factors

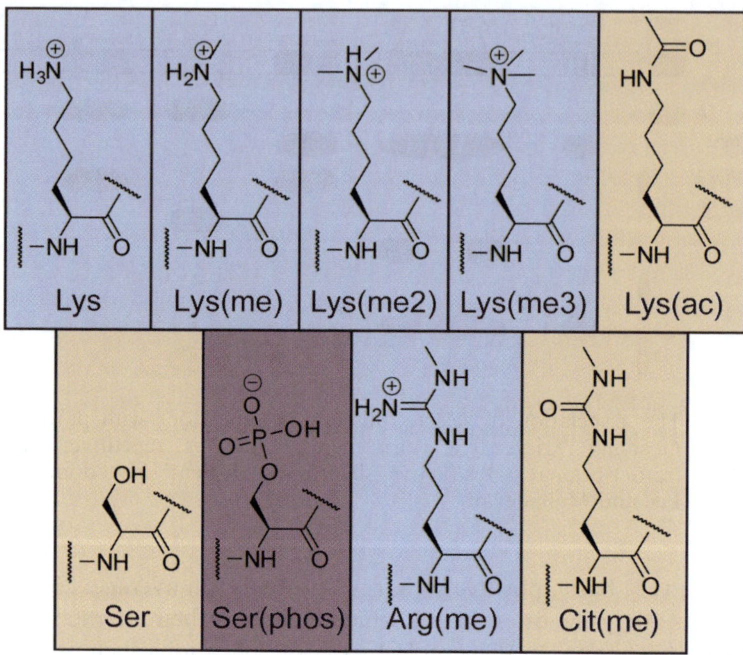



**Figure 5.7** Charge states associated with various histone residues modifications. Methylation does not have a significant effect on the overall charge of lysine/arginine sidechains. Acetylation and phosphorylation both reduce the overall charge on histone tails, leading to weakened binding of histones to DNA. Arginine deimination (to citrulline) also leads to a reduction in overall histone charge.

(reviewed in Segal and Widom[9]). The replacement of the four "canonical" histones with a variety of histone variants (reviewed in Talbot and Henikoff[10]) also has significant implications for transcription and for chromosome structure during meiosis. Nucleosome remodelling (the moving, restructuring, destabilising or ejection of nucleosomes) is also regulated to some extent by histone and DNA modifications (reviewed in Clapier and Cairns[11]). Finally, dsRNA has also been implicated in the silencing of gene-poor chromosomal regions through the RNAi pathway.[6]

## 5.1.5 Epigenetics

Epigenetics can be defined as the study of how heritable changes in phenotype are transmitted without changes in the underlying DNA sequence, as defined by the four canonical bases, A, T, C and G.[12,13] There is evidence that epigenetic transfer of information between generations can be mediated by 5-methylcytosine methylation of DNA and by some modifications to histones.[14,15]

Although modifications to DNA and histones are commonly referred to as being epigenetic, in many cases there is little or no evidence that they are involved in the epigenetic transfer of information, though they may be involved in the regulation of transcription/expression in individual cells or organisms.

Although we divide the remainder of the review into particular types of covalent histone modification, it is important to emphasise that it is likely that, in most cases, combinations of modifications, together with non-covalent mechanisms, enable biological regulation. The possibility of specific combinations of modifications leading to a specific output has led to the idea of a histone code regulating transcription.[16] While this remains a possibility, the available evidence is that such a code, if it exists, may be complex. In regard to potential complexity it is perhaps worth considering modifications to lysine residues in isolation. In addition to the unmodified forms, histone lysines can exist in acetylated, mono-, di- and trimethylated, ubiquitylated, SUMOylated, biotinylated, and possibly hydroxylated forms. Some of these (potential) modifications are mutually exclusive on chemical grounds, *e.g.* acetylation, SUMOylation/ubiquitylation and di-/trimethylation, whilst others are not (*e.g.* hydroxylation and acetylation/methylation). Despite the mutual exclusivity of some modifications, based on known modifications, the number of potential combinations of known lysine modifications to histone H3 alone is in excess of $4 \times 10^5$; this increases to $> 3.5 \times 10^7$ if other modifications on H3 are also counted. It is also worth noting that analyses of histone modifications have, for the most part, employed antibodies that are specific (or at least selective) for particular modifications. The use of such antibodies has been highly productive and enabled major scientific advances in our understanding of the biochemistry of genetics. However, selective antibodies are necessarily biased to particular modifications. There is thus a need for the development of less-biased analytical methodologies. Mass-spectrometric analyses of histones are of particular interest, though there are considerable challenges, not least because of the redundancies in mass and complexity.

In the following sections, we outline the features of the enzymes responsible for catalysing the placement and removal of post-translational modifications of histones, with particular focus on structure, mechanism and the potential for pharmacologically relevant inhibition. We also examine some of the proteins that recognise these modifications by selectively binding to them. We discuss, where applicable, the biological roles of some of these proteins, and the links to human disease.


## 5.2 The Lysine Acetylation System

Acetylation of histone lysines is carried out by histone acetyltransferases (HATs), while histone deacetylases serve to remove acetyl groups by hydrolysis. Histone lysine acetylation results in a decrease in positive charge on histones, leading to reduced affinity for DNA, which correlates with transcriptional activation. Conversely, histone deacetylation results in an increase in the electrostatic attraction between DNA and histones, resulting in tighter binding and thereby repression of transcription. However, histone acetylation also influences the recruitment of transcriptional regulators (including other histone modifying enzymes); the most important class of acetyl lysine binding proteins identified are bromodomain proteins. These three protein classes, *viz.* HATs, HDACs and bromodomains, regulate the lysine acetylation system of histone modifications, and are described in more detail below.

## 5.2.1   Histone Acetyltransferases

Histone acetyl transferases (HATs, EC 2.3.1.48; Table 5.1) catalyse the transfer of an acetyl group to the ε-amino group of lysine residues in the tails and/or cores of histones, although, as for HDACs, some HATs also modify non-histone substrates (*e.g.* transcription factors). HATs generally contain multiple subunits, and the functions of the catalytic subunit depend on the presence of other subunits. Domains that cooperate to recruit the HAT to the appropriate location in the genome include bromodomains, chromodomains, WD40 repeats, Tudor domains and PHD fingers.[17]

HATs can be classified into three broad families depending on their catalytic domains: the GNAT (GCN5-related *N*-acetyltransferase)) family, including SAGA and PCAF, with more than 50 members in the human genome; the MYST (MOZ, Ybf2/Sas3, Sas2, Tip60) family, including HBO1; and "orphan" HATs with greater sequence diversity from classical HAT domains, including HAT1, p300 and CBP.

The gene specificity of HAT-mediated transcriptional activation is driven in part by the specificity of complex formation, and in part by specific cell signalling pathways. In addition, HATs acetylate other proteins including transcription factors, thereby playing a key role in complex regulatory networks.

p300 and its close homolog CBP are among the most widely studied histone acetyltransferases. Both enzymes acetylate a broad range of substrates, including all four core histones and a range of non-histone proteins including the tumour suppressor p53. These acetylations regulate important cellular functions including DNA repair, cell cycle, differentiation, and establishment of retroviral pathogenesis.[18]

The potential clinical impact of HAT modulation covers a wide range of indications including cancer, HIV, diabetes mellitus, cardiac disease and neurodegeneration.[19]

Many HATs are involved in cancer pathogenesis and several are proposed as cancer biomarkers due to correlations of expression levels with outcomes, including a positive correlation between p300 expression levels and prostate cancer recurrence, and a negative correlation between hMOF and primary breast carcinomas.[20] A series of isothiazolone PCAF-p300 inhibitors has shown inhibition of growth of a panel of colon and ovarian tumour cell lines.[21]

**Table 5.1** Human histone acetyltransferases (HATs).

| Subfamily | HAT | Known/preferred substrates | Biological role | Disease association |
| --- | --- | --- | --- | --- |
| GNAT (GCN5-like *N*-acetyl-transferases) | hGCN5 | H3 (K9, K14, K18)/H2B | Transcription activation | Oncology |
| | PCAF | H3 (K9, K14, K18)/H2B | Transcription activation | Oncology, memory |
| | TAF1 | H3 > H4 | Transcription activation | Oncology |
| | ELP3 | H3 | Transcription elongation | Motor neuron disease |
| MYST | MOZ/MYST3 | H3 (K14) | Transcription activation | — |
| | MORF/MYST4 | H3 (K14) | Transcription activation | — |
| | HBO1/MYST2 | H4 (K5, K8, K12) > H3 | Transcription, DNA replication | — |
| | HMOF/MYST1 | H4 (K16) | Chromatin boundaries, dosage compensation, DNA repair | Breast cancer |
| | TIP60/PLIP | H4 (K5, K8, K12, K16) | Transcription activation, DNA repair | — |
| Other | HAT1 | H4 (K5, K12) | Histone deposition, DNA repair | Oncology |
| | P300 | H2A (K5); H2B (K12, K15) | Transcription activation | Oncology, immunology, virology |
| | CBP | H2A (K5); H2B (K12, K15) | Transcription activation | Oncology, diabetes, neurodegeneration |
| | TFIIIC90 | H3 (K9, K14, K18) | Pol III transcription | — |
| | SRC1 | H3/H4 | Transcription activation | — |
| | ACTR | H3/H4 | Transcription activation | — |
| | CLOCK | H3/H4 | Transcription activation | Circadian rhythm |

Modulation of HATs has multiple effects on immune cells. Conditional knockouts of CBP and p300 in mice indicate key roles in T-cell development and differentiation.[22] A series of selective natural product-derived inhibitors of p300-HAT were nontoxic to T-cells, but inhibited histone acetylation of HIV infected cells, and thereby inhibited the multiplication of HIV.[18] A role of CBP in diabetes mellitus has been suggested by the observation that CBP-mediated gene activation increased pancreatic beta-cell proliferation.[23] CBP has also been implicated in various stages of neuronal development, learning and, in later life, neurodegeneration.[19,24]

### 5.2.1.1  Histone Acetyltransferase Mechanism

The currently accepted mechanism for both the Gcn5 and MYST HAT families involves a process in which the $N^\varepsilon$-lysine, deprotonated by a conserved glutamate serving as a general base, nucleophilically attacks the enzyme-bound acetyl-CoA (Figure 5.8b). This forms a putative tetrahedral intermediate, which fragments to generate the acetamide and release CoASH.[25] In the p300 HAT domain there is no obvious catalytic base, and a mechanism that does not involve tight-knit substrate interactions has been proposed, consistent with the broad sequence selectivity of p300.[26]

### 5.2.1.2  Histone Acetyltransferase Structures

Crystal structures of HATs have been solved, including RTT109 (PDB ID 3D35), p300 (3BIY), MYST3 (2RC4, 2OZU), HAT1 (2P0W), and GCN5 (1Z4R). The structure of human GCN5 complexed with co-enzyme A (Figure 5.8a) showed the relative positions of the glutamate side chain that acts as a general base to deprotonate the substrate lysine in proximity to the acetyl group of CoA.[27] A series of ternary co-structures of *Tetrahymena* GCN5 with CoA and histone peptides from histone H3 phosphorylated at Ser10 or Thr11 have provided structural insights into how combinations of histone modifications are related.[28] The structure of p300/CBP shows structural conservation with other HATs within the central core region, despite the low overall sequence homology with the GCN5 and MYST families.[26]

### 5.2.1.3  Histone Acetyltransferase Inhibitors

At the time of writing there is a paucity of drug-like HAT inhibitors in the public domain. Structure-based design has been used to generate bisubstrate analogues of inhibitors that link CoA and substrate peptide to mimic the Ac-CoA-lysine intermediate. The simplest analogue, LysCoA (Figure 5.8c), showed a $K_i$ of 16 nM against p300. This compound, and a number of LysCoA-containing histone peptides, have been used as inhibitor probes for a range of HATs.[29] A number of natural products including curcumin and garcinol show non-selective inhibition of HATs.[30,31] Synthetic modification of garcinol led to a series of analogues with varying selectivity profiles across HATs, with some

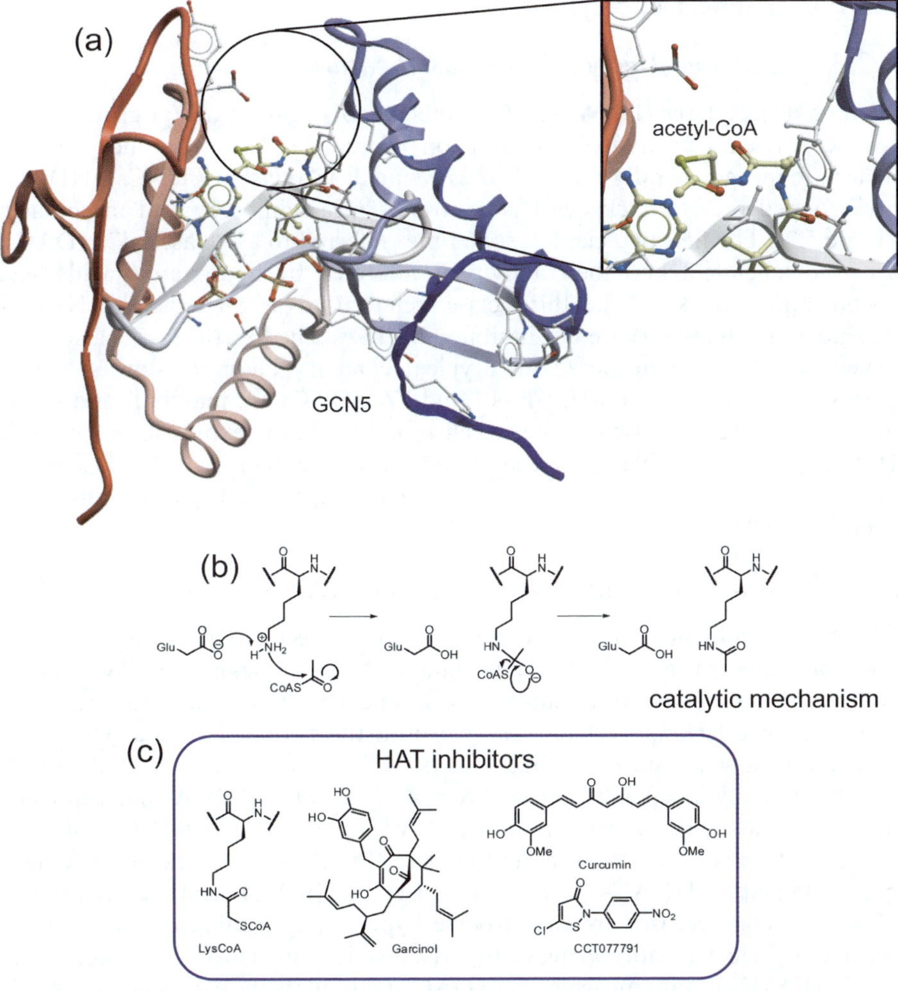



**Figure 5.8** (a) View from a crystal structure of human GCN5 complexed with co-enzyme A (PDB code: 1Z4R); (b) outline catalytic mechanism; and (c) representative inhibitors of histone acetyl transferases.

compounds showing selective inhibition of p300 in the micromolar range. These compounds were used to probe the role of p300 in T cells.[18] A series of isothiazolones (represented by CCT077791, Figure 5.8c) that inhibit p300 and PCAF was discovered by high-throughput screening; these reduced total acetylation of histones H3 and H4, levels of specific acetylated lysine marks, and acetylation of α-tubulin. These compounds were shown to be covalent modifiers of HATs.[21]

## 5.2.2 Histone Deacetylases

### 5.2.2.1 Histone Deacetylase Classification

Histone deacetylases (HDACs, EC number 3.5.1) remove acetyl groups from $N^\varepsilon$-acetyl lysines by hydrolysis, both on histones and non-histone proteins, hence are more generally referred to as lysine deacetylases (KDACs). HDACs are grouped into four classes based on sequence homology and mechanism (Table 5.2). The first two classes, sometimes referred to as "classical" HDACs, are zinc-dependent and their activity is inhibited by hydroxamic acids, *e.g.* trichostatin A (TSA). The third class, referred to as Sirtuins, are $NAD^+$-dependent proteins and are not inhibited by TSA. The fourth class is also zinc-dependent, but is considered an atypical category based on low sequence homology to classes I and II. Class I and IV HDACs are mainly found in the nucleus and are expressed in many cell types, while the expression of class II HDACs, which are able to shuttle in and out of the nucleus, is tissue specific. Sirtuin localisation depends on the particular isoform (cytoplasm, mitochondria and nucleus).[32]

### 5.2.2.2 Biological Functions of Histone Deacetylases

The roles of individual HDACs in mammals have been studied using knockout and transgenic mice, and RNAi techniques.[32] Mice lacking HDAC1 die at embryo stage due to proliferation defects and developmental retardation. In contrast, mice lacking HDAC2 survive until the perinatal period when they succumb to a spectrum of cardiac defects. HDAC3 has been reported to be a transcriptionally independent critical regulator of mitosis. HDAC8 is expressed in smooth muscle cells, where it associates with the actin cytoskeleton and regulates contractile capacity. Among the class IIa HDACs, knockout mice have been obtained for HDAC4, 5 and 9. Mice lacking HDAC4 display a remarkable phenotype characterised by chondrocyte hypertrophy leading to ectopic bone formation, while in normal neurons HDAC4 is a mediator of neuronal cell death. HDAC7 is a thymus-specific HDAC and inhibits the expression of Nur77 involved in apoptosis and negative selection during developing thymocytes. HDAC5 and 9 knockout mice display cardiac hypertrophy.

Class IIb subtypes 6 and 10 are characterised by a duplication of their catalytic domain; the best-characterised isoform is HDAC6, which deacetylates tubulin, microtubules,[33] and HSP90.[34] Little is known about HDAC11.

Seven human subtypes of sirtuins have been identified. SIRT1 deacetylates proteins such as p53 or BCL6.[35] SIRT2 deacetylates tubulin and its inhibition results in neuroprotection.[36,37] Elevated levels of SIRT3 have been observed in breast cancer.[38] A number of sirtuins promote extended lifespan in various organisms.[39]

### 5.2.2.3 Histone Deacetylase Mechanisms

The proposed mechanism for zinc-containing HDACs involves Lewis acid activation of the acetyl carbonyl oxygen by Zn(II) coordination, with

**Table 5.2** Human histone deacetylases.

| *Zinc-dependent enzymes* | | | *Localisation* | *Biological role* |
|---|---|---|---|---|
| Class I (RPD3 homologue) | | HDAC1 | Nucleus | Transcriptional repression |
| | | HDAC2 | Nucleus | Transcriptional repression |
| | | HDAC3 | Nucleus, cytoplasm | Transcriptional repression |
| | | HDAC8 | Nucleus, cytoplasm | Smooth muscle cell contractility |
| Class II (HDA1 homologue) | IIa | HDAC4 | Nucleus, cytoplasm | Transcriptional repression |
| | | | Muscle differentiation block | |
| | | | Retinal protection | |
| | | HDAC5, 7 and 9 | Nucleus, cytoplasm | Transcriptional repression |
| | | | Muscle differentiation block | |
| | IIb | HDAC6 | Cytoplasm | Regulation of microtubule stability and function |
| | | | Regulation of molecular chaperone function | |
| | | | Regulation of aggresome function | |
| | | HDAC10 | Cytoplasm | Unknown |
| Class IV | | HDAC11 | Nucleus | Transcriptional repression |

| *NAD$^+$-dependent enzymes* | | | *Localisation* | *Biological role* |
|---|---|---|---|---|
| Class III (Sir 2 homologue) | | SIRT1 | Nucleus | Functional regulation of p53 |
| | | | Transcriptional repression | |
| | | SIRT2 | Nucleus, cytoplasm | Cell cycle regulation |
| | | SIRT3 | Mitochondria | Activation of mitochondrial functions |
| | | | Thermogenesis in brown adipose | |
| | | | Activation of acetyl-CoA synthethase 2 | |
| | | SIRT4 | Mitochondria | Insulin secretion |
| | | SIRT5 | Mitochondria | Unknown |
| | | SIRT6 | Nucleus | DNA repair and genomic stability |
| | | | Regulation of telomere metabolism and function | |

hydrolysis mediated by general base catalysis, involving a histidine residue (Figure 5.9b). This generates a tetrahedral intermediate which fragments to give the products of hydrolysis, deacetylated lysine and acetate.

The proposed mechanism for the $NAD^+$-utilising HDACs (Sirtuins) is more complex, involving in the first step the formation of an *O*-glycosyl iminoyl ether and release of nicotinamide (Figure 5.10b). The iminoyl ether is then hydrolysed by a molecule of water, assisted anchimerically by the 2′-OH group of the ribosyl moiety, to give 2-*O*-acetyl ADP-ribose and the deacetylated lysine.

### 5.2.2.4   *Histone Deacetylase Structures*

Structures for HDAC7 (PDB ID 3C0Z), HDAC4 (2VQM), and HDAC8 (1T69) have been reported with inhibitors and/or substrates bound. The structure of HDAC8 bound to the clinically approved inhibitor SAHA (vorinostat, Figure 5.9a) shows how the catalytic zinc residue, held in place by a cluster of two histidine, glutamic acid and methionine residues, is coordinated in a bidentate fashion by the hydroxamic acid moiety of SAHA.

Structures have been determined for the sirtuins SIRT2 (PDB ID 3JR3), SIRT5 (2NYR), SIRT3 (3GLR) and SIRT6 (3K35) with peptide substrates or inhibitors bound. The ternary complex of yeast HST2 (the homologue of human SIRT2) with carba-NAD and a histone H4 acetyllysine peptide (PDB ID 1SZD, Figure 5.10a) provides insights into the mechanism of the sirtuin-catalysed deacetylation reaction.[40]

### 5.2.2.5   *Histone Deacetylase Inhibitors*

HDAC silencing or inhibition regulates cell cycle, cell growth, chromatin decondensation, cell differentiation, apoptosis, and angiogenesis in several cancer cell types. HDAC-dependent aberrant transcriptional repression is implicated in non-Hodgkin's lymphomas and in M2 subtype of myeloid leukaemia. HDACs are associated (by siRNA knockdown or over-expression) with gastric, human renal cell carcinoma, and promyelocytic leukaemia. Two HDAC inhibitors are currently marketed: the hydroxamic acid derivative vorinostat/SAHA, and the natural product depsipeptide romidepsin (Figure 5.10c), both for the treatment of cutaneous T-cell lymphoma. At the time of writing, these and at least 11 other HDAC inhibitors are in clinical trials for Hodgkin's lymphoma, melanoma, pancreatic, nasopharyngeal and prostate cancers.[41]

The success of vorinostat has encouraged the development of many HDAC inhibitors with diverse selectivity profiles (for recent reviews on the subject, see Paris *et al.*,[32] Tan *et al.*,[41] Grayson *et al.*[42] and Hahene *et al.*[43]). Many, but not all, HDAC inhibitors conform to a "cap–linker–chelator" pharmacophore. Those that have entered clinical trials are variants of one of two zinc-chelating groups: the hydroxamic acid, present in vorinostat and TSA, or the aminophenylamide moiety present in MS-275. Other zinc-chelating groups under investigation include mercaptoacetamides, α-keto oxazoles, trifluoromethyl ketones, and silane-diols (Figure 5.9c).

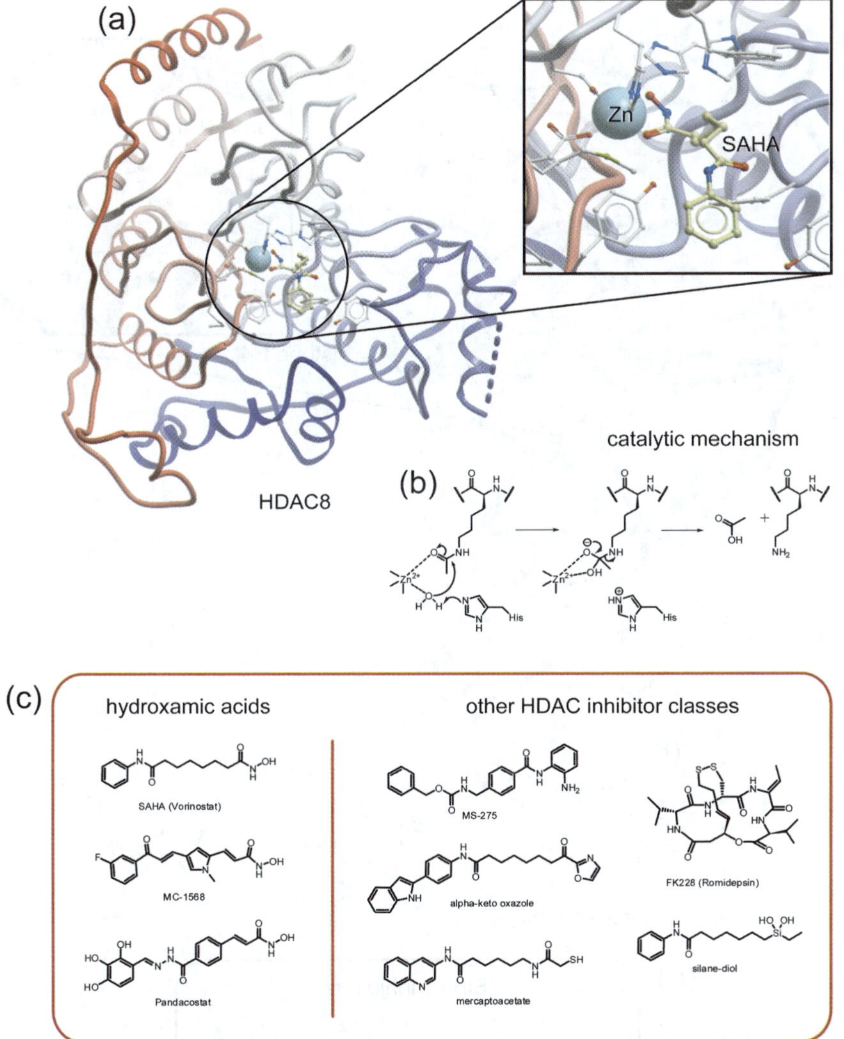



**Figure 5.9** (a) View from a crystal structure of HDAC8 in complex with vorinostat (PDB code: 1T69); (b) outline catalytic mechanism of hydrolytic lysine deacetylation by Class I, II and IV HDACs; (c) representative HDAC inhibitor chemotypes, including two clinically approved inhibitors, the hydroxamic acid SAHA/Vorinostat, and the depsipeptide FK228/ Romidepsin.

In general, the hydroxamic acids in clinical trials are pan-active across HDAC isoforms, while the aminophenylamides show some selectivity for class I over class II isoforms. The nature of the linker and cap moieties are also important in determining isoform selectivity; for example, within the cinnamic hydroxamic acids, the pyrrole-linked MC-1568 shows unusual selectivity in

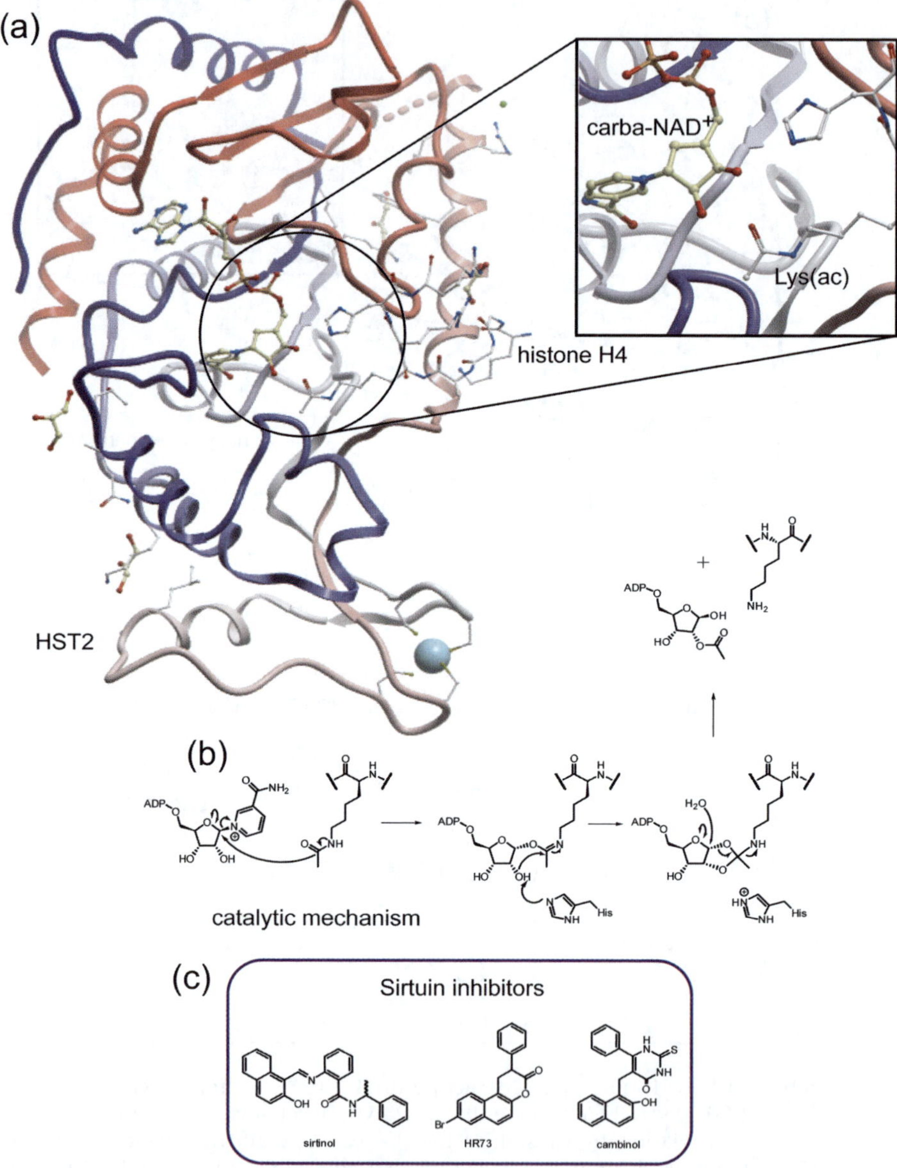



**Figure 5.10**    (a) View from a crystal structure of yeast HST2 (homologue of human SIRT2) in complex with carba-NAD and a histone H4 acetyllysine peptide (PDB code: 1SZD); (b) outline catalytic mechanism of lysine deacetylation by SIRT deacetylases, which proceeds by activation of the acetyl group by *O*-glycosylation; (c) representative SIRT inhibitor chemotypes.

favour of class II isoforms, whilst the phenyl-hydrazine linked pandacostat is almost completely pan-active.[44] It is notable that hydroxamic acids such as vorinostat may inhibit other metalloenzymes, including histone demethylases.[45]

Modulators of HDAC activity may be of interest in therapeutic areas beyond cancer. Epigenetic abnormalities, possibly introduced during either embryogenesis, puberty or adulthood, have also been noted in several psychiatric disorders, including schizophrenia, depression and drug addiction.[42] The potential of HDAC inhibitors to activate the expression of mRNAs that are down-regulated in these and other conditions has led to interest in embarking on clinical trials with CNS-penetrant HDAC inhibitors such as MS-275.[42]

Perturbation of histone acetylation homeostasis has been shown to be intimately involved in the neurodegenerative pathomechanisms of Huntington's, Parkinson's and Kennedy disease, amyotropic lateral sclerosis, Rubinstein–Taybi syndrome as well as stroke.[43] Based on promising *in vitro* and *in vivo* data, clinical trials have been initiated to evaluate the safety and efficacy of HDAC inhibitors for the treatment of Huntington's disease, amyotropic lateral sclerosis and spinal muscular atrophy.[43] Inhibition experiments showing that HDAC2 negatively regulates memory formation and synaptic plasticity created excitement that the development of specific inhibitors of this isoform could be used as cognitive enhancers, particularly in neurodegenerative patients.[46]

HDAC inhibitors have effects on the acetylation of key factors that regulate immune cell function *via* modulation of inflammatory transcriptions: for example HDAC3 and 6 have been shown to control acetylation of the NF-κB subunit p65. This ability to suppress the release of proinflammatory cytokines has triggered studies on HDAC inhibitors for treatment of immuno-inflammatory conditions such as arthritis, Crohn's disease and airway inflammation.[47]

Finally, inhibition of protein deacetylation with histone deacetylase (HDAC) inhibitors has shown promise in preclinical models of cardiovascular and renal disease, including myocyte hypertrophy, fibrosis, inflammation and epithelial-to-mesenchymal transition.[48]

SIRT inhibitors (Figure 5.10c) include sirtinol,[49] splitomycins including the derivative HR73 which shows activity on human SIRT,[50] and cambinol, which inhibited growth of Burkitt lymphoma xenografts in mice.[51]


### 5.2.3 Bromodomains

Bromodomains, which are small ($\sim$110 amino acids) conserved protein modules, were originally identified using sequence alignment as a common motif among *Drosophila* Brahma, female-sterile homeotic (fsh) and four other potential transcription regulators.[52–54] Bromodomains recognise acetylated lysines in proteins including histones.[55–59] The human genome encodes for 56 distinct bromodomains contained in 42 proteins, with some proteins containing two or more distinct bromodomains.[60]

### 5.2.3.1   Bromodomain Classification

Bromodomain-containing proteins have been classified (Table 5.3) into several distinct subgroups based on their function: (i) histone acetyltransferases (HATs), including GCN5, PCAF and TAFII250; (ii) ATP-dependent chromatin-remodelling complexes, including Brahma, Swi2, Snf2 and Brg1; (iii) the BET (bromodomain and extra C-terminal domain) family, a class of transcriptional regulators carrying two tandem bromodomains and an ET domain.[61] The BET family proteins are retained on chromosomes during mitosis,[62] distinguishing them from other bromodomain-containing proteins which are displaced from chromosomes.[63]

### 5.2.3.2   Bromodomains and Disease

Although some bromodomain-containing proteins have been studied in detail, allowing some association with specific diseases, the roles of many bromodomains are unknown.

Within the BET family, BRD2 is genetically associated with juvenile myoclonic epilepsy.[64] Although complete knockout in mice results in embryonic lethality (with neural tube defects as a major pathology),[65] targeted reduction in expression by promoter disruption in adult mice causes an unusual obese phenotype without type 2 diabetes.[66]

RNAi-mediated knockdown of BRD4 has been shown to arrest mitosis, whilst in the clinic expression levels of BRD4 correlate with breast cancer survival rates.[67] Knockdown experiments have also implicated BRD4 in regulation of viral transcription (HIV) and degradation (HPV), as well as transcriptional co-activation of NF-κB.[68] The testis-specific bromodomain-containing protein BRDT has been shown by *in vivo* mutation to be essential for spermatogenesis, indicating a likely role in fertility.[69] Both BRD3 and BRD7 are associated with nasopharyngeal cancer. BRD7 shows reduced expression in NPC biopsies and cell lines, while over-expression inhibits cell growth and cell cycle progression from G1 to S phase of NPC cells.[70]

BRD8 is an accessory subunit of human NuA4 HAT, and is over-expressed in human metastatic colorectal cancer cell lines, while BRD8 knockdown induced cell death or growth delay in colorectal cancer cell lines, and surviving BRD8-knockdown cells were particularly sensitive to spindle poisons. This suggests that a BRD8 inhibitor could be effective as an adjunct to anti-mitotic treatments.[71] Other bromodomain-containing HATs including CREBBP, PCAF and GCN5 have been extensively studied, but conclusions drawn from gene knockdown studies do not allow the individual contributions of the different domains to be elucidated. Disease association for these proteins is discussed in the HAT chapter section.

PB1 (polybromo-1 protein) is a subunit of the PBAF (polybromo, Brg1-associated factors) chromatin-remodelling complex required for kinetochore localisation during mitosis and the transcription of estrogen-responsive genes.

**Table 5.3** Human bromodomains.

| Subfamily | Bromodomain | Cognate sequence | Biological role/disease association |
| --- | --- | --- | --- |
| BET (BRD2, BRD3, BRD4, BRD8, BRDT) | BRD2 | H4K5/8/12/16 | Epilepsy |
| | BRD3 | — | Oncology |
| | BRD4 | H4K5, K12 | Cell cycle/mitosis, viral gene expression, NF-κB transcription |
| | BRDT | H4K5/8/12/16 | Spermatogenesis |
| | BRD8 | No data | Colorectal cancer |
| CREBBP, EP300, BAZ1A, BAZ1B, BRD1, BRD7, BRD9, BRPF1, BRPF3, KIAA1240, ATAD2, WDR9_2, PHIP_2, BRWD3_2 | CREBBP | H3K27, K36, H4K20, p53K382 | Neurological diseases, Alzheimer's |
| | BRD7 | H3K14, H4K8 | Nasopharyngeal cancer, cell cycle |
| PCAF, FALZ, GCN5L, PRKCBP1, CECR2, TAF1_1, TAF1_2, TAF1L_1, TAF1L_2 | PCAF | H3K9, H3K14, H3K36, | |
| | | H4K16, H4K20, HIV-TatK50 | — |
| | GCN5L | H4K8, K16 | — |
| | TAF1_1 | H4K8, K12, K16, H3K14 | — |
| TIF1, SP100, SP110, SP140, LOC93349, MLL, TIF1, TRIM28, TRIM33, TRIM66 | — | No data | — |
| BAZ2A, BAZ2B, ZMYND11 | — | No data | — |
| PB1, SMARCA2, SMARCA4 | PB1_1 | H3K4, H2AK5 | Mitosis, cardiac development, tumour suppression |
| | PB1_2 | H3K9 | |
| | PB1_3 | H3K9 | |
| | PB1_4 | H3K23, H2BK20 | |
| | PB1_5 | Various | |
| | PB1_6 | Various | |
| | SMARCA2 | H4K8/12, H3K9/14 | — |
| | SMARCA4 | H3K14, H4K8, H4K12 | — |

PB1 is mutated in some breast cancers and has been proposed as playing a role as a tumour suppressor.[72]

### 5.2.3.3   *Bromodomain Structure*

Crystal structures have been solved for over 20 bromodomains. The structure of the GCN5 bromodomain complexed with an acetylated histone H4 peptide shows the characteristic fold and features of histone recognition (Figure 5.11a).[58] The overall structure consists of a bundle of four α-helices (αZ, αA, αB and αC) joined by two loops αZ–αA and αB–αC, forming a hydrophobic cleft which contains the recognition motif for acetyl lysine. This cleft contains a highly conserved asparagine residue, the side chain of which donates a hydrogen bond to the carbonyl group of acetyl lysine. The remainder of the cleft is lined with hydrophobic aromatic and aliphatic side chains that contribute to the overall affinity of acetyl lysine. The lack of side chains in the cleft capable of stabilising a positive charge probably explains the basis for selective binding of acetylated lysine over unmodified or methylated lysine.

Of particular interest is the way in which the side chains that line the acetyl lysine binding pocket and neighbouring regions dictate the selectivity of

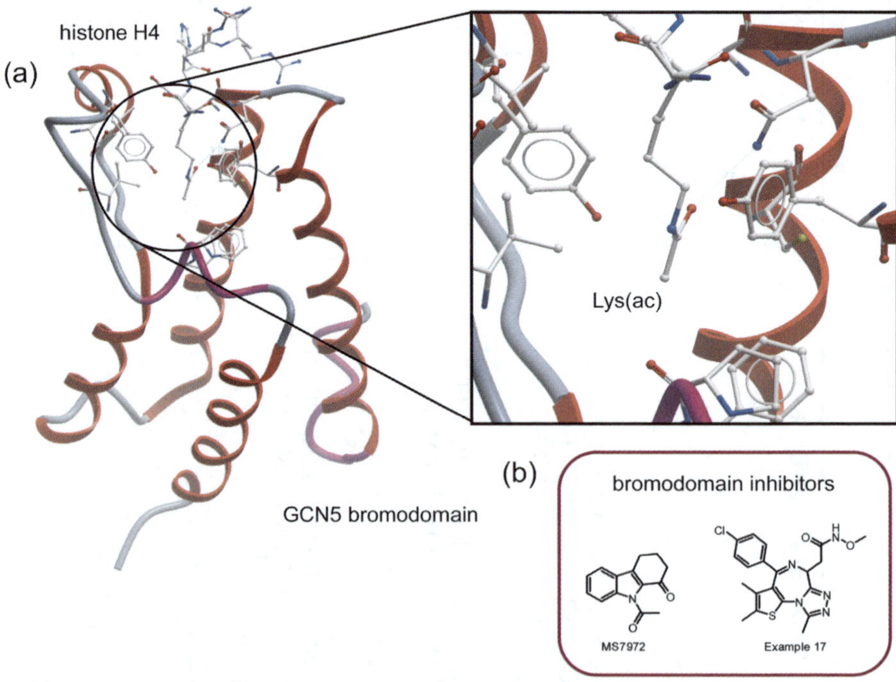


**Figure 5.11**   (a) View from a crystal structure of the GCN5 bromodomain in complex with an acetylated histone H4 peptide; (b) ligands for **CREBBP** (MS7972) and **BRD4** (Example 17) bromodomains.

binding to specific flanking sequences on histones. An NMR study of PCAF and CREBBP bromodomains bound to histone peptides concluded that for PCAF, the preferred acetyl lysine-binding site should have a hydrophobic residue at $(K(ac)+2)$ and an aromatic or positively charged residue at $(K(ac)+3)$, as observed for example at H3K14 with the sequence TGGK (ac)APR; while for CREBBP, the consensus target sequence should contain hydrophobic residues at $(K(ac)+1)$ and $(K(ac)+2)$ and a positively charged residue at $(K(ac)-1)$, as well as an aromatic residue at $(K(ac)-2)$, as observed at H4K20 with the sequence RHRK(ac)VLR.[73]

### 5.2.3.4  Bromodomain Inhibitors

At the time of writing, few small molecule ligands for bromodomains have been described in the academic literature. The carbazole MS7972 (Figure 5.11b), discovered by NMR screening of commercially available acetamide-containing small molecules against CREBBP, showed a $K_d$ of $\sim 20\,\mu M$ as determined by titration in tryptophan fluorescence experiments. In U2OS cells, MS7972 at $200\,\mu M$ modulated p53 stability and function in response to DNA damage.[74]

The triazolothienodiazepine (Example 17, Figure 5.11b) is reported to compete with acetylated histone H4 peptides for binding to BRD 4 with an $IC_{50}$ of 13 nM using a FRET assay with FLAG-tagged BRD4 and an H4 peptide. This compound showed antiproliferative effects at submicromolar concentrations against a panel of cancer cell lines, in particular a $GI_{50}$ of 14 nM against MV4-11 leukaemic cells.[75]


# 5.3  The Lysine and Arginine Methylation System

Histone lysine and arginine methylation sites serve as markers and initiators of different chromatin states. In electrostatic terms, methylation may be predicted to have less effect on the target residue than does acetylation, because the positive charge on the lysine or arginine can be maintained. In biological terms this difference is apparently reflected in the complex functions of methylation/ demethylation which can be either transcriptionally activating or deactivating. In contrast, acetylation is a predominantly activating modification. Methylation at specific residues (*e.g.* H3K4) is associated with the promoters of actively transcribed genes, whereas methylation at others (*e.g.* H3K9, H3K27) generally leads to transcriptional silencing and heterochromatinisation. The known functions of different methylation events are too complex to be described in detail here (see Lee and Mahadevan[5] and Ng *et al.*[76] for recent reviews). Thus methylation/demethylation of lysines and arginines by histone methyltransferases (HMTs) and demethylases (HDMs), in combination with the action of other chromatin modifying enzymes (including HDACs/HATs, DNMTs and binding domains), has the effect of changing transcriptional states, depending on which lysines are methylated. Different lysine methylation marks serve as binding sites for methyl binding domains (MBDs), which may be part of other histone-modifying enzymes or protein complexes involved in transcriptional regulation. These three protein classes, *viz.* HMTs, HDMs and MBDs, regulate the lysine and arginine methylation system of histone modifications, and are described in more detail below.

### 5.3.1 Histone Methyltransferases

#### 5.3.1.1 Histone Lysine Methyltransferases

Methylation of histone lysine residues is catalysed by site- and methylation-state specific (*i.e.* tri-/di-/monomethylation states) histone methyltransferases (EC number 2.1.1.43, Table 5.4). With the exception of DOT1L, which methylates H3K79, all identified KMTs catalyse methylation using a SET domain (SET is an abbreviation for Su(var)3-9, Enhancer of zeste and Trithorax; and refers to the *Drosophila* protein suppressor of variegation where this domain was first identified).[77]

#### 5.3.1.2 Arginine Methyltransferases

Arginine methylation is catalysed by protein arginine methyltransferases (PRMTs, EC number 2.1.1.125), and thus far eight human PRMTs have been identified, with several more putative enzymes predicted.[78] Two subclasses of PRMTs exist: Class I PRMTs catalyse asymmetric dimethylation, while Class II PRMTs catalyse symmetric dimethylation of arginine. All three methyltransferase subfamilies (SET-domain KMTs, DOT1L and PRMTs) utilise *S*-adenosylmethionine as an electrophilic methyl source.

#### 5.3.1.3 Histone Methyltransferases and Disease

Methyltransferases have been implicated in diseases (recently reviewed by Copeland *et al.*[79]), with the most well-documented examples being those of translocations in the MLL methyltransferases in various leukaemias (reviewed by Krivtsov and Armstrong[80]), and over-expression of EZH2, which is part of the polycomb PRC2 silencing complex, in a variety of cancers (reviewed by Simon and Lange[81]). DOT1L mistargeting has also been implicated in the formation of acute leukaemias.[82,83] Arginine methyltransferases such as PRMT4 have been linked to spinal muscular atrophy and NF-κB-related inflammatory diseases.[84,85] Several methyltransferases thus represent promising targets for pharmaceuticals.

#### 5.3.1.4 Lysine/Arginine Methyltransferase Mechanisms

Methylation of lysine or arginine sidechains is accomplished by $S_N2$ transfer of a methyl group from *S*-adenosylmethionine (SAM) to nucleophilic lysine/arginine sidechains (Figure 5.12b). The mechanism of deprotonation of the lysine sidechains has not yet been elucidated fully; however, a combination of

**Table 5.4** Histone methyltransferase enzymes.

| Subfamily | Methyltransferase | Known/preferred substrates |
|---|---|---|
| SET-domain lysine methyltransferases | MLL1 | H3K4 |
| | MLL2 | |
| | MLL3 | |
| | MLL4 | |
| | MLL5 | |
| | SET1A | |
| | SET1B | |
| | SET7/9 | |
| | ASH1 | |
| | SUV39H1 | H3K9 |
| | SUV39H2 | |
| | SETDB1/ESET | |
| | EuHMTase/GLP | |
| | G9a | |
| | CLL8 | |
| | RIZ1 | |
| | EZH2 | H3K27, H1K26 |
| | NSD1 | H3K36 |
| | SMYD2 | |
| | SET2 | |
| | SUV4 | H4K20 |
| | SUV4 | |
| | PR-SET7/8 | |
| Non-SET-domain lysine methyltransferases | DOT1L | H3K79 |
| Class I arginine methyltransferases | PRMT1 | H4R3 |
| | PRMT4/CARM1 | H3R2, H3R17, H3R26, H4R3 |
| | PRMT6 | H3R2 |
| Class II arginine methyltransferases | PRMT5 | H3R8, H4R3 |

experimental and theoretical approaches suggests that the presence of a water channel between the protonated lysine sidechain and the bulk solvent is important for deprotonation.[86] Successive methylation steps may then have the effect of obstructing the water channel, thus hindering further methylation cycles. Steric factors may affect the degree of methylation (mono-, di- or trimethylation); further, single amino acid variations in the lysine-binding pocket have been shown to act as a switch between different degrees of methylation.[87–89] Some degree of substrate sequence selectivity has been observed (reviewed by Rathert *et al.*[90]); however, the observation that many lysine methyltransferases also have non-histone substrates[91] suggests that these enzymes may be more promiscuous than once thought.

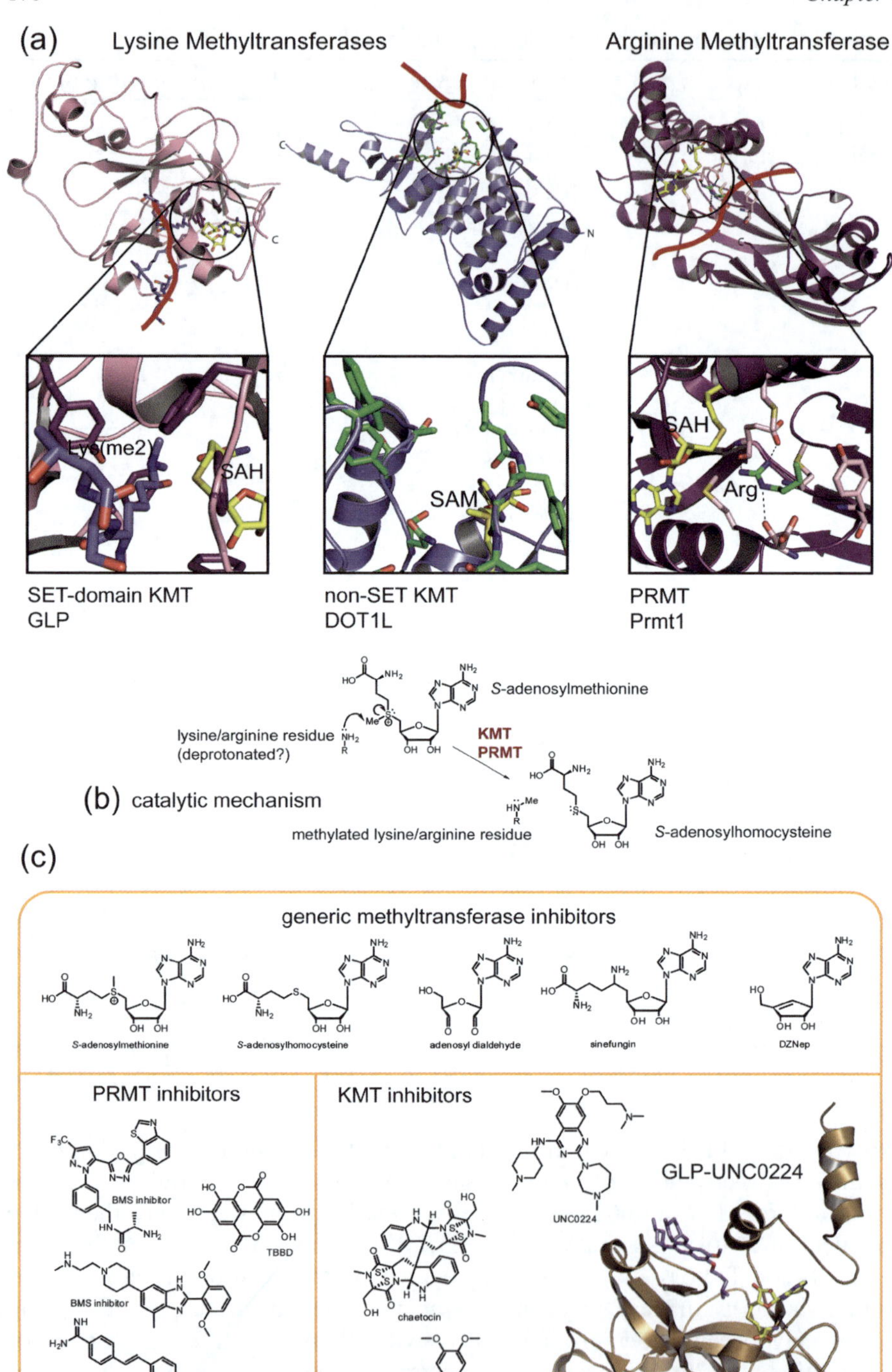

(a)
Lysine Methyltransferases
Arginine Methyltransferase
Lys(me2)
SAH
SET-domain KMT
GLP
SAM
non-SET KMT
DOT1L
SAH
Arg
PRMT
Prmt1
HO
NH2
NH2
S-adenosylmethionine
Me
lysine/arginine residue
(deprotonated?)
NH2
R
OH OH
KMT
PRMT
HO
NH2
NH2
S
N
Me
R
OH OH
S-adenosylhomocysteine
methylated lysine/arginine residue
(b) catalytic mechanism
(c)
generic methyltransferase inhibitors
S-adenosylmethionine
S-adenosylhomocysteine
adenosyl dialdehyde
sinefungin
DZNep
PRMT inhibitors
KMT inhibitors
GLP-UNC0224
BMS inhibitor
TBBD
BMS inhibitor
stilbamidine
AMI-1
UNC0224
chaetocin
BIX-01294

## 5.3.1.5    SET-domain Lysine Methyltransferase Structures

X-ray crystal and NMR structures have been reported for several SET-domain histone lysine methyltransferases (see Qian and Zhou[92] for a review of structural characteristics), in complex with SAM/SAH, peptide substrates and inhibitors.[87,89,93–101] In general, the SAM binding site and the lysine substrate binding groove are located on opposite faces of the enzyme, and are linked by a relatively narrow hydrophobic channel in which methyl transfer occurs (Figure 5.12a). Mutations in the lysine binding site have been shown to control the degree of methylation,[87–89] though, as discussed above, this may be more complex than a simple steric effect, and may employ a water channel to enable deprotonation of the target lysine.[86]

## 5.3.1.6    Non-SET-domain Lysine Methyltransferase Structures

A crystal structure for the non-SET domain KMT DOT1L has also been solved.[102] The structure (Figure 5.12a), which utilised a truncated DOT1L construct, shows an α-helical N-terminal domain, an open central α/β structure, and an unstructured *C*-terminal region (not observed in the crystal structure), which was shown to be essential for nucleosome binding and catalytic activity (DOT1L methylates H3K79 only in the nucleosomal context, and cannot methylate free histone H3). The SAM binding site is located adjacent to a putative lysine binding channel which is located suitably close to the *C*-terminal nucleosome binding region for methylation to occur.

## 5.3.1.7    Arginine Methyltransferase Structures

X-ray crystal structures have been reported for three Class I PRMTs, rat PRMT1 (Figure 5.12a) and PRMT3 and mouse CARM1/PRMT4.[103–106] Like the non-SET domain lysine methyltransferase, the substrate-binding and SAM-binding sites are located on the same face of the enzyme. SAM-binding residues are strictly conserved within the PRMT family and, while the general

---

**Figure 5.12**    Lysine and arginine methyltransferases. (a) Views from X-ray crystal structures of human SET-domain (GLP, PDB ID 2RFI) and non-SET domain (DOT1L, PDB ID 1NW3) lysine methyltransferases, as well as the rat Class I arginine methyltransferase Prmt1 (PDB ID 1OR8), are illustrated. Also indicated (or suggested) are substrate-binding grooves, in red; *S*-adenosylmethionine or its product, *S*-adenosylhomocysteine, are in yellow, and peptide substrates (blue for Lys substrate, green for Arg substrate). (b) Outline catalytic mechanism of methyltransferases. (c) Methyltransferase inhibitors; first, *S*-adenosylmethionine analogues which are generic inhibitors of methyltransferases; second, PRMT inhibitors discovered as a result of various screening efforts, and finally, KMT inhibitors, including the natural product chaetocin. The crystal structure of the methyltransferase GLP in complex with the potent inhibitor UNC0224 is also illustrated (PDB ID 3K5K).

architecture is similar for the different PRMTs, variations within the substrate-binding grooves appear to be responsible for substrate selectivity; dimerisation of the enzyme was also found to be essential for catalytic activity.

### 5.3.1.8   Generic Methyltransferase Inhibition by S-Adenosylmethionine Analogues

Inhibition of lysine/arginine methyltransferases might be achieved by competition either with the peptide substrate, with the *S*-adenosylmethionine cofactor, or a combination of both (Figure 5.12c). However, for the SET-domain KMTs, SAM and the substrate peptide bind on opposite faces of the enzyme,[107] suggesting that it may not be possible bind at both sites with a single pharmaceutically useful inhibitor. Further, specificity may be achieved more readily if the substrate binding site is targeted, rather than the relatively well-conserved cofactor binding site. However, differences in SAM-binding conformations are apparent in structural studies on several methyltransferases;[95,102,104] these difference suggest that selectivity for a subfamily of methyltransferases might be achieved by targeting the SAM binding site.[79]

### 5.3.1.9   Lysine Methyltransferase Inhibitors

Lysine methyltransferases are inhibited by SAM analogues, as are the PRMTs.[108] The *Drosophila* lysine methyltransferase SU(VAR)3-9 is selectively inhibited by chaetocin, a fungal mycotoxin of the epidithiodiketopiperazine class (Figure 5.12c). High-throughput screens have also been useful in identifying inhibitors of KMTs, including the related compounds BIX-01294[109] and its derivative UNC0224,[101] which selectively inhibit GLP and G9a. Crystal structures of these methyltransferases in complex with the BIX-01294/UNC0224 inhibitors have been solved,[94,101] and these imply that inhibition occurs by competition with peptide substrate.

### 5.3.1.10   Arginine Methyltransferase Inhibitors

Arginine methyltransferase inhibition by SAM analogues has been reported; these include the product of methylation reactions, *S*-adenosylhomocysteine, as well as sinefungin, adenosine dialdehyde and DZNep.[78,110] Aromatic sulfonates, including compound AMI-1 (Figure 5.12c), have been shown to selectively inhibit the PRMTs (but not the SET-domain KMTs);[111] carboxy analogues of these are also selective for PRMT inhibition. Phenol and bromophenol derivatives are also effective PRMT inhibitors,[111] though some also inhibit a wide range of other epigenetic regulatory enzymes, including HATs, KMTs and sirtuins.[112] High-throughput screening efforts against PRMT4 led to the identification of sulfone,[113] pyrazole[114–117] ("BMS inhibitor 1" in Figure 5.12c) and benzo[*d*]imidazole[118] lead compounds ("BMS inhibitor 2" in

Figure 5.12c) with good activity *in vitro*. Some inhibitors have also been reported which act by binding to the peptide substrate (TBBD, Figure 5.12c), thus preventing its binding to the enzyme; this results in inhibitors which inhibit methylation of only particular arginine residues.[119]


## 5.3.2 Histone Demethylases

Prior to the identification (in 2005) of LSD1 as a histone demethylase, histone methylation was thought to be a stable, irreversible modification. The only means of removing methyl marks on lysines or arginines was considered to be *via* degradation of the methylated histone followed by its replacement with an unmethylated histone, or by proteolytic "clipping" of the methylated histone tails. However, by 2005 cell biological analyses had implied that histone methylation is a dynamic process; methylation of certain lysine residues at silenced genes disappeared when these genes were activated. The timescale of the disappearance was not consistent with it being mediated by protein synthesis. These observations prompted the search for histone demethylases,[120] which led first to the discovery of LSD1 and subsequently to the extended family of JmjC histone demethylases (Table 5.5). Arginine methylation is antagonised by conversion of methylarginine to methylcitrulline by peptidyl-arginine deiminase 4 (PADI4),[121] though this is not equivalent to formal demethylation of arginine, as the resulting methylcitrulline is not positively charged, and is thus likely to have different properties to unmethylated arginine. Deimination also blocks further methylation by arginine methyl-transferases such as CARM1. However, deimination by PADI4 was shown to block hormone-induced transcriptional activation. Thus, although PADI4 is not a true demethylase, it does prevent methylated arginine dependent transcriptional activation, as a true arginine demethylase would be expected to do. PADI4 has also been shown to associate with HDAC1, thus providing another link between arginine deimination and transcriptional repression.[122] JMJD6, which is part of the JmjC subfamily of 2OG oxygenases but which is more closely related to the asparagine hydroxylase factor inhibiting hypoxia inducible factor (FIH), has been reported to be a histone arginine demethylase.[123] However, a different report has provided evidence that JMJD6 is a lysine hydroxylase, regulating splicing-regulatory proteins.[124]

The substrate specificities and known biological roles of the histone demethylases (discovered between 2005 and 2009) are summarised below.


### 5.3.2.1 *Histone Lysine Demethylase Subfamilies*

Two main families of histone demethylases have been identified. The LSD family (EC number 1.14.11.B1) employs a monoamine oxidase mechanism to oxidise methyllysine to an imine (with concomitant reduction of FAD to $FADH_2$), followed by addition of water to give a hemiaminal intermediate, which fragments to achieve demethylation. The JmjC family (EC number

**Table 5.5**   Histone demethylases.

| Subfamily | Demethylase | Known/preferred substrates | Biological role/disease associations |
| --- | --- | --- | --- |
| LSD | LSD1 | H3K4me2/me1<br>H3K9me2/me1<br>p53 K370me2 | Associates with Co-REST to repress neuronal-specific genes<br>Demethylates p53; possible role in tumourigenesis |
|  | LSD2 | H3K4me2/me1 | Required for *de novo* DNA methylation of imprinted genes |
| FBXL | FBXL11 | H3K36me2/me1 | Avoidance of senescence by silencing $p15^{Ink4b}$ locus[190] |
|  | FBXL10 | H3K36me2/me1 | Avoidance of senescence by silencing $p15^{Ink4b}$ locus[190,191]<br>Represses *c-Jun*[192]<br>Over-expressed in lymphomas and adenocarcinomas |
|  | PHF8 | H3K9me2/me1 | Mutations associated with cleft lip/palate and mental retardation[154] |
|  | KIAA1718 | H3K27me2/me1 | — |
| JMJD1 | JMJD1A | H3K9me2/me1 | Necessary for androgen-receptor dependent transcription[193]<br>Regulation of self-renewal in ES cells[194]<br>Spermatogenesis[195]<br>Energy homeostasis[196,197] |
| JMJD2 | JMJD2A | H3K9me3/me2<br>H3K36me3/me2<br>H1K26me3/me2 | Necessary for androgen-receptor dependent transcription[198,199] |

| | JMJD2B | H3K9me3/me2<br>H3K36me3/me2 | — |
| | JMJD2C | H3K9me3/me2<br>H3K36me3/me2 | Necessary for androgen-receptor dependent transcription[198,199]<br>Upregulated in prostate and oesophageal cancers[200]<br>Regulation of self-renewal[194] and of pluripotency genes[201] in ES cells |
| | JMJD2D | H3K9me3/me2 | — |
| | JMJD2E | H3K9me3/me2 | — |
| JMJD3 | JMJD3 | H3K27me3/me2 | Upregulated by NF-κB, responds to inflammatory stimuli, regulates genes involved in bone healing[202]<br>Induces differentiation of pluripotent ES cells[203,204] |
| | UTX | H3K27me3/me2 | Tumour suppressor activity by activating *INK4a/ARF* locus[205]<br>Controls genes involved in morphogenesis[206,207] |
| JARID1 | JARID1A | H3K4me3/me2/me1 | Associates with pRB, controls differentiation[208–210]<br>Recruited to PRC2 target genes to induce silencing[211] |
| | JARID1B | H3K4me3/me2/me1 | Over-expressed in breast cancer, facilitates G1 to S progression[212]<br>Associates with androgen receptor, upregulated in prostate cancer[213]<br>Regulates genes controlling cell cycle, differentiation and lineage[214] |
| | JARID1C | H3K4me3/me2/me1 | Mutations associated with XLMR[215–220] |
| | JARID1D | H3K4me3/me2/me1 | — |

1.14.11.27, part of the larger 2OG oxygenase superfamily, with members involved in oxidation of a variety of protein, nucleotide and small molecule substrates) utilises molecular oxygen, Fe(II) and 2-oxoglutarate (2OG) to hydroxylate methyllysine to the hemiaminal, which similarly fragments to give formaldehyde and the demethylated lysine. Work on individual members of each family, together with their substrate selectivities, biological roles/functions, and putative disease links, are outlined in Table 5.5.

### 5.3.2.2   Non-histone Substrates for Histone Demethylases

The only non-histone substrate identified (to date) for histone demethylases is the tumour suppressor p53. LSD1 demethylates K370 and K372 of p53.[125] These methylation sites are necessary for binding to 53BP1 (p53-binding protein 1), which is critical for binding of p53 to DNA, and subsequent tumour suppression; thus, LSD1-mediated demethylation of p53 may contribute to tumourigenesis. Work with peptides reveals that the JmjC demethylases also have the potential to accept other protein substrates.[126]

### 5.3.2.3   Structure and Mechanism of Amine Oxidase Histone Demethylases

LSD1 is a flavin adenine dinucleotide (FAD) dependent oxidase which catalyses demethylation of H3K4me2/me1[127] and, in association with the androgen receptor, of H3K9me2/me1.[128] Flavin dependent oxygenases oxidise methylamines to imines (with concomitant reduction of FAD to $FADH_2$), which are then hydrolysed non-enzymatically to yield formaldehyde and the demethylated amine (Figure 5.13a). This mechanism is incapable of demethylating trimethylated lysine residues, because a protonated amine is required (*i.e.* imine formation cannot take place from the quaternary ammonium species). $FADH_2$ is oxidised by molecular oxygen to generate hydrogen peroxide, which results in the regeneration of FAD.[129]

Crystal structures of LSD1 have been reported both with and without histone fragments. These reveal that LSD1 consists of three domains, the SWIRM (Swi3, Rsc8, and Moira), Tower and amine oxidase domains.[130–132] The amine oxidase domain binds the cofactor FAD, and both the amine oxidase and Tower domains are involved in substrate recognition and binding. The Tower domain also interacts with Co-REST, a key protein partner in LSD1's regulation of neuronal genes. The SWIRM domain is essential for protein stability *in vitro*, and, while SWIRM domains in other proteins have been shown to bind dsDNA, this has not observed in the case of LSD1. Discrimination between di- and trimethyllysine substrates occurs entirely on chemical grounds, as the methyllysine binding pocket in LSD1 is apparently sufficiently large to accommodate even a trimethylated lysine residue (Figure 5.13a).

### 5.3.2.4   Structure and Mechanism of 2-Oxoglutarate Dioxygenase Histone Demethylases

The JmjC domain containing histone demethylases are Fe(II) and 2-oxoglutarate dependent dioxygenases that can catalyse demethylation of all three possible $N$-methylation states of lysine residues. These enzymes utilise enzyme-bound Fe(II) to carry out oxidation of 2OG by molecular oxygen, with subsequent decarboxylation to yield succinate and carbon dioxide. This produces a reactive Fe(IV) = O intermediate, which hydroxylates the $N^{\varepsilon}$-methyl groups to yield unstable hemiaminal intermediates (Figure 5.13b). These decompose spontaneously to produce formaldehyde and the demethylated lysine residue. Substrate selectivity appears to occur on steric grounds, with trimethyl demethylases (*e.g.* JMJD2A[133]) having larger methyllysine binding pockets than dimethyl demethylases (*e.g.* FBXL11,[134] PHF8[135]).

Crystal structures have been reported for the 2OG dependent histone demethylases JMJD2A,[133] FBXL11[134] and PHF8.[135] All display the double-stranded β-helix (DSBH) fold which is typical of 2OG oxygenases, and which supports the Fe(II)-binding facial triad of one glutamate or aspartate and two histidine residues. The co-substrate 2OG binds Fe(II) in a bidentate manner through the ketone and carboxylate moieties at C-1 and C-2, while the C-5 carboxylate is tethered by forming a salt bridge to a lysine residue at the other end of the co-substrate binding site (Figure 5.13b). The methylated lysine residue of the substrate is accommodated in a relatively spacious binding cleft bounded by several aromatic residues, with the methylammonium moiety suitably close to the Fe(II) for hydroxylation. Variations in the size of this binding cleft are proposed to be responsible for selectivity between tri- and dimethylated lysine residues, while selectivity for different substrate sequences is achieved by a combination of factors. The catalytic domains of the 2OG dependent demethylases exert some degree of control over which sequences are recognised as substrates. Additional substrate selectivity is likely also provided by other domains within the enzymes, as has been demonstrated for PHF8; the PHD finger, which binds H3K4me3, confers selectivity for, and increases activity towards, H3K9me2. Whether such binding domain control exists in the substrate selectivity of other demethylase subfamilies (which also contain PHD and Tudor domains, both of which bind methylated lysines), remains to be determined.

### 5.3.2.5   Amine Oxidase Histone Demethylase Inhibitors

Several types of LSD1 inhibitors have been reported (Figure 5.13c): these include known non-specific monoamine oxidase inhibitors, including tranylcypromine[136–138] and pargyline,[128] selective inhibition by substrate peptide-based analogues of pargyline,[139,140] selective inhibition by FAD analogues,[141] and inhibition by polyamines[142] and thalidomide derivatives;[143] further, effects on histone H3K9 and H3K4 methylation levels in cells have been ascribed to inhibition of LSD1-catalysed demethylation (by tranylcypromine, pargyline and polyamines).[138,142]

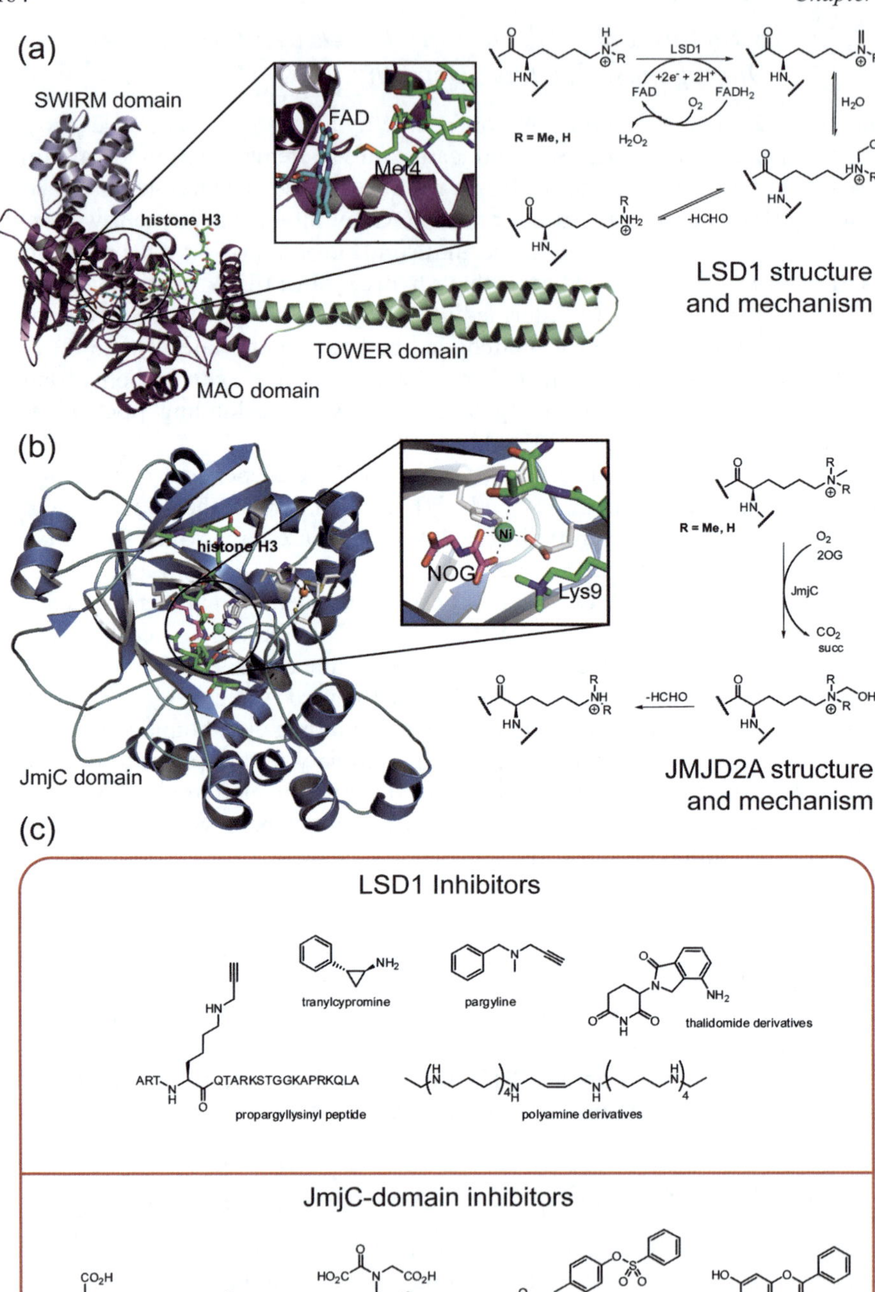

(a)
SWIRM domain
FAD
Met4
histone H3
TOWER domain
MAO domain
LSD1
FAD
FADH2
O2
H2O2
+2e- + 2H+
R = Me, H
H2O
-HCHO
LSD1 structure
and mechanism
(b)
histone H3
NOG
Ni
Lys9
JmjC domain
R = Me, H
O2
2OG
JmjC
CO2
succ
-HCHO
JMJD2A structure
and mechanism
(c)
LSD1 Inhibitors
HN
tranylcypromine
NH2
pargyline
thalidomide derivatives
NH2
O
O
N
O
ART
N
H
O
QTARKSTGGKAPRKQLA
propargyllysinyl peptide
N
H
N
H
4
N
H
N
H
4
polyamine derivatives
JmjC-domain inhibitors
CO2H
N
CO2H
pyridine dicarboxylates
HO2C
N
H
O
O
CO2H
N-oxalylglycine (NOG)
HO2C
N
O
CO2H
NMe2
N-oxalyl amino acids
HO2C
N
H
O
CO2H
O
S
O
O
N-oxalyl-D-tyrosine derivatives
HO
HO
O
O
OH
catechols
N
H
O
O
N
H
OH
HDAC inhibitors, e.g. SAHA
HO
O
O
OH
succinic acid
N
S
S
S
S
N
disulfiram
N
O
Se
ebselen

## 5.3.2.6   2OG Dioxygenase Histone Demethylase Inhibitors

Most inhibitors of 2OG oxygenases described in the literature are 2OG competitors, and bind to Fe(II) at the active site (Figure 5.13c). A variety of Fe(II)-chelating moieties have been shown to inhibit 2OG oxygenases, including the JMJD2 histone demethylases.[144] However, these strategies face the challenge of achieving selectivity. 2OG oxygenases catalyse a wide variety of reactions in diverse biological processes, including the hypoxic response, fatty acid metabolism, epigenetics, RNA splicing, collagen synthesis, *etc.* Further, all possess highly similar active sites, with almost identical iron-binding motifs, and all rely on 2OG as a cosubstrate. Thus, inhibition strategies relying on 2OG competition may be challenging.

Inhibition of the JMJD2 demethylases by a variety of 2OG analogues/competitors has been reported.[144–146] These include *N*-oxalylamino acids, pyridine dicarboxylates and related bipyridyl derivatives, catechols, hydroxamic acids (including the clinically used HDAC inhibitor SAHA/Vorinostat), and TCA cycle intermediates, such as succinate and fumarate. High-throughput screening methods have been reported for the JMJD2 demethylases,[146] and high-throughput screens on this and other JmjC demethylase subfamilies may be useful in identifying novel inhibitor scaffolds.

In an effort to avoid inhibiting the demethylases with 2OG competitors, we recently described the inhibition of the JMJD2 histone demethylases by ejection of the structural Zn(II) ion by known Zn-ejecting compounds such as disulfiram and ebselen.[147] This strategy presents a means of achieving selectivity for a particular demethylase family by targeting a non-active site part of the protein. It is possible that similar approaches may be useful in the development of demethylase inhibitors with clinical relevance.

### 5.3.3   Methyl Binding Domains

Changes in chromatin structure (and resulting changes in transcriptional activity) as a result of histone modifications are thought to occur in one of two ways: modifications may result in steric or electrostatic changes to histone tails which change the nature of the histones' interactions with DNA, or, alternatively, modifications may recruit modification-specific binding domains

---

**Figure 5.13**   Histone demethylases. Structures and mechanisms are illustrated for the two subclasses of histone lysine demethylases. (a) Views from an X-ray crystal structure of LSD1 (PDB ID 2V1D) in complex with cofactor FAD and a peptide substrate analogue, where methionine replaces methylated lysine, and outline mechanism of the LSD1-catalysed demethylation reaction. (b) Views from an X-ray crystal structure of JMJD2A (PDB ID 2OQ6) in complex with co-factor analogue *N*-oxalylglycine and histone substrate trimethylated at H3K9 (note that Ni(II) replaces Fe(II) for crystallography), and outline mechanism of JmjC-domain catalysed lysine demethylation. (c) Representative inhibitors of LSD1 and JmjC-domain demethylases.

which then target other histone-modifying or nucleosome-remodelling proteins, leading to changes in chromatin structure. Unlike acetylation or citrullination, methylation of lysine and arginine residues does not result in a change in electrostatic charge, suggesting that the effects of histone methylation are mediated by interactions with methyl-binding domains.

Six major subfamilies of MBDs have been identified (Table 5.6). These comprise the $C_4HC_3$-Zn-finger plant homeodomains (**PHD**), Malignant brain tumour (**MBT**) repeats, chromatin organisation modifier (**Chromo**) domains, double/tandem **Tudor** domains, **ankyrin** repeat domains, and **WD-40** (conserved tryptophan/aspartate-40) domains. The chromo, Tudor and MBT domains are suggested to have evolved from a common methyl-binding ancestor, and have consequently been grouped as the so-called "royal family" proteins.[148] In some cases, these exist as independent proteins, though often they are part of larger proteins (for example, the histone demethylase JMJD2A also contains a double Tudor domain and two PHD domains[149]) or protein complexes (*e.g.* the WD40 domain EED forms part of the polycomb repressive complex 2 (PRC2)).[150] The proteins to which these MBDs are attached include a wide variety of histone-modifying enzymes, transcription factors, chromatin binding and remodelling factors, *etc.*[151] However, the general characteristics of the methyllysine or methyl arginine recognition motif are conserved across all subfamilies, though variations exist, permitting selectivity for different sequences or methylation states.

### 5.3.3.1 Histone-modifying Enzymes are Targeted by Methyl Binding Domains

Some MBDs are reported to target the actions of histone-modifying enzymes. Notable examples include Suv39h-HP1 (where Suv39h methylates H3K9, and HP1 targets its activity by binding to H3K9me3[152]), LSD1-BHC80 (where LSD1 demethylates K9me2 and BHC80 binds to H3K4me0, targeting its activity[153]), and the PRC2 complex, where EZH2 methylates K27, while EED WD40 repeat binds to the same mark, facilitating the spreading of the K27me mark.[150] Each of these examples occurs in the context of multi-protein complexes. Some proteins contain domains that bind to one methylated amino acid while reacting with another. Examples include the histone demethylase PHF8, where the relatively unspecific demethylase activity of the catalytic domain (towards K9me2, K27me2 and K36me2[154,155]) is targeted to K9me2 by simultaneous binding of the PHF8 PHD finger to K4me3;[155] in a similar fashion, the histone methyltransferase GLP methylates H3K9, and also binds to this mark through its ankyrin repeat domain,[156] suggesting a means of spreading this modification through chromatin.

### 5.3.3.2 Methyl Binding Domains and Disease

The contributions of MBDs to disease have been recently reviewed,[157] suggesting that some MBDs may present therapeutic targets. Several examples are

**Table 5.6**  Selected human methyl binding domains (MBDs).

| Binding domain subfamily | Protein | Cognate sequence | Disease associations |
|---|---|---|---|
| PHD | BHC80 | H3K4me0 | — |
| | RAG2 | H3K4me3 | Immunodeficiency if RAG recombinase activity (V(D)J recombination) diminished |
| | ING | H3K4me3/me2 | Various cancers |
| | BPTF | H3K4me3/me2 | — |
| | PYGO | H3K4me2 | — |
| | DNMT3L | H3K4me0 | — |
| | PHF8, KIAA1718 | H3K4me3 | Cleft lip/palate/MR |
| | JMJD2A-C | — | Prostate/oesophageal cancers |
| | JARID1A-D | — | Leukaemogenesis when fused to NUP98 |
| | MLL | — | Leukaemias |
| MBT | L3MBT1L | H1bK26me2/me1 H4K20me2/me1 | Malignant brain tumour formation in *Drosophila* |
| Chromo | HP1 | H3K9me3 | — |
| | CHD1 | H3K4me3/me2 | — |
| | Polycomb | H3K27me3/me2 | — |
| Tudor | 53BP1 | H3K79me2, H4K20me2 | — |
| | Crb2 | H4K20me2 | — |
| | SMN | sDMA | — |
| | JMJD2A | H3K4me3/ H4K20me3 | Prostate/oesophageal cancers |
| | TDRD3 | sDMA | — |
| | C20orf104 | H3K9me2, H3K4me2, H4K20me2 | — |
| Ankyrin repeat | G9a/GLP | K9me2/me1 | — |
| WD40 | WDR5 | H3R2me0 | — |
| | EED | Kme3 | — |

outlined here. Mutations in the RAG2 PHD finger lead to reduced recombinase activity and are linked to immunodeficiency disorders,[158–160] while mutations in the ING PHD fingers are associated with the susceptibility to and progression of several cancers. In one suggested model, ING PHD fingers bind to promoters of active genes after DNA damage, recruiting HDACs and other silencing factors to silence these genes (reviewed by Baker *et al.*[157]). Finally, translocations which fuse the nuclear pore protein NUP98 to PHD fingers from a variety of other proteins (JARID1A, PHF23,

NSD1) produce a potent oncoprotein which is leukaemogenic.[161–164] In contrast, the PHD fingers of MLL were shown to be necessary to prevent MLL-driven leukaemogenesis.[165,166]


### 5.3.3.3    *Methyl Binding Domain Structure and Function*

Although MBDs are structurally diverse (with the family including Zn-fingers, propeller-like repeat structures, *etc*), the common methyllysine binding motif consists of a box-like structure of two to four aromatic residues, sometimes together with an acidic residue (Asp or Glu), which binds to methylated lysine tails (Figure 5.14).[167] An important contributor to binding is the cation–π interaction between the quaternary ammonium lysine sidechain and the delocalised π-electrons of aromatic sidechains. Hydrophobic interactions likely have some contribution to binding, though they are apparently not the major contributor. This was demonstrated in the case of the binding of trimethyllysine and its uncharged carbon analogue, *t*-butylnorleucine, to the chromodomain HP1; while a histone fragment containing trimethyllysine bound efficiently, the carbon analogue did not bind at all.[168]

Two main binding modes have been observed for MBDs; for those that preferentially bind higher methylation states (me3), the methylated lysine binds to a hydrophobic box on the surface of the MBD (*e.g.* PHD, chromo, double Tudor and WD40 MBDs), whereas for those that preferentially bind lower methylation states (*e.g.* MBT, tandem Tudor and ankyrin repeat domains), the hydrophobic box forms a shallow cleft into which the methylated lysine binds. Further steric constraints are added by the presence of an acidic residue in the cleft, which is proposed to hinder binding of the trimethyl state, but forms electrostatic and hydrogen bonding interactions with the dimethyl state.[167] Finally, some methyllysine sequences are capable of forming a β-strand, which "docks" onto the existing antiparallel β-sheet structure of some MBDs, thus providing some sequence selectivity.[169]

Interestingly, a second, related class of PHD Zn-finger proteins recognises unmethylated lysine exclusively. The BHC80 and DNMT3L PHDs bind to unmethylated lysine using a salt bridge to an acidic residue; steric hindrance prevents the binding of methylated lysines.[153,170,171] Similarly, the WD40 protein WDR5 preferentially binds unmethylated H3R2, to the exclusion of the methylated forms.[172]

---

**Figure 5.14**    Views from X-ray crystal structures of methyl binding domains. For each MBD is shown: the class of MBD (*e.g.* PHD finger), the specific protein name (*e.g.* PHF8), the PDB ID (*e.g.* 2PUY), the mode of recognition (surface or cleft), the methylation state preference (*e.g.* Kme3) and the residues comprising the recognition motif (*e.g.* three aromatic, one acidic (Asp or Glu)). In each case, the lysine residue which is recognised by the MBD is shown in orange, while the recognition motif residues are shown in blue.

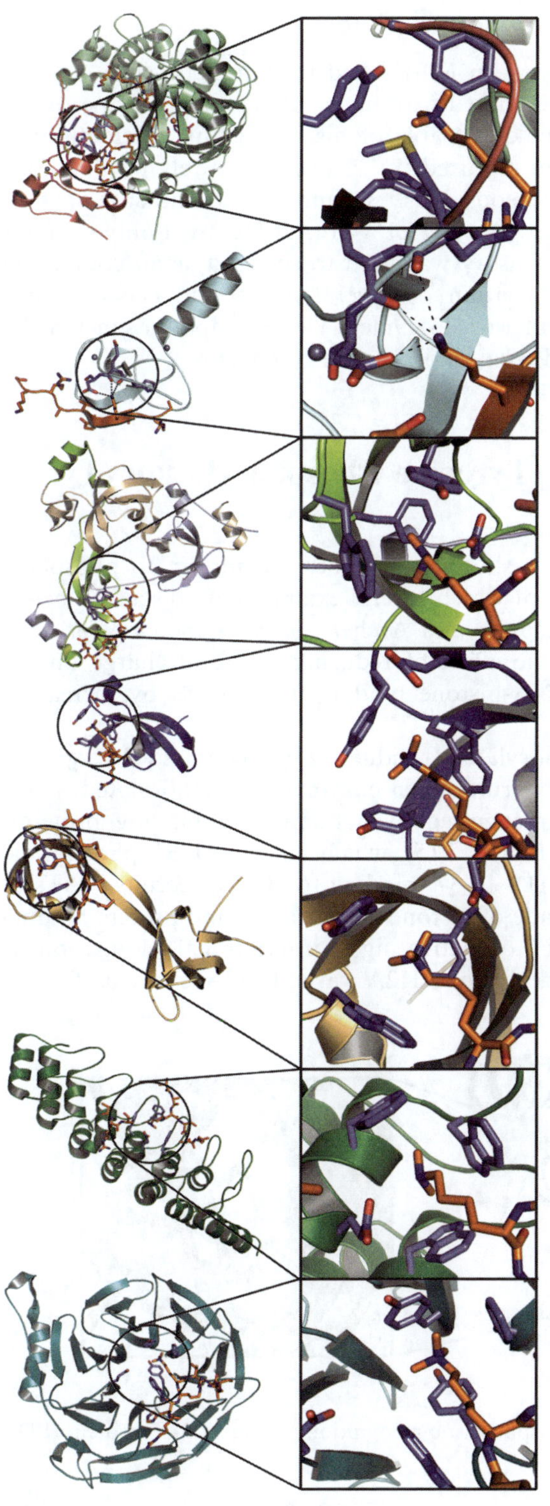

PHD finger (PHF8)
3KV4
Surface recognition
Kme3 preference
3 aromatic

PHD finger (BHC80)
2PUY
Surface recognition
Kme0 preference
1 acidic + 2 backbone

MBT domain (L3MBTL1)
2PQW
Cleft recognition
Kme2 preference
3 aromatic + 1 acidic

Chromodomain (HP1)
1KNE
Surface recognition
Kme3 preference
3 aromatic + 1 acidic

Tandem Tudor domain
(JMJD2A) 2GFA
Surface recognition
Kme3 preference
3 aromatic + 1 acidic

Ankyrin repeat domain (GLP)
3B95
Cleft recognition
Kme2 preference
3 aromatic + 1 acidic

WD40 domain (EED)
3IJ1
Surface recognition
Kme3 preference
4 aromatic + 1 acidic

### 5.3.3.4   *Methyl Binding Domain Inhibitors*

Inhibitors of the interaction between MBDs and their target lysines may be useful in combating diseases associated with misregulated methyllysine recognition. Assays for small molecules capable of disrupting this interaction by binding to the MBD have been reported.[173,174] The assays rely on amplified luminescent proximity homogeneous assay technology to measure loss of methylated histone binding to L3MBTL1[173] or JMJD2A double Tudor domains,[174] and the L3MBTL1 assay was used to screen a small compound library, resulting in the identification of several compound classes which inhibited the binding of L3MBTL1 to H3K9me1. These compounds included a cephalosporin, Cefsulodin, and a catechol, I-OMe-tyrphostin.[173]

## 5.4   Serine/Threonine/Tyrosine Phosphorylation of Histones

Serine, threonine and tyrosine residues in histones are subject to phosphorylation by a variety of kinases, of which several examples are discussed below (reviewed in more detail by Keppler and Archer[175]). As is the case for acetylation, phosphorylation has the effect of reducing the total charge on the histones, and may disrupt DNA-histone binding interactions by increasing electrostatic repulsion.

Additionally, some phosphorylated residues are recognised by specific binding proteins which then recruit other enzymes. Phosphorylated serine residues on histone tails are recognised by a number of binding proteins, including the 14-3-3 proteins (Figure 5.15), which bind to H3S10phos (phosphorylated in this case by PIM1). 14-3-3 binding to H3S10 recruits HATs to nucleosomes, leading to a cascade of histone modification and protein binding events, with the overall effect of activating transcriptional elongation.[176] Phosphorylation of S139 in the histone H2A variant H2AX signals for the

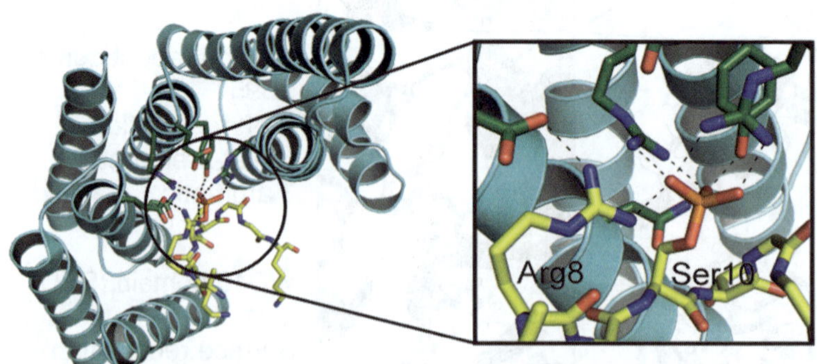


**Figure 5.15**   Histone H3S10 phosphoserine recognition by the 14-3-3ζ protein (PDB ID 2C1N).

recruitment of repair enzymes to DNA double-stranded breaks; this is mediated through the binding of MDC1 to phosphorylated H2AX.[177]

Some phosphorylation events may disrupt or enhance binding of other histone-binding and histone-modifying proteins. For example, phosphorylation of H3S10 (by the kinase auroraB) causes dissociation of heterochromatin protein 1 (HP1) from H3K9me3 in heterochromatin.[178] AuroraB-catalysed phosphorylation of H3S10 and H3S28 is required for chromosome condensation, and its inhibition or down-regulation leads to a variety of errors in mitosis, causing cell death. Similarly, the tyrosine kinase JAK2 phosphorylates H3Y41, which also has the effect of disrupting HP1 binding, leading to activation of HP1-silenced genes.[179] Phosphorylation of H3T11 (by the kinase PRK1) apparently accelerates demethylation of H3K9me3 by the demethylase JMJD2C;[180] in contrast, phosphorylation of H3S10 prevents demethylation of H3K9 by the JMJD2 demethylases.[133] Phosphorylation levels on histones are controlled by interplay between kinases and phosphatases, which include protein phosphatase type 1 (PP1).

Inhibitors of the histone serine kinase AuroraB have been developed as anticancer drugs that cause disruption of mitosis, leading to cell death (reviewed by Perez Fidalgo *et al.*[181]). Similarly, a variety of JAK2 tyrosine kinase inhibitors are in clinical trials for the treatment of myeloproliferative disorders.[182] It has also been suggested that inhibition of the H3T11 kinase, PRK1, may provide a means of therapeutically targeting prostate cancer cells.[180] Finally, though it has not been investigated, there is the possibility of inhibiting the interactions between phosphoserine binding proteins and their target phosphorylated serines, which might be useful in preventing the activation of 14-3-3 target genes.

## 5.5 Lysine Ubiquitylation/SUMOylation/Biotinylation of Histones

Histone lysine residues are subject to ubiquitylation and sumoylation, and biotinylation of lysine residues in all four histones has been reported.[183] Histone ubiquitylation may contribute to transcriptional activation due to the sheer size of the ubiquitin protein; its conjugation to histones may result in increased spacing between nucleosomes, thus possibly facilitating transcriptional activation. There is also some evidence for crosstalk between histone ubiquitylation and other modifications; for example, ubiquitylation of H2BK123 by the ubiquitin conjugating enzyme Rad6 is reported to be essential for methylation of H3K4[184] and for efficient methylation of H3K79 by DOT1L.[185] Further, it appears that deubiquitylation of H2B after methylation is also required for transcriptional activation,[186] suggesting that the roles of ubiquitylation may be more complex than merely increasing nucleosome spacing. H2A is also known to be ubiquitylated, with roles in heritable gene silencing and X-inactivation.[187]

Histones are also modified by the small ubiquitin-like modifier (SUMO). SUMO-specific activating, conjugating and ligating enzymes catalyse the ligation of SUMO proteins to lysines in histones, particularly in histone H4, where SUMOylation is associated with transcriptional repression, through the recruitment of HDACs and HP1.[188]

At present it is unclear whether any of the histone ubiquitin or SUMO ligation complexes may prove useful as therapeutic targets; however, there is some evidence for upregulation of the SUMO conjugating enzyme Ubc9 in several cancers (reviewed by Keppler and Archer[175]), suggesting that therapeutic inhibition of SUMOylation may be useful. Although beyond the scope of this review, inhibitors of ubiquitylation/deubiquitylation are being developed and may be useful in therapeutic applications aimed at modulating histone modifications.[189]

## 5.6   Appendix: Abbreviations Used in this Chapter

Histone modifications are identified throughout using the standard nomenclature, where, *e.g.*, H3K9me3 refers to histone H3, residue lysine-9, trimethylated. One-letter amino acid codes are used for histone residues. Other abbreviations used include: 2OG, 2-oxoglutarate; 5-meC, 5-methylcytosine; BRD, bromodomain; CNS, Central nervous system; CpG, C-phospho-G; DNMT, DNA methyltransferase; FAD, flavin adenine dinucleotide; HAT, histone acetyltransferase; HDAC, histone deacetylase; HDM, histone demethylase; HP1, heterochromatin protein 1; JmjC, Jumonji-C; KDM, lysine demethylase; KMT, Lysine methyltransferase; LSD1, lysine-specific demethylase 1; MBT, malignant brain tumour; $NAD^+$, nicotinamide adenine dinucleotide; PDB, protein data bank; PHD, plant homeodomain; PRC, polycomb repressive complex; PRMT, protein arginine methyltransferase; SAH, *S*-adenosylhomocysteine; SAHA, suberoylanilide hydroxamic acid; SAM, *S*-adenosylmethionine; SET, Su(var), Enhancer of zeste, trithorax; SIRT, sirtuin; SUMO, small ubiquitin-like modifier; TSA, trichostatin A.

## Acknowledgements

We thank the Wellcome Trust, the Biotechnology and Biological Sciences Research Council, and the European Union for funding our work (N.R.R., C.J.S.). The Structural Genomics Consortium is a registered charity (number 1097737) that receives funds from the Canadian Institutes for Health Research, the Canadian Foundation for Innovation, Genome Canada through the Ontario Genomics Institute, GlaxoSmithKline, Karolinska Institutet, the Knut and Alice Wallenberg Foundation, the Ontario Innovation Trust, the Ontario Ministry for Research and Innovation, Merck & Co., Inc., the Novartis Research Foundation, the Swedish Agency for Innovation Systems, the Swedish Foundation for Strategic Research, and the Wellcome Trust. We thank our co-workers for their efforts in the field of epigenetic regulators and apologise for the lack of complete citations due to space constraints.

# References


1. N. Happel and D. Doenecke, *Gene*, 2009, **431**, 1.
2. S. Kriaucionis and N. Heintz, *Science*, 2009, **324**, 929.
3. M. Tahiliani, K. P. Koh, Y. Shen, W. A. Pastor, H. Bandukwala, Y. Brudno, S. Agarwal, L. M. Iyer, D. R. Liu, L. Aravind and A. Rao, *Science*, 2009, **324**, 930.
4. C. D. Allis, T. Jenuwein, D. Reinberg and M.-L. Caparros, *Epigenetics*, Cold Spring Harbor Laboratory Press, 2007.
5. B. M. Lee and L. C. Mahadevan, *J. Cell. Biochem.*, 2009, **108**, 22.
6. S. I. S. Grewal and S. C. R. Elgin, *Nature*, 2007, **447**, 399.
7. J. A. Simon and R. E. Kingston, *Nat. Rev. Mol. Cell Biol.*, 2009, **10**, 697.
8. H. Cedar and Y. Bergman, *Nat. Rev. Genet.*, 2009, **10**, 295.
9. E. Segal and J. Widom, *Trends Genet.*, 2009, **25**, 335.
10. P. B. Talbert and S. Henikoff, *Nat. Rev. Mol. Cell Biol.*, 2010, **11**, 264.
11. C. R. Clapier and B. R. Cairns, *Annu. Rev. Biochem.*, 2009, **78**, 273.
12. S. L. Berger, T. Kouzarides, R. Shiekhattar and A. Shilatifard, *Genes Dev.*, 2009, **23**, 781.
13. A. Bird, *Nature*, 2007, **447**, 396.
14. C. Martin and Y. Zhang, *Curr. Opin. Cell Biol.*, 2007, **19**, 266.
15. A. V. Probst, E. Dunleavy and G. Almouzni, *Nat. Rev. Mol. Cell Biol.*, 2009, **10**, 192.
16. T. Jenuwein and C. D. Allis, *Science*, 2001, **293**, 1074.
17. K. K. Lee and J. L. Workman, *Nat. Rev. Mol. Cell Biol.*, 2007, **8**, 284.
18. K. Mantelingu, B. A. Reddy, V. Swaminathan, A. H. Kishore, N. B. Siddappa, G. V. Kumar, G. Nagashankar, N. Natesh, S. Roy, P. P. Sadhale, U. Ranga, C. Narayana and T. K. Kundu, *Chem. Biol.*, 2007, **14**, 645.
19. F. Manzo, F. P. Tambaro, A. Mai and L. Altucci, *Expert Opin. Ther. Pat.*, 2009, **19**, 761.
20. S. Isharwal, M. C. Miller, C. Marlow, D. V. Makarov, A. W. Partin and R. W. Veltri, *Prostate*, 2008, **68**, 1097.
21. L. Stimson, M. G. Rowlands, Y. M. Newbatt, N. F. Smith, F. I. Raynaud, P. Rogers, V. Bavetsias, S. Gorsuch, M. Jarman, A. Bannister, T. Kouzarides, E. McDonald, P. Workman and G. W. Aherne, *Mol. Cancer Ther.*, 2005, **4**, 1521.
22. L. H. Kasper, T. Fukuyama, M. A. Biesen, F. Boussouar, C. Tong, A. de Pauw, P. J. Murray, J. M. van Deursen and P. K. Brindle, *Mol. Cell Biol.*, 2006, **26**, 789.
23. M. A. Hussain, D. L. Porras, M. H. Rowe, J. R. West, W. J. Song, W. E. Schreiber and F. E. Wondisford, *Mol. Cell Biol.*, 2006, **26**, 7747.
24. Y. Tanaka, I. Naruse, T. Hongo, M. Xu, T. Nakahata, T. Maekawa and S. Ishii, *Mech. Dev.*, 2000, **95**, 133.

25. C. E. Berndsen, B. N. Albaugh, S. Tan and J. M. Denu, *Biochemistry*, 2007, **46**, 623.
26. X. Liu, L. Wang, K. Zhao, P. R. Thompson, Y. Hwang, R. Marmorstein and P. A. Cole, *Nature*, 2008, **451**, 846.
27. A. Schuetz, G. Bernstein, A. Dong, T. Antoshenko, H. Wu, P. Loppnau, A. Bochkarev and A. N. Plotnikov, *Proteins*, 2007, **68**, 403.
28. A. Clements, A. N. Poux, W. S. Lo, L. Pillus, S. L. Berger and R. Marmorstein, *Mol. Cell*, 2003, **12**, 461.
29. O. D. Lau, T. K. Kundu, R. E. Soccio, S. Ait-Si-Ali, E. M. Khalil, A. Vassilev, A. P. Wolffe, Y. Nakatani, R. G. Roeder and P. A. Cole, *Mol. Cell*, 2000, **5**, 589.
30. K. Balasubramanyam, M. Altaf, R. A. Varier, V. Swaminathan, A. Ravindran, P. P. Sadhale and T. K. Kundu, *J. Biol. Chem.*, 2004, **279**, 33716.
31. K. Balasubramanyam, R. A. Varier, M. Altaf, V. Swaminathan, N. B. Siddappa, U. Ranga and T. K. Kundu, *J. Biol. Chem.*, 2004, **279**, 51163.
32. M. Paris, M. Porcelloni, M. Binaschi and D. Fattori, *J. Med. Chem.*, 2008, **51**, 1505.
33. C. Hubbert, A. Guardiola, R. Shao, Y. Kawaguchi, A. Ito, A. Nixon, M. Yoshida, X. F. Wang and T. P. Yao, *Nature*, 2002, **417**, 455.
34. J. J. Kovacs, P. J. Murphy, S. Gaillard, X. Zhao, J. T. Wu, C. V. Nicchitta, M. Yoshida, D. O. Toft, W. B. Pratt and T. P. Yao, *Mol. Cell*, 2005, **18**, 601.
35. M. T. Borra, B. C. Smith and J. M. Denu, *J. Biol. Chem.*, 2005, **280**, 17187.
36. T. F. Outeiro, E. Kontopoulos, S. M. Altmann, I. Kufareva, K. E. Strathearn, A. M. Amore, C. B. Volk, M. M. Maxwell, J. C. Rochet, P. J. McLean, A. B. Young, R. Abagyan, M. B. Feany, B. T. Hyman and A. G. Kazantsev, *Science*, 2007, **317**, 516.
37. K. Suzuki and T. Koike, *Neuroscience*, 2007, **147**, 599.
38. N. Ashraf, S. Zino, A. Macintyre, D. Kingsmore, A. P. Payne, W. D. George and P. G. Shiels, *Br. J. Cancer*, 2006, **95**, 1056.
39. W. Dang, K. K. Steffen, R. Perry, J. A. Dorsey, F. B. Johnson, A. Shilatifard, M. Kaeberlein, B. K. Kennedy and S. L. Berger, *Nature*, 2009, **459**, 802.
40. K. Zhao, R. Harshaw, X. Chai and R. Marmorstein, *Proc. Natl. Acad. Sci. U. S. A.*, 2004, **101**, 8563.
41. J. Tan, S. Cang, Y. Ma, R. L. Petrillo and D. Liu, *J. Hematol. Oncol.*, 2010, **3**, 5.
42. D. R. Grayson, M. Kundakovic and R. P. Sharma, *Mol. Pharmacol.*, 2010, **77**, 126.
43. E. Hahnen, J. Hauke, C. Trankle, I. Y. Eyupoglu, B. Wirth and I. Blumcke, *Expert Opin. Invest. Drugs*, 2008, **17**, 169.
44. J. E. Bradner, N. West, M. L. Grachan, E. F. Greenberg, S. J. Haggarty, T. Warnow and R. Mazitschek, *Nat. Chem. Biol.*, **6**, 238.
45. N. R. Rose, S. S. Ng, J. Mecinovic, B. M. Lienard, S. H. Bello, Z. Sun, M. A. McDonough, U. Oppermann and C. J. Schofield, *J. Med. Chem.*, 2008, **51**, 7053.

46. J. S. Guan, S. J. Haggarty, E. Giacometti, J. H. Dannenberg, N. Joseph, J. Gao, T. J. Nieland, Y. Zhou, X. Wang, R. Mazitschek, J. E. Bradner, R. A. DePinho, R. Jaenisch and L. H. Tsai, *Nature*, 2009, **459**, 55.
47. I. M. Adcock, *Br. J. Pharmacol.*, 2007, **150**, 829.
48. E. W. Bush and T. A. McKinsey, *Circ. Res.*, **106**, 272.
49. C. M. Grozinger, E. D. Chao, H. E. Blackwell, D. Moazed and S. L. Schreiber, *J. Biol. Chem.*, 2001, **276**, 38837.
50. S. Pagans, A. Pedal, B. J. North, K. Kaehlcke, B. L. Marshall, A. Dorr, C. Hetzer-Egger, P. Henklein, R. Frye, M. W. McBurney, H. Hruby, M. Jung, E. Verdin and M. Ott, *PLoS Biol.*, 2005, **3**, e41.
51. B. Heltweg, T. Gatbonton, A. D. Schuler, J. Posakony, H. Li, S. Goehle, R. Kollipara, R. A. Depinho, Y. Gu, J. A. Simon and A. Bedalov, *Cancer Res.*, 2006, **66**, 4368.
52. S. R. Haynes, C. Dollard, F. Winston, S. Beck, J. Trowsdale and I. B. Dawid, *Nucleic Acids Res.*, 1992, **20**, 2603.
53. J. W. Tamkun, R. Deuring, M. P. Scott, M. Kissinger, A. M. Pattatucci, T. C. Kaufman and J. A. Kennison, *Cell*, 1992, **68**, 561.
54. F. Jeanmougin, J. M. Wurtz, B. Le Douarin, P. Chambon and R. Losson, *Trends Biochem. Sci.*, 1997, **22**, 151.
55. C. Dhalluin, J. E. Carlson, L. Zeng, C. He, A. K. Aggarwal and M. M. Zhou, *Nature*, 1999, **399**, 491.
56. P. Ornaghi, P. Ballario, A. M. Lena, A. Gonzalez and P. Filetici, *J. Mol. Biol.*, 1999, **287**, 1.
57. R. H. Jacobson, A. G. Ladurner, D. S. King and R. Tjian, *Science*, 2000, **288**, 1422.
58. D. J. Owen, P. Ornaghi, J. C. Yang, N. Lowe, P. R. Evans, P. Ballario, D. Neuhaus, P. Filetici and A. A. Travers, *EMBO J.*, 2000, **19**, 6141.
59. B. D. Strahl and C. D. Allis, *Nature*, 2000, **403**, 41.
60. In http://pfam.sanger.ac.uk/family?acc = PF00439.
61. B. Florence and D. V. Faller, *Front. Biosci.*, 2001, **6**, D1008.
62. T. Kanno, Y. Kanno, R. M. Siegel, M. K. Jang, M. J. Lenardo and K. Ozato, *Mol. Cell*, 2004, **13**, 33.
63. M. J. Kruhlak, M. J. Hendzel, W. Fischle, N. R. Bertos, S. Hameed, X. J. Yang, E. Verdin and D. P. Bazett-Jones, *J. Biol. Chem.*, 2001, **276**, 38307.
64. G. L. Cavalleri, N. M. Walley, N. Soranzo, J. Mulley, C. P. Doherty, A. Kapoor, C. Depondt, J. M. Lynch, I. E. Scheffer, A. Heils, A. Gehrmann, P. Kinirons, S. Gandhi, P. Satishchandra, N. W. Wood, A. Anand, T. Sander, S. F. Berkovic, N. Delanty, D. B. Goldstein and S. M. Sisodiya, *Epilepsia*, 2007, **48**, 706.
65. E. Shang, X. Wang, D. Wen, D. A. Greenberg and D. J. Wolgemuth, *Dev. Dyn.*, 2009, **238**, 908.
66. F. Wang, H. Liu, W. P. Blanton, A. Belkina, N. K. Lebrasseur and G. V. Denis, *Biochem. J.*, **425**, 71.
67. Z. Yang, N. He and Q. Zhou, *Mol. Cell Biol.*, 2008, **28**, 967.
68. M. Zhou, K. Huang, K. J. Jung, W. K. Cho, Z. Klase, F. Kashanchi, C. A. Pise-Masison and J. N. Brady, *J. Virol.*, 2009, **83**, 1036.

69. E. Shang, H. D. Nickerson, D. Wen, X. Wang and D. J. Wolgemuth, *Development*, 2007, **134**, 3507.
70. S. G. Z. Y. Yu, B. C. Zhang, M. Zhou, J. J. Xiang and G. Y. Li, *Chin. J. Cancer*, 2001, **20**, 569.
71. H. Y. Yamada and C. V. Rao, *Int. J. Oncol.*, 2009, **35**, 1101.
72. M. Thompson, *Biochimie*, 2009, **91**, 309.
73. L. Zeng, Q. Zhang, G. Gerona-Navarro, N. Moshkina and M. M. Zhou, *Structure*, 2008, **16**, 643.
74. Sachchidanand, L. Resnick-Silverman, S. Yan, S. Mutjaba, W. J. Liu, L. Zeng, J. J. Manfredi and M. M. Zhou, *Chem. Biol.*, 2006, **13**, 81.
75. Mitsubishi/Tanabe, in "Int. Pat. Appl. WO09/084693".
76. S. Ng, W. Yue, U. Oppermann and R. Klose, *Cell. Mol. Life Sci.*, 2009, **66**, 407.
77. C. Qian and M. M. Zhou, *Cell. Mol. Life Sci.*, 2006, **63**, 2755.
78. C. D. Krause, Z. H. Yang, Y. S. Kim, J. H. Lee, J. R. Cook and S. Pestka, *Pharmacol. Ther.*, 2007, **113**, 50.
79. R. A. Copeland, M. E. Solomon and V. M. Richon, *Nat. Rev. Drug Discovery*, 2009, **8**, 724.
80. A. V. Krivtsov and S. A. Armstrong, *Nat. Rev. Cancer*, 2007, **7**, 823.
81. J. A. Simon and C. A. Lange, *Mutat. Res./Fundam. Mol. Mech. Mutagenesis*, 2008, **647**, 21.
82. Y. Okada, Q. Feng, Y. Lin, Q. Jiang, Y. Li, V. M. Coffield, L. Su, G. Xu and Y. Zhang, *Cell*, 2005, **121**, 167.
83. Y. Okada, Q. Jiang, M. Lemieux, L. Jeannotte, L. Su and Y. Zhang, *Nat. Cell Biol.*, 2006, **8**, 1017.
84. D. Cheng, J. Cote, S. Shaaban and M. T. Bedford, *Mol. Cell*, 2007, **25**, 71.
85. M. Covic, P. O. Hassa, S. Saccani, C. Buerki, N. I. Meier, C. Lombardi, R. Imhof, M. T. Bedford, G. Natoli and M. O. Hottiger, *EMBO J*, 2005, **24**, 85.
86. X. Zhang and T. C. Bruice, *Proc. Natl. Acad. Sci. U. S. A.*, 2008, **105**, 5728.
87. B. Xiao, C. Jing, J. R. Wilson, P. A. Walker, N. Vasisht, G. Kelly, S. Howell, I. A. Taylor, G. M. Blackburn and S. J. Gamblin, *Nature*, 2003, **421**, 652.
88. X. Cheng, R. E. Collins and X. Zhang, *Annu. Rev. Biophys. Biomol. Struct.*, 2005, **34**, 267.
89. J.-F. Couture, L. M. A. Dirk, J. S. Brunzelle, R. L. Houtz and R. C. Trievel, *Proc. Natl. Acad. Sci. U. S. A.*, 2008, **105**, 20659.
90. P. Rathert, A. Dhayalan, H. Ma and A. Jeltsch, *Mol. BioSyst.*, 2008, **4**, 1186.
91. P. Rathert, A. Dhayalan, M. Murakami, X. Zhang, R. Tamas, R. Jurkowska, Y. Komatsu, Y. Shinkai, X. Cheng and A. Jeltsch, *Nat. Chem. Biol.*, 2008, **4**, 344.
92. C. Qian and M. Zhou, *Cell. Mol. Life Sci.*, 2006, **63**, 2755.
93. K. Briknarová, X. Zhou, A. Satterthwait, D. W. Hoyt, K. R. Ely and S. Huang, *Biochem. Biophys. Res. Commun.*, 2008, **366**, 807.

94. Y. Chang, X. Zhang, J. R. Horton, A. K. Upadhyay, A. Spannhoff, J. Liu, J. P. Snyder, M. T. Bedford and X. Cheng, *Nat. Struct. Mol. Biol.*, 2009, **16**, 312.

95. J. F. Couture, E. Collazo, J. S. Brunzelle and R. C. Trievel, *Genes Dev.*, 2005, **19**, 1455.

96. J.-F. Couture, E. Collazo, G. Hauk and R. C. Trievel, *Nat. Struct. Mol. Biol.*, 2006, **13**, 140.

97. T. Kwon, J. H. Chang, E. Kwak, C. W. Lee, A. Joachimiak, Y. C. Kim, J. W. Lee and Y. Cho, *EMBO J.*, 2003, **22**, 292.

98. S. M. Southall, P.-S. Wong, Z. Odho, S. M. Roe and J. R. Wilson, *Mol. Cell*, 2009, **33**, 181.

99. J. R. Wilson, C. Jing, P. A. Walker, S. R. Martin, S. A. Howell, G. M. Blackburn, S. J. Gamblin and B. Xiao, *Cell*, 2002, **111**, 105.

100. X. Zhang, Z. Yang, S. I. Khan, J. R. Horton, H. Tamaru, E. U. Selker and X. Cheng, *Mol. Cell*, 2003, **12**, 177.

101. F. Liu, X. Chen, A. Allali-Hassani, A. M. Quinn, G. A. Wasney, A. Dong, D. Barsyte, I. Kozieradzki, G. Senisterra, I. Chau, A. Siarheyeva, D. B. Kireev, A. Jadhav, J. M. Herold, S. V. Frye, C. H. Arrowsmith, P. J. Brown, A. Simeonov, M. Vedadi and J. Jin, *J. Med. Chem.*, 2009, **52**, 7950.

102. J. Min, Q. Feng, Z. Li, Y. Zhang and R. M. Xu, *Cell*, 2003, **112**, 711.

103. X. Zhang, L. Zhou and X. Cheng, *EMBO J.*, 2000, **19**, 3509.

104. N. Troffer-Charlier, V. Cura, P. Hassenboehler, D. Moras and J. Cavarelli, *EMBO J.*, 2007, **26**, 4391.

105. W. W. Yue, M. Hassler, S. M. Roe, V. Thompson-Vale and L. H. Pearl, *EMBO J.*, 2007, **26**, 4402.

106. X. Zhang and X. Cheng, *Structure*, 2003, **11**, 509.

107. K. Subramanian, D. Jia, P. Kapoor-Vazirani, D. R. Powell, R. E. Collins, D. Sharma, J. Peng, X. Cheng and P. M. Vertino, *Mol. Cell*, 2008, **30**, 336.

108. S. Huang, *Nat. Rev. Cancer*, 2002, **2**, 469.

109. S. Kubicek, R. J. O'sullivan, E. M. August, E. R. Hickey, Q. Zhang, L. Miguel Teodoro, S. Rea, K. Mechtler, J. A. Kowalski, C. A. Homon, T. A. Kelly and T. Jenuwein, *Mol. Cell*, 2007, **25**, 473.

110. T. B. Miranda, C. C. Cortez, C. B. Yoo, G. Liang, M. Abe, T. K. Kelly, V. E. Marquez and P. A. Jones, *Mol. Cancer Ther.*, 2009, **8**, 1579.

111. D. Cheng, N. Yadav, R. W. King, M. S. Swanson, E. J. Weinstein and M. T. Bedford, *J. Biol. Chem.*, 2004, **279**, 23892.

112. A. Mai, D. Cheng, M. T. Bedford, S. Valente, A. Nebbioso, A. Perrone, G. Brosch, G. Sbardella, F. De Bellis, M. Miceli and L. Altucci, *J. Med. Chem.*, 2008, **51**, 2279.

113. A. Spannhoff, R. Heinke, I. Bauer, P. Trojer, E. Metzger, R. Gust, R. Schule, G. Brosch, W. Sippl and M. Jung, *J. Med. Chem.*, 2007, **50**, 2319.

114. T. Huynh, Z. Chen, S. Pang, J. Geng, T. Bandiera, S. Bindi, P. Vianello, F. Roletto, S. Thieffine, A. Galvani, W. Vaccaro, M. A. Poss, G. L. Trainor, M. V. Lorenzi, M. Gottardis, L. Jayaraman and A. V. Purandare, *Bioorg. Med. Chem. Lett.*, 2009, **19**, 2924.

115. A. V. Purandare, Z. Chen, T. Huynh, S. Pang, J. Geng, W. Vaccaro, M. A. Poss, J. Oconnell, K. Nowak and L. Jayaraman, *Bioorg. Med. Chem. Lett.*, 2008, **18**, 4438.
116. M. Allan, S. Manku, E. Therrien, N. Nguyen, S. Styhler, M.-F. Robert, A.-C. Goulet, A. J. Petschner, G. Rahil, A. Robert MacLeod, R. Déziel, J. M. Besterman, H. Nguyen and A. Wahhab, *Bioorg. Med. Chem. Lett.*, 2009, **19**, 1218.
117. E. Therrien, G. Larouche, S. Manku, M. Allan, N. Nguyen, S. Styhler, M.-F. Robert, A.-C. Goulet, J. M. Besterman, H. Nguyen and A. Wahhab, *Bioorg. Med. Chem. Lett.*, 2009, **19**, 6725.
118. H. Wan, T. Huynh, S. Pang, J. Geng, W. Vaccaro, M. A. Poss, G. L. Trainor, M. V. Lorenzi, M. Gottardis, L. Jayaraman and A. V. Purandare, *Bioorg. Med. Chem. Lett.*, 2009, **19**, 5063.
119. B. R. Selvi, K. Batta, A. H. Kishore, K. Mantelingu, R. A. Varier, K. Balasubramanyam, S. K. Pradhan, D. Dasgupta, S. Sriram, S. Agrawal and T. K. Kundu, *J. Biol. Chem.*, 2010, **285**, 7143.
120. S. C. Trewick, P. J. McLaughlin and R. C. Allshire, *EMBO Rep.*, 2005, **6**, 315.
121. G. L. Cuthbert, S. Daujat, A. W. Snowden, H. Erdjument-Bromage, T. Hagiwara, M. Yamada, R. Schneider, P. D. Gregory, P. Tempst, A. J. Bannister and T. Kouzarides, *Cell*, 2004, **118**, 545.
122. H. Denis, R. Deplus, P. Putmans, M. Yamada, R. Metivier and F. Fuks, *Mol. Cell. Biol.*, 2009, **29**, 4982.
123. B. Chang, Y. Chen, Y. Zhao and R. K. Bruick, *Science*, 2007, **318**, 444.
124. C. J. Webby, A. Wolf, N. Gromak, M. Dreger, H. Kramer, B. Kessler, M. L. Nielsen, C. Schmitz, D. S. Butler, J. R. Yates 3rd, C. M. Delahunty, P. Hahn, A. Lengeling, M. Mann, N. J. Proudfoot, C. J. Schofield and A. Bottger, *Science*, 2009, **325**, 90.
125. J. Huang, R. Sengupta, A. B. Espejo, M. G. Lee, J. A. Dorsey, M. Richter, S. Opravil, R. Shiekhattar, M. T. Bedford, T. Jenuwein and S. L. Berger, *Nature*, 2007, **449**, 105.
126. V. K. C. Ponnaluri, D. T. Vavilala, S. Putty, W. G. Gutheil and M. Mukherji, *Biochem. Biophys. Res. Commun.*, 2009, **390**, 280.
127. Y. Shi, F. Lan, C. Matson, P. Mulligan, J. R. Whetstine, P. A. Cole, R. A. Casero and Y. Shi, *Cell*, 2004, **119**, 941.
128. E. Metzger, M. Wissmann, N. Yin, J. M. Muller, R. Schneider, A. H. F. M. Peters, T. Gunther, R. Buettner and R. Schule, *Nature*, 2005, **437**, 436.
129. F. Forneris, C. Binda, M. A. Vanoni, A. Mattevi and E. Battaglioli, *FEBS Lett.*, 2005, **579**, 2203.
130. Y. Chen, Y. Yang, F. Wang, K. Wan, K. Yamane, Y. Zhang and M. Lei, *Proc. Natl. Acad. Sci. U. S. A.*, 2006, **103**, 13956.
131. F. Forneris, C. Binda, A. Adamo, E. Battaglioli and A. Mattevi, *J. Biol. Chem.*, 2007, **282**, 20070.
132. P. Stavropoulos, G. Blobel and A. Hoelz, *Nat. Struct. Mol. Biol.*, 2006, **13**, 626.

133. S. S. Ng, K. L. Kavanagh, M. A. McDonough, D. Butler, E. S. Pilka, B. M. R. Lienard, J. E. Bray, P. Savitsky, O. Gileadi, F. von Delft, N. R. Rose, J. Offer, J. C. Scheinost, T. Borowski, M. Sundstrom, C. J. Schofield and U. Oppermann, *Nature*, 2007, **448**, 87.

134. Z. Han, P. Liu, L. Gu, Y. Zhang, H. Li, S. Chen and J. Chai, *Front. Sci.*, 2007, **1**, 52.

135. J. R. Horton, A. K. Upadhyay, H. H. Qi, X. Zhang, Y. Shi and X. Cheng, *Nat. Struct. Mol. Biol.*, 2010, **17**, 38.

136. M. Yang, J. C. Culhane, L. M. Szewczuk, P. Jalili, H. L. Ball, M. Machius, P. A. Cole and H. Yu, *Biochemistry*, 2007, **46**, 8058.

137. D. M. Z. Schmidt and D. G. McCafferty, *Biochemistry*, 2007, **46**, 4408.

138. M. G. Lee, C. Wynder, D. M. Schmidt, D. G. McCafferty and R. Shiekhattar, *Chem. Biol.*, 2006, **13**, 563.

139. J. C. Culhane, L. M. Szewczuk, X. Liu, G. Da, R. Marmorstein and P. A. Cole, *J. Am. Chem. Soc.*, 2006, **128**, 4536.

140. L. M. Szewczuk, J. C. Culhane, M. Yang, A. Majumdar, H. Yu and P. A. Cole, *Biochemistry*, 2007, **46**, 6892.

141. R. Ueda, T. Suzuki, K. Mino, H. Tsumoto, H. Nakagawa, M. Hasegawa, R. Sasaki, T. Mizukami and N. Miyata, *J. Am. Chem. Soc.*, 2009, **131**, 17536.

142. Y. Huang, T. M. Stewart, Y. Wu, S. B. Baylin, L. J. Marton, B. Perkins, R. J. Jones, P. M. Woster and R. A. Casero, *Clin. Cancer Res.*, 2009, **15**, 7217.

143. L. Escoubet-Lozach, I. L. Lin, K. Jensen-Pergakes, H. A. Brady, A. K. Gandhi, P. H. Schafer, G. W. Muller, P. J. Worland, K. W. H. Chan and D. Verhelle, *Cancer Res.*, 2009, **69**, 7347.

144. N. R. Rose, S. S. Ng, J. Mecinović, B. M. R. Liénard, S. H. Bello, Z. Sun, M. A. McDonough, U. Oppermann and C. J. Schofield, *J. Med. Chem.*, 2008, **51**, 7053.

145. S. Hamada, T. D. Kim, T. Suzuki, Y. Itoh, H. Tsumoto, H. Nakagawa, R. Janknecht and N. Miyata, *Bioorg. Med. Chem. Lett.*, 2009, **19**, 2852.

146. M. Sakurai, N. R. Rose, L. Schultz, A. M. Quinn, A. Jadhav, S. S. Ng, U. Oppermann, C. J. Schofield and A. Simeonov, *Mol. BioSyst.*, 2010, **6**, 357.

147. R. Sekirnik, N. R. Rose, A. Thalhammer, P. T. Seden, J. Mecinović and C. J. Schofield, *Chem. Commun. (Cambridge, UK)*, 2009, 6376.

148. S. Maurer-Stroh, N. J. Dickens, L. Hughes-Davies, T. Kouzarides, F. Eisenhaber and C. P. Ponting, *Trends Biochem. Sci.*, 2003, **28**, 69.

149. R. J. Klose, K. Yamane, Y. Bae, D. Zhang, H. Erdjument-Bromage, P. Tempst, J. Wong and Y. Zhang, *Nature*, 2006, **442**, 312.

150. R. Margueron, N. Justin, K. Ohno, M. L. Sharpe, J. Son, W. J. Drury Iii, P. Voigt, S. R. Martin, W. R. Taylor, V. De Marco, V. Pirrotta, D. Reinberg and S. J. Gamblin, *Nature*, 2009, **461**, 762.

151. G. Lomberk, L. Wallrath and R. Urrutia, *Genome Biol.*, 2006, **7**, 228.

152. S. I. Grewal and S. Jia, *Nat. Rev. Genet.*, 2007, **8**, 35.

153. F. Lan, R. E. Collins, R. De Cegli, R. Alpatov, J. R. Horton, X. Shi, O. Gozani, X. Cheng and Y. Shi, *Nature*, 2007, **448**, 718.

154. C. Loenarz, W. Ge, M. L. Coleman, N. R. Rose, C. D. Cooper, R. J. Klose, P. J. Ratcliffe and C. J. Schofield, *Hum. Mol. Genet.*, 2010, **19**, 217.

155. J. R. Horton, A. K. Upadhyay, H. H. Qi, X. Zhang, Y. Shi and X. Cheng, *Nat. Struct. Mol. Biol.*, 2010, **17**, 38.

156. R. E. Collins, J. P. Northrop, J. R. Horton, D. Y. Lee, X. Zhang, M. R. Stallcup and X. Cheng, *Nat. Struct. Mol. Biol.*, 2008, **15**, 245.

157. L. A. Baker, C. D. Allis and G. G. Wang, *Mutat. Res./Fundam. Mol. Mech. Mutagenesis*, 2008, **647**, 3.

158. A. G. Matthews, A. J. Kuo, S. Ramon-Maiques, S. Han, K. S. Champagne, D. Ivanov, M. Gallardo, D. Carney, P. Cheung, D. N. Ciccone, K. L. Walter, P. J. Utz, Y. Shi, T. G. Kutateladze, W. Yang, O. Gozani and M. A. Oettinger, *Nature*, 2007, **450**, 1106.

159. S. Ramon-Maiques, A. J. Kuo, D. Carney, A. G. Matthews, M. A. Oettinger, O. Gozani and W. Yang, *Proc. Natl. Acad. Sci. U. S. A.*, 2007, **104**, 18993.

160. C. Sobacch, V. Marella, F. Rucci, P. Vezzoni and A. Villa, *Hum. Mutat.*, 2006, **27**, 1174.

161. L. H. Kasper, P. K. Brindle, C. A. Schnabel, C. E. Pritchard, M. L. Cleary and J. M. van Deursen, *Mol. Cell Biol.*, 1999, **19**, 764.

162. M. A. Moore, K. Y. Chung, M. Plasilova, J. J. Schuringa, J. H. Shieh, P. Zhou and G. Morrone, *Ann. N. Y. Acad. Sci.*, 2007, **1106**, 114.

163. G. G. Wang, L. Cai, M. P. Pasillas and M. P. Kamps, *Nat. Cell Biol.*, 2007, **9**, 804.

164. G. G. Wang, J. Song, Z. Wang, H. L. Dormann, F. Casadio, H. Li, J.-L. Luo, D. J. Patel and C. D. Allis, *Nature*, 2009, **459**, 847.

165. J. Chen, D. A. Santillan, M. Koonce, W. Wei, R. Luo, M. J. Thirman, N. J. Zeleznik-Le and M. O. Diaz, *Cancer Res.*, 2008, **68**, 6199.

166. A. G. Muntean, D. Giannola, A. M. Udager and J. L. Hess, *Blood*, 2008, **112**, 4690.

167. S. D. Taverna, H. Li, A. J. Ruthenburg, C. D. Allis and D. J. Patel, *Nat. Struct. Mol. Biol.*, 2007, **14**, 1025.

168. R. M. Hughes, K. R. Wiggins, S. Khorasanizadeh and M. L. Waters, *Proc. Natl. Acad. Sci. U. S. A.*, 2007, **104**, 11184.

169. A. J. Ruthenburg, C. D. Allis and J. Wysocka, *Mol. Cell*, 2007, **25**, 15.

170. S. K. T. Ooi, C. Qiu, E. Bernstein, K. Li, D. Jia, Z. Yang, H. Erdjument-Bromage, P. Tempst, S.-P. Lin, C. D. Allis, X. Cheng and T. H. Bestor, *Nature*, 2007, **448**, 714.

171. J.-L. Hu, B. O. Zhou, R.-R. Zhang, K.-L. Zhang, J.-Q. Zhou and G.-L. Xu, *Proc. Natl. Acad. Sci. U. S. A.*, 2009, **106**, 22187.

172. J. F. Couture, E. Collazo and R. C. Trievel, *Nat. Struct. Mol. Biol.*, 2006, **13**, 698.

173. T. J. Wigle, J. M. Herold, G. A. Senisterra, M. Vedadi, D. B. Kireev, C. H. Arrowsmith, S. V. Frye and W. P. Janzen, *J. Biomol. Screen.*, 2010, **15**, 62.

174. A. M. Quinn, M. T. Bedford, A. Espejo, A. Spannhoff, C. P. Austin, U. Oppermann and A. Simeonov, *Nucl. Acids Res.*, 2010, **38**, e11.
175. B. R. Keppler and T. K. Archer, *Expert Opin. Ther. Target.*, 2008, **12**, 1457.
176. A. Zippo, R. Serafini, M. Rocchigiani, S. Pennacchini, A. Krepelova and S. Oliviero, *Cell*, 2009, **138**, 1122.
177. M. Stucki, J. A. Clapperton, D. Mohammad, M. B. Yaffe, S. J. Smerdon and S. P. Jackson, *Cell*, 2005, **123**, 1213.
178. T. Hirota, J. J. Lipp, B.-H. Toh and J.-M. Peters, *Nature*, 2005, **438**, 1176.
179. M. A. Dawson, A. J. Bannister, B. Gottgens, S. D. Foster, T. Bartke, A. R. Green and T. Kouzarides, *Nature*, 2009, **461**, 819.
180. E. Metzger, N. Yin, M. Wissmann, N. Kunowska, K. Fischer, N. Friedrichs, D. Patnaik, J. M. G. Higgins, N. Potier, K.-H. Scheidtmann, R. Buettner and R. Schule, *Nat. Cell Biol.*, 2008, **10**, 53.
181. J. A. Perez Fidalgo, D. Roda, S. Rosello, E. Rodriguez-Braun and A. Cervantes, *Clin. Trans. Oncol.*, 2009, **11**, 787.
182. A. Pardanani, *Leukemia*, 2007, **22**, 23.
183. Y. I. Hassan and J. Zempleni, *J. Nutr.*, 2006, **136**, 1763.
184. J. Kim, M. Guermah, R. K. McGinty, J.-S. Lee, Z. Tang, T. A. Milne, A. Shilatifard, T. W. Muir and R. G. Roeder, *Cell*, 2009, **137**, 459.
185. R. K. McGinty, J. Kim, C. Chatterjee, R. G. Roeder and T. W. Muir, *Nature*, 2008, **453**, 812.
186. K. W. Henry, A. Wyce, W. S. Lo, L. J. Duggan, N. C. Emre, C. F. Kao, L. Pillus, A. Shilatifard, M. A. Osley and S. L. Berger, *Genes Dev.*, 2003, **17**, 2648.
187. M. de Napoles, J. E. Mermoud, R. Wakao, Y. A. Tang, M. Endoh, R. Appanah, T. B. Nesterova, J. Silva, A. P. Otte, M. Vidal, H. Koseki and N. Brockdorff, *Dev. Cell*, 2004, **7**, 663.
188. Y. Shiio and R. N. Eisenman, *Proc. Natl. Acad. Sci. U. S. A.*, 2003, **100**, 13225.
189. S. R. Ande, J. Chen and S. Maddika, *Eur. J. Pharmacol.*, 2009, **625**, 199.
190. R. Pfau, A. Tzatsos, S. C. Kampranis, O. B. Serebrennikova, S. E. Bear and P. N. Tsichlis, *Proc. Natl. Acad. Sci. U. S. A.*, 2008, **105**, 1907.
191. J. He, E. M. Kallin, Y. Tsukada and Y. Zhang, *Nat. Struct. Mol. Biol.*, 2008, **15**, 1169.
192. R. Koyama-Nasu, G. David and N. Tanese, *Nat. Cell Biol.*, 2007, **9**, 1074.
193. K. Yamane, C. Toumazou, Y.-i. Tsukada, H. Erdjument-Bromage, P. Tempst, J. Wong and Y. Zhang, *Cell*, 2006, **125**, 483.
194. Y. H. Loh, W. Zhang, X. Chen, J. George and H. H. Ng, *Genes Dev.*, 2007, **21**, 2545.
195. Y. Okada, G. Scott, M. K. Ray, Y. Mishina and Y. Zhang, *Nature*, 2007, **450**, 119.
196. T. Inagaki, M. Tachibana, K. Magoori, H. Kudo, T. Tanaka, M. Okamura, M. Naito, T. Kodama, Y. Shinkai and J. Sakai, *Genes Cells*, 2009, **14**, 991.
197. K. Tateishi, Y. Okada, E. M. Kallin and Y. Zhang, *Nature*, 2009, **458**, 757.
198. S. Shin and R. Janknecht, *Biochem. Biophys. Res. Commun.*, 2007, **359**, 742.

199. M. Wissmann, N. Yin, J. M. Muller, H. Greschik, B. D. Fodor, T. Jenuwein, C. Vogler, R. Schneider, T. Gunther, R. Buettner, E. Metzger and R. Schule, *Nat. Cell Biol.*, 2007, **9**, 347.
200. P. A. C. Cloos, J. Christensen, K. Agger, A. Maiolica, J. Rappsilber, T. Antal, K. H. Hansen and K. Helin, *Nature*, 2006, **442**, 307.
201. J. Wang, M. Zhang, Y. Zhang, Z. Kou, Z. Han, D. Y. Chen, Q. Y. Sun and S. Gao, *Biol. Reprod.*, 2010, **82**, 105.
202. F. De Santa, M. G. Totaro, E. Prosperini, S. Notarbartolo, G. Testa and G. Natoli, *Cell*, 2007, **130**, 1083.
203. T. Burgold, F. Spreafico, F. De Santa, M. G. Totaro, E. Prosperini, G. Natoli and G. Testa, *PLoS One*, 2008, **3**, e3034.
204. G. L. Sen, D. E. Webster, D. I. Barragan, H. Y. Chang and P. A. Khavari, *Genes Dev.*, 2008, **22**, 1865.
205. M. Barradas, E. Anderton, J. C. Acosta, S. Li, A. Banito, M. Rodriguez-Niedenfuhr, G. Maertens, M. Banck, M. M. Zhou, M. J. Walsh, G. Peters and J. Gil, *Genes Dev.*, 2009, **23**, 1177.
206. K. Agger, P. A. C. Cloos, J. Christensen, D. Pasini, S. Rose, J. Rappsilber, I. Issaeva, E. Canaani, A. E. Salcini and K. Helin, *Nature*, 2007, **449**, 731.
207. F. Lan, P. E. Bayliss, J. L. Rinn, J. R. Whetstine, J. K. Wang, S. Chen, S. Iwase, R. Alpatov, I. Issaeva, E. Canaani, T. M. Roberts, H. Y. Chang and Y. Shi, *Nature*, 2007, **449**, 689.
208. E. V. Benevolenskaya, H. L. Murray, P. Branton, R. A. Young and W. G. Kaelin Jr., *Mol. Cell*, 2005, **18**, 623.
209. G. M. Gutierrez, E. Kong and P. W. Hinds, *Cancer Cell*, 2005, **7**, 501.
210. N. Lopez-Bigas, T. A. Kisiel, D. C. DeWaal, K. B. Holmes, T. L. Volkert, S. Gupta, J. Love, H. L. Murray, R. A. Young and E. V. Benevolenskaya, *Mol. Cell*, 2008, **31**, 520.
211. D. Pasini, K. H. Hansen, J. Christensen, K. Agger, P. A. Cloos and K. Helin, *Genes Dev.*, 2008, **22**, 1345.
212. K. Yamane, K. Tateishi, R. J. Klose, J. Fang, L. A. Fabrizio, H. Erdjument-Bromage, J. Taylor-Papadimitriou, P. Tempst and Y. Zhang, *Mol. Cell*, 2007, **25**, 801.
213. Y. Xiang, Z. Zhu, G. Han, X. Ye, B. Xu, Z. Peng, Y. Ma, Y. Yu, H. Lin, A. P. Chen and C. D. Chen, *Proc. Natl. Acad. Sci. U. S. A.*, 2007, **104**, 19226.
214. B. K. Dey, L. Stalker, A. Schnerch, M. Bhatia, J. Taylor-Papidimitriou and C. Wynder, *Mol. Cell. Biol.*, 2008, **28**, 5312.
215. S. Iwase, F. Lan, P. Bayliss, L. de la Torre-Ubieta, M. Huarte, H. H. Qi, J. R. Whetstine, A. Bonni, T. M. Roberts and Y. Shi, *Cell*, 2007, **128**, 1077.
216. F. E. Abidi, L. Holloway, C. A. Moore, D. D. Weaver, R. J. Simensen, R. E. Stevenson, R. C. Rogers and C. E. Schwartz, *J. Med. Genet.*, 2008, **45**, 787.
217. L. R. Jensen, M. Amende, U. Gurok, B. Moser, V. Gimmel, A. Tzschach, A. R. Janecke, G. Tariverdian, J. Chelly, J. P. Fryns, H. Van Esch,

T. Kleefstra, B. Hamel, C. Moraine, J. Gecz, G. Turner, R. Reinhardt, V. M. Kalscheuer, H. H. Ropers and S. Lenzner, *Am. J. Hum. Genet.*, 2005, **76**, 227.
218. C. Santos, L. Rodriguez-Revenga, I. Madrigal, C. Badenas, M. Pineda and M. Mila, *Eur. J. Hum. Genet.*, 2006, **14**, 583.
219. A. Tzschach, S. Lenzner, B. Moser, R. Reinhardt, J. Chelly, J. P. Fryns, T. Kleefstra, M. Raynaud, G. Turner, H. H. Ropers, A. Kuss and L. R. Jensen, *Hum. Mutat.*, 2006, **27**, 389.
220. A. Adegbola, H. Gao, S. Sommer and M. Browning, *Am. J. Med. Genet. A*, 2008, **146A**, 505.

# *Chemologics*

LYN H. JONES

Pfizer Inc., Sandwich Laboratories ipc 432, Sandwich, Kent, CT13 9NJ, UK

## 6.1   Introduction

The successful application of synthetic organic chemistry to the biotherapeutic arena has created an exciting paradigm for drug discovery and new opportunities for medicinal chemistry design. Chemistry is no longer restricted to the discovery of small molecule organic drugs since the workable opportunity space is arguably the entire spectrum of chemical space, inclusive of inorganics and biomolecules. Moreover, the emphasis for biotherapeutics is increasingly becoming the alignment of structure with function, and therefore methods that enable the delineation of structure-function relationships will become powerful tools in the design of improved therapeutics, thus avoiding traditionally empirical approaches. Chemistry can significantly modify the structure of a biological, improving its therapeutic effectiveness or imparting totally new properties to the designed molecule, thereby increasing the functionality of the construct. Here, the recent advances made in the area of chemistry-enabled biotherapeutics, or chemologics, are reviewed with a focus on the opportunities being unearthed through close partnerships between biology and chemistry.

## 6.2   Synthetic Vaccines

Vaccines are traditionally prepared from the inactivated or attenuated form of a microorganism that can stimulate the immune system into recognising a "foreign body" which results in an immunological response that removes that

RSC Drug Discovery Series No. 5
New Frontiers in Chemical Biology: Enabling Drug Discovery
Edited by Mark E. Bunnage
© Royal Society of Chemistry 2011
Published by the Royal Society of Chemistry, www.rsc.org

pathogen from the body. Dendritic cells (antigen-presenting cells) possess cell surface receptors that are responsible for the recognition of antigens which are then processed by specialised intracellular compartments into molecular determinants (epitopes) that are presented by the major histocompatibility complex (MHC) in order to elicit the appropriate T-cell response. This includes the activation of *helper* T cells ($T_H$), resulting in the release of cytokines that regulate/augment the immune response, and *cytotoxic* T cells ($T_c$) that destroy virally infected or cancer cells. The vaccine also imparts to the immune system of the host a memory of the epitope through *memory* T cells that enables a rapid and specific immunological response to future exposure to the pathogen. Additionally, B cells recognise antigens *via* the B-cell receptor, and together with a signal from a helper T cell, they become activated and initiate the antibody-mediated immune response.

The intrinsic heterogeneity of many existing vaccine approaches, that are formed from a complex mixture of biologically active proteins and polysaccharides (the fragments of the microorganism), can lead to a variety of effects in different individuals, resulting in inter-patient variability regarding their therapeutic effectiveness and off-target/toxicological outcomes. A well-defined synthetic vaccine is an attractive alternative to traditional approaches as it confers improved reproducibility in the production, efficacy and safety of the therapy. Additionally, during the development of a new vaccine, it can be important to generate relationships between the vaccine chemical structure and its therapeutic activity, as these facilitate the design of new and improved structures in an iterative cycle. Complex mixtures would complicate the elucidation and interpretation of these structure–activity relationships.

The immune system has not evolved to develop a response to molecules with a molecular weight less than about 10 000. In order to create a synthetic vaccine, any molecule below this limit must be covalently linked (bioconjugated) to a carrier protein in order to elicit a reproducible immune response. Therefore, synthetic chemistry expertise is called upon to optimise these conjugations, exploring a variety of different reaction conditions and/or disconnections to provide the desired construct in optimal and reproducible yield. Analytical and purification methods have also been developed to work with these large, and highly complex, molecules. A variety of biophysical methods can also be applied to the characterisation of the resulting constructs, additional to the commonly used technique of gel electrophoresis that separates biomolecules using an electric field applied to a polymer matrix. Of particular utility is the widely used technology of matrix-assisted laser desorption/ionisation (MALDI), which is a very "soft" ionisation technique used in mass spectrometry.[1] The matrix is used to protect the relatively fragile biomolecular construct from being destroyed by the laser and facilitates vapourisation and ionisation of the conjugate. Nuclear magnetic resonance (NMR) techniques are also being used increasingly to characterise not only epitope structure and conformation (particularly useful for complex polysaccharide[2] or constrained peptide epitopes[3]) but also to assess the extent of bioconjugation to a carrier protein.[4]

## 6.2.1  Carbohydrate-based Vaccines

One of the best known conjugate vaccines is the *Haemophilus influenzae* type B (Hib) vaccine, which has been shown to significantly reduce cases of childhood meningitis to the extent that this disease is no longer a public health problem in Europe or the US. Early versions of the vaccine simply used the purified bacterial polysaccharide polyribosylribitol as an antigen, but this was removed from the market due to a lack of efficacy in children less than 18 months of age. These issues were resolved when the polysaccharide was conjugated to a carrier protein, resulting in a reproducible immunogenic response in young children.[5]

In recent years, there has been a focus on developing carbohydrate-based antitumour synthetic vaccines. These rely on the fact that cancer cells express unusual glycosylation patterns on their surface (tumour-associated carbohydrate antigens) and therefore a vaccine that is able to present these aberrant sugars effectively to the immune system should be able to generate an immune response to these tumours. This work has been pioneered by Danishefsky at the Sloan–Kettering Institute in New York and over several years his group has developed significant advances in the area, particularly regarding the construction of highly synthetically complex vaccine candidates.[6] This work led to the Globo-H-KLH synthetic vaccine (KLH = keyhole limpet haemocyanin, a carrier protein), currently being evaluated in clinical trials for the treatment of prostate cancer (Figure 6.1). Globo-H is a sugar expressed on many different tumour cells but this carbohydrate can only be isolated in very small quantities from these cells (unlike the Hib vaccine where polyribosylribitol can be expressed and isolated from cultured bacteria). The ability to prepare this complex carbohydrate using differentially protected saccharide monomers is a significant synthetic achievement in its own right, and the synthetic oligosaccharide confirmed the structure originally assigned to the natural Globo-H antigen.[7]

Danishefsky has also designed and synthesised multivalent vaccine candidates that express multiple carbohydrate-associated cancer antigens on a single oligopeptide backbone to mimic the intrinsic heterogeneity of carbohydrate surface antigens present on cancer cells (Figure 6.2).[8] These highly complex synthetic vaccines are constructed using solid-phase peptide synthesis; each sugar is tethered to an amino acid (bearing a protecting group on the amino function) that can be linked to a polymeric resin bead. The amino group can be

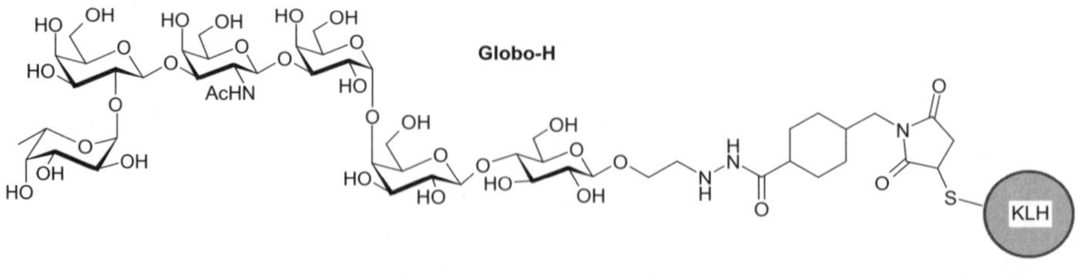

**Figure 6.1**   Globo-H-KLH synthetic vaccine.

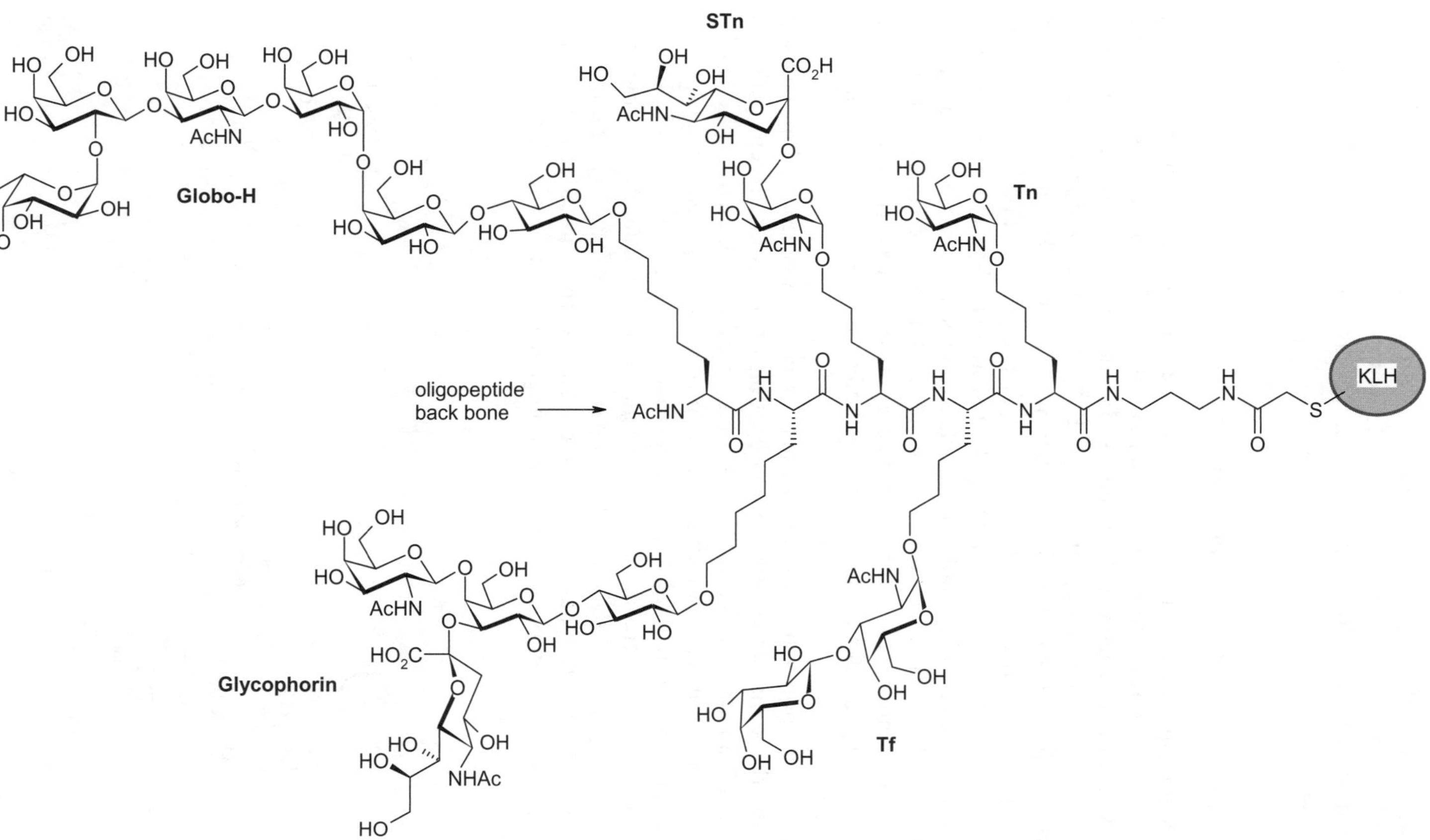


**Figure 6.2** Heterogeneous carbohydrate cancer vaccine.

deprotected, ready for peptide formation with another sugar-linked amino acid, and the process repeated until the desired peptide sequence is achieved, which can then be cleaved off the resin and conjugated to the carrier protein.[9]

## 6.2.2 Purely Synthetic Vaccines

The bioconjugation chemistries described above still provide a certain degree of heterogeneity *via* random protein coupling that could be eliminated using advanced synthetic chemistry techniques. In the area of carbohydrate-based vaccines for example, a key finding by Kasper and co-workers found that zwitterionic polysaccharides (ZPSs) can invoke an MHC-mediated T-cell response in the absence of protein[10,11] and then Andreana applied this observation to the creation of an entirely carbohydrate-based vaccine.[12] In this case the ZPS contained a vicinal diol that was cleaved by sodium periodate to provide an aldehyde that was subsequently conjugated to a hydroxylamine-containing carbohydrate antigen in a totally chemoselective manner (Figure 6.3). The resulting conjugate elicited high titre antibodies in the absence of standard adjuvants.

Yet another approach to totally synthetic vaccines was reported by Jackson and co-workers that relies on the conjugation of a toll-like receptor (TLR) ligand (that can activate dendritic cells) to a Tc epitope and a $T_H$ epitope (the target epitope) to furnish a vaccine that can protect mice against viral infections and mediate prophylactic and therapeutic activity against tumours.[13] In this study, the epitopes were linked by a single lysine residue that enables the conjugation of the TLR-2 targeting lipid moiety *S*-[2,3-bis(palmitoyloxy) propyl]cysteine, *via* two additional serine residues, that provides a branched conformation that improves the immunogenicity and solubility of the construct.

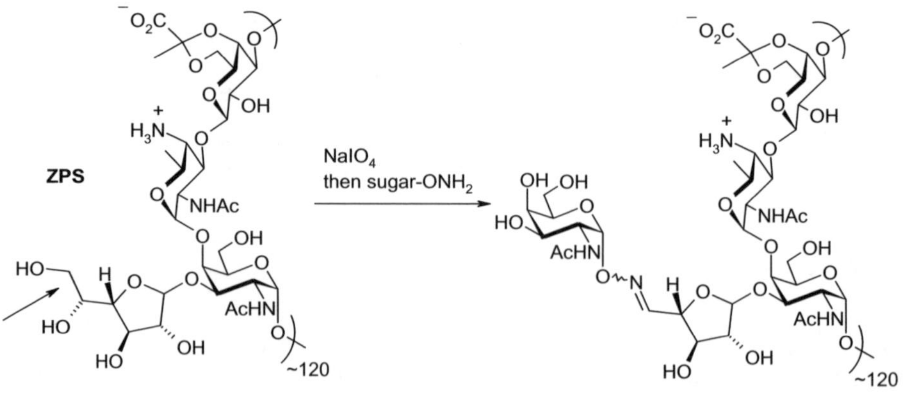

**Figure 6.3**  Zwitterionic polysaccharide conjugation to a carbohydrate antigen.

## 6.2.3 Synthetic Virus-like Particles

Virus-like particles are typically composed of recombinant proteins that form the outer virus shell, but which lack the viral nucleic acid, thereby preventing infection, but which elicit strong T-cell and B-cell responses. These particles can be exploited to display multiple antigenic units via traditional conjugation chemistries, thus presenting a multivalent array with enhanced immunogenicity. Robinson at the University of Zurich has recently developed an approach to synthetic virus-like particles that can be harnessed to provide a myriad of different antigen-presenting constructs in a well-controlled manner.[14] In this work, a self-assembling lipopeptide relies on a coiled-coil association in the peptide component into a parallel helical bundle (consisting of three monomers) and the resulting lipid cluster drives self-assembly of multiple coiled coil bundles into discreet nanosize VLPs. Antigens can be coupled to the C-terminal lipopeptide resulting in the creation of well-defined SVLPs that can induce antibody responses *in vivo* without the need for adjuvant (Figure 6.4). Various biophysical techniques were applied to characterise the structure of the VLP. Hydrodynamic studies using analytical centrifugation indicated monomer association, and sedimentation equilibrium provided an average mass of 431 kDa (equating to about 72 monomers). Electron microscopy confirmed a spherical shape with an average diameter of 20 nm for the VLP, and the far-UV circular dichroism spectrum was indicative of a random coil/α-helix structure.

## 6.2.4 Expanding the Genetic Code and Breaking Self-tolerance

Another mechanism to create a well-defined chemical vaccine utilises a synthetic biology approach recently developed by Schultz and Lerner at The Scripps Research Institute that created a nitrophenylalanine derivative of TNF-α that breaks self-tolerance.[15] This work results from the pioneering work of Schultz that has expanded the genetic code by generating a unique codon-tRNA pair and the corresponding aminoacyl-tRNA synthetase.[16] Using this technique, specific genetic substitution of Tyr-86 for the highly immunogenic *p*-nitrophenylalanine amino acid created a T-cell epitope that elicits a strong

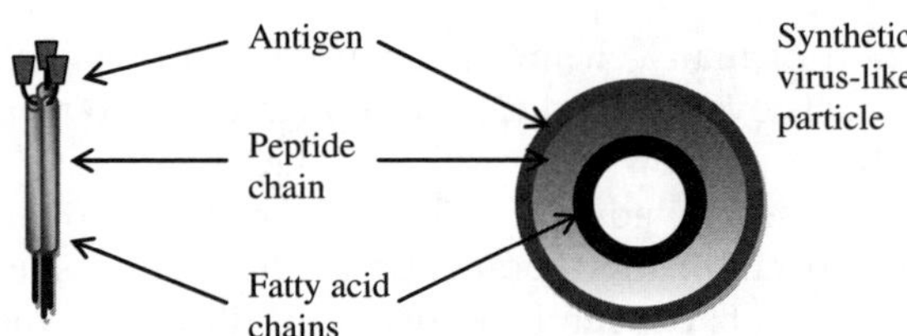


**Figure 6.4** Diagrammatic representations of the synthetic virus-like particles.

cross-reactive immune response to the inflammatory cytokine protein TNF-α that protects mice from LPS challenge.

Using these chemical methods, Schultz has expanded the genetic code to approximately 50 amino acids to provide proteins with a variety of functions *via* the site-selective incorporation of unnatural residues, including redox and photosensitive groups, fluorophores, metal-binding groups for catalysis and chemical tags that facilitate bioconjugations.[16] A number of drugs are in development that utilise this technology to selectively conjugate various groups to therapeutically relevant proteins. For example, ARX201 is a recombinant form of human growth hormone that has been modified to achieve precise spatial positioning of the site of poly(ethylene) glycol (PEG) attachment, to provide a long-acting therapeutic protein candidate that is currently in Phase II clinical trials for growth deficiency.[17]

## 6.2.5   Chemically Programmed Vaccination

Reactive immunisation was developed originally as a method to generate catalytic antibodies by using reactive haptens that were designed to elicit covalent antibodies, a concept first described by Janda at The Scripps Research Institute.[18] Barbas subsequently applied this concept to the generation of aldol catalytic antibodies *via* reactive immunisation with a 1,3-diketone antigen (see Section 6.3).[19] Recently, Barbas was able to extend this idea to the elicitation of a polyclonal response *in vivo* that can be triggered by a small molecule that can protect the treated animals with "instant immunity". The approach involves immunising mice with the immunogen **1** (Figure 6.5) that generates a reactive polylonal response.[20] Subsequent treatment with ligands for tumour surface integrins ($\alpha_v\beta_3$ and $\alpha_v\beta_5$), with an appended di-ketone moiety, induced an immune response that was targeted to the tumour cell surface. The therapeutic potential of this approach was illustrated in colon and melanoma tumour models, whereby significant reductions in tumour growth were observed in treated mice.

## 6.3   Antibodies

### 6.3.1   Antibody-directed Therapies

The mode of action of many chemotherapeutics relies on the greater rate of division of cancer cells over normal cells, although this also means that rapidly dividing cells in bone marrow, hair follicles and the digestive tract are also targeted, leading to severe adverse effects. A method to ameliorate these cytocidal effects is to target the therapeutic to the desired site of action, and Mylotarg® (gemtuzumab ozogamicin) perfectly illustrates the value of this approach by combining the advantages of a potent small molecule anti-cancer drug with the exquisitely selective binding capabilities of an antibody.

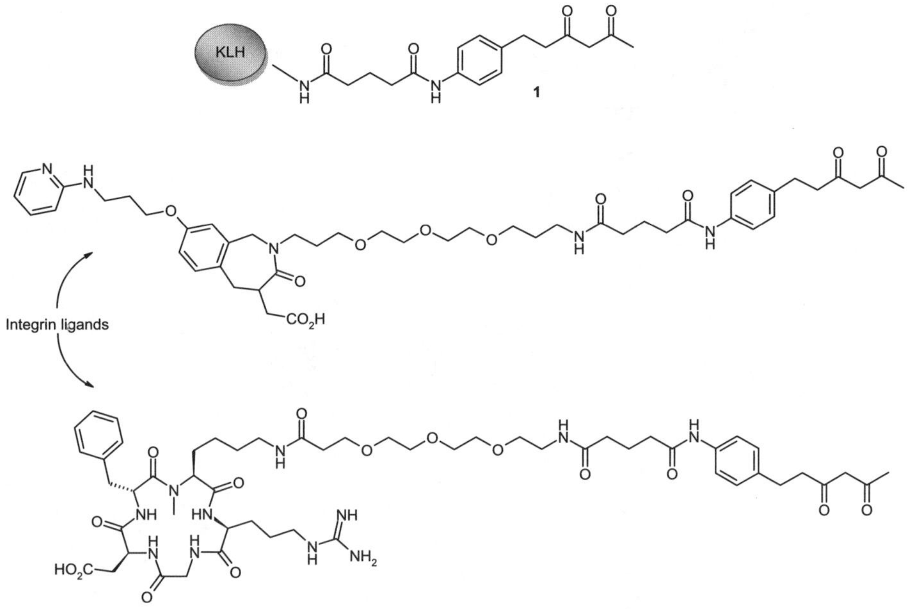

Figure 6.5 Reactive immunogen **1** and tumour surface integrins.

Mylotarg is a chemotherapeutic composed of a recombinant humanised IgG antibody conjugated *via* a bifunctional linker to a cytotoxic antitumour agent called calicheamicin $\gamma^{I}_{1}$.[21] Calicheamycin is an enediyne antibiotic that binds to the minor groove of DNA and then undergoes a Bergman cycloaromatisation[22] to generate a 1,4-benzyne diradical that abstracts hydrogen atoms from the carbohydrate backbone of the oligonucleotide resulting in strand cleavage (Figure 6.6).[23] The fascinating mechanism of action of the enediyne natural products combined with their intriguingly elaborate structure has resulted in significant interest within the synthetic chemistry community and several members of this family have succumbed to total synthesis.[24]

Calicheamicin itself is therefore highly cytotoxic and a method to direct this therapeutic to the desired site of action is required. The antibody portion of Mylotarg binds specifically to the CD33 antigen, a sialic acid-dependent adhesion protein found on the surface of leukaemic blast cells and immature normal cells of myelomonocytic lineage, but not on normal haematopoietic stem cells.[25] Binding of the antibody with the CD33 antigen results in the formation of a complex that is internalised *via* receptor-mediated endocytosis. As the endosome matures into a lysosome, the acidity increases to approximately pH 5, causing the hydrazone moiety to be cleaved. Intracellular glutathione also reduces the disulfide, releasing the calicheamicin inside the myeloid cell resulting in death. This effectively increases the concentration of the drug within the target area, thus increasing the therapeutic index of calicheamicin.

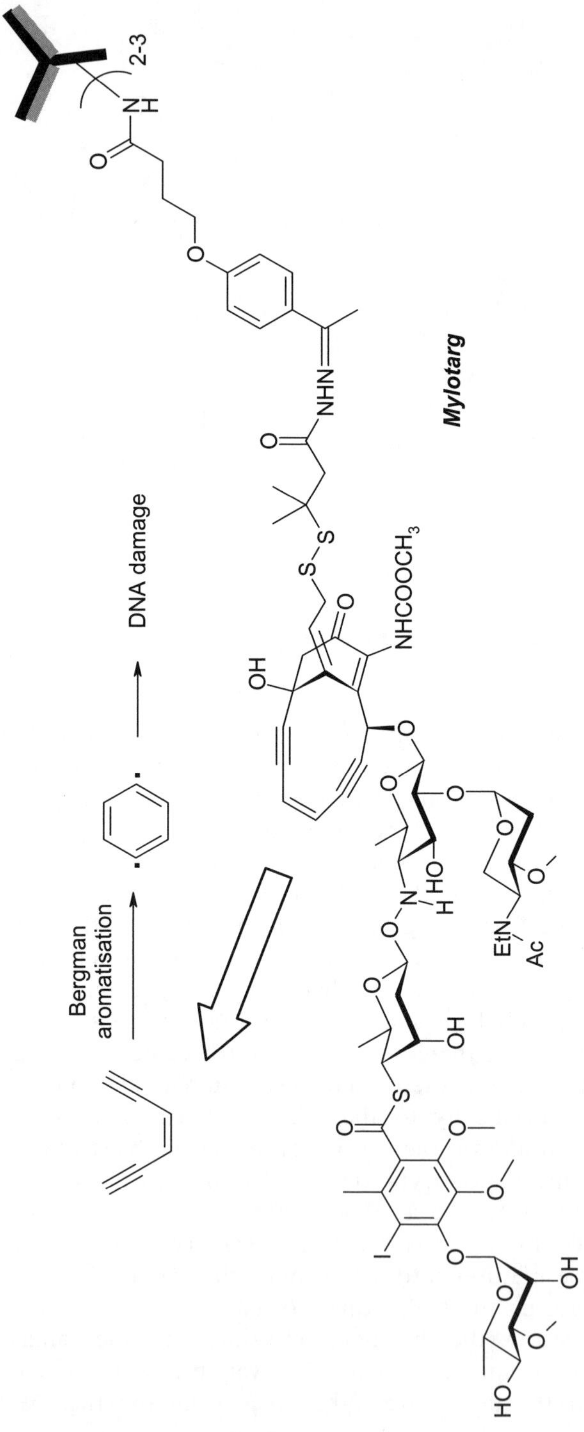

**Figure 6.6**   Bergman aromatisation and the structure of Mylotarg.

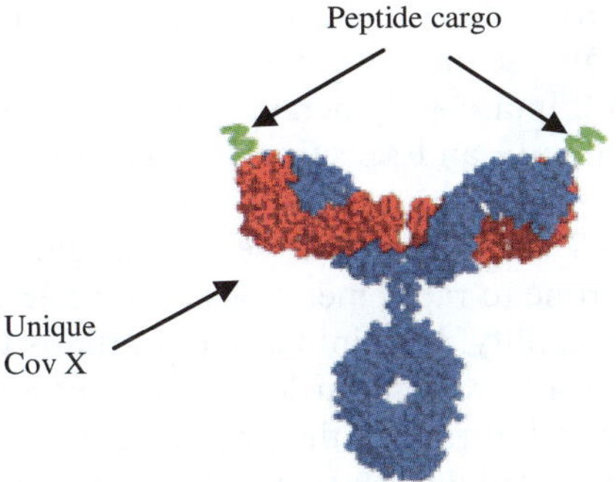



**Figure 6.7** CovX body.

## 6.3.2 Improved Pharmacokinetics

Additional to the highly selective nature of antibody binding, the pharmaco-kinetics of antibodies can result in extended half-lives since these macro-molecules can be protected from catabolism through binding to the Brambell receptor (FcRn). Peptides can possess highly effective pharmacodynamics, but are frequently poor therapeutics due to rapid degradation in the body and frequency of administration. Therefore, the half lives of peptides can be improved significantly through conjugation to antibodies.

Lerner and Barbas at The Scripps Research Institute have applied the specific chemistry of the aldolase catalytic antibody 38C2 to append targeting modules to the antibody *via* a specially designed linker. 38C2 was generated using a reactive immunisation strategy, which uses a hapten containing a moderately reactive functional group (a 1,3-diketone) that induces the aldol reaction *via* an enamine mechanism that uses a functionalised catalytic lysine residue in the antibody binding site (see Section 6.2.5). The unusually low $pK_a$ of $\sim 6.0$ is essential for the catalytic mechanism of the antibody, that is able to catalyse the aldol reaction of a large variety of substrates.[26] The specificity of the reaction of these lysine residues with β-lactam or 1,3-diketone modified substrates enables efficient and selective programming of the antibody with two functionalised units, which are termed CovX bodies™ (Figure 6.7).[27] There are a number of these conjugates in clinical trials for various therapeutic indications.

# 6.4 Oligonucleotides

## 6.4.1 Aptamers – Macugen

Aptamers are highly structured, single-stranded RNA or DNA oligonucleo-tides that have been selected for their ability to bind target proteins with high affinity and selectivity, and which therefore possess numerous therapeutic applications. The process for selection is called SELEX (systematic evolution of ligands by exponential enrichment) and usually relies on an immobilised protein target being able to select against a randomised library of oligonu-cleotides ($\sim 10^{14}$ unique sequences), and an enrichment process that uses standard reverse transcriptase–polymerase chain reaction (RT-PCR) methods to build the desired affinity and selectivity.[28] Using this technique, aptamers with affinities in the region of picomolar to the low-nanomolar range can be identified following $\sim 7$ to 15 rounds of selection. Unfortunately, unmodified oligonucleotides are prone to rapid metabolism by nucleases in the body which limits their therapeutic utility, but simple modifications to the oligonucleotide backbone can dramatically reduce nuclease turnover and are often incorpo-rated into the randomised library in the first stage of the process.

Macugen (pegaptanib sodium) is an anti-vascular endothelial growth factor (anti-VEGF) RNA aptamer that was discovered using the technology described above.[29] In particular, the incorporation of 2′-F and 2′-OMe groups through the oligonucleotide strand increased nuclease stability significantly, resulting in picomolar affinity and highly selective aptamers. Finally, pegylation of the 5′-end of the oligonucleotide further improved the pharmacokinetics, yielding pegaptanib as a highly chemically modified, complex RNA aptamer, 28 nucleotides in length, with a $K_D$ of 49 pM for VEGF and exquisite selectivity over related proteins (Figure 6.8). Macugen is approved for the treatment of all types of neovascular age-related macular degeneration, a leading form of blindness. There are currently six other aptamers in clinical trials for a variety of indications.[30]

## 6.4.2 siRNA

RNA interference (RNAi) holds significant promise as a therapeutic approach to silence disease-causing genes, particularly those that are not amenable to con-ventional therapeutics, such as small molecules, proteins or monoclonal anti-bodies. RNAi is an evolutionary conserved process by which double-stranded

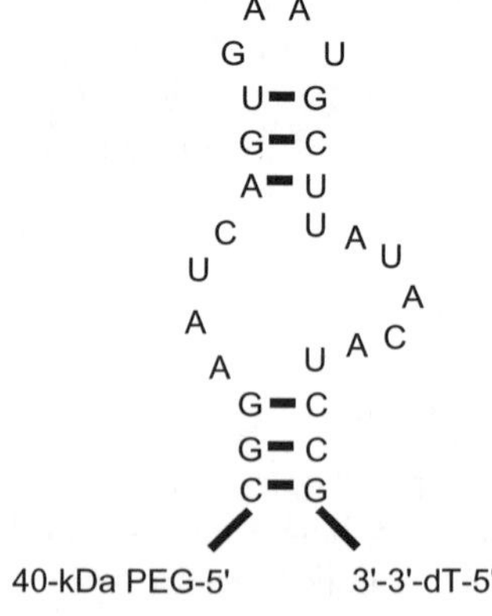


**Figure 6.8** Pegylated RNA aptamer pegaptanib.

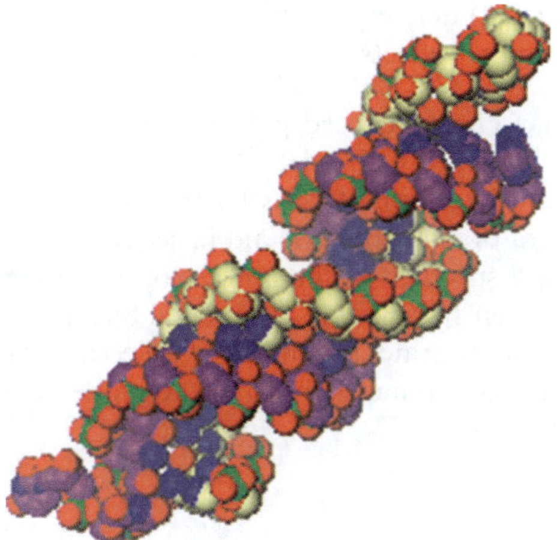

**Figure 6.9**  Representation of a double-stranded siRNA molecule.

small interfering RNA (siRNA; Figure 6.9) induces sequence-specific, post-transcriptional gene silencing. The endogenous process is initiated with the cleavage of long double-stranded RNA (dsRNA) into siRNA (19–21 base pairs) by the Dicer enzyme complex. These siRNAs are then incorporated into the RNAi-induced silencing complex (RISC) where the duplex unwinds to reveal the guide (antisense) strand that recognises target sites to direct specific mRNA cleavage, thus effecting gene silencing. The complex is then able to re-use the guide strand and therefore silencing is catalytic in nature. RNAi is considered to be a self-defence mechanism of eukaryotic cells to combat infections by RNA viruses and transposons and it is also assumed to regulate protein expression levels in response to certain environmental stimuli. Additionally, gene silencing induced by siRNA has become a powerful functional genomics tool for reverse genetics studies.

It was quickly realised that this highly sequence-specific mechanism of gene silencing could be harnessed to develop a new therapeutic approach that interferes with disease-causing genes.[31] Particularly attractive features of siRNA are very high potency combined with a long duration of silencing effect (their catalytic nature is usually limited by the rate at which the targeted cells divide). Notably, the first *in vivo* example of synthetic siRNA displaying efficacy in an animal disease model was reported by Song and co-workers in 2003.[32] Since then, a number of siRNA drug candidates have entered clinical trials, including approaches towards age-related macular degeneration and respiratory syncytial viral infection.

Despite the promise siRNA holds as a therapeutic approach, there are currently a number of hurdles that need to be overcome for their potential to be realised. They are:

- Poor cell penetration/uptake
- Nuclease instability

- Achieving (targeted) delivery to the relevant tissue/cells
- Minimising off-target effects

The issue of off-target effects, caused predominantly by the silencing of genes sharing partial homology with the siRNA and immune stimulation resulting from the recognition of certain siRNAs by the immune system, can be minimised using a plethora of algorithms to aid in sequence design, thus improving both functionality and specificity. The process of designing functional, specific siRNAs *in silico* has been successfully automated by a number of available web applications which incorporate hundreds of selection rules (the underlying science behind these programmes is beyond the scope of this chapter and has been reviewed elsewhere).[33]

### 6.4.2.1 Stability

Chemical modifications made to siRNA and lipid-based formulations have been used to address the issues concerning effective delivery of the oligonucleotide. Initial work in the area of RNAi focussed on diseases such as macular degeneration and RSV infection as this allowed for the direct administration of unmodified siRNA to the desired target (through local injection and inhalation respectively).

Nuclease stability is a problem associated with all oligonucleotides, and various backbone modifications have been used to successfully, and significantly, improve the stability of siRNA *in vivo* whilst still retaining silencing activity. Phosphorothioate and 2′-base sugar modifications (e.g. 2′-OMe, 2′-F) are illustrative of nuclease stabilising chemistries (see Section 6.3.1). Generally, to achieve adequate stabilisation (to allow systemic delivery) a significant fraction of the RNA bases need to be modified (sometimes > 50%) and it can be difficult not to compromise the silencing effect. By studying the degradation fragments of the siRNA in plasma it is possible to design the minimal number of modifications required to improve stability, thus minimising changes that could also reduce activity.[34] In some cases, the unwinding of the siRNA during knockdown may be inhibited due to these chemical modifications and therefore the duplex must be designed to have higher degrees of flexibility to compensate.[33]

### 6.4.2.2 Delivery

**6.4.2.2.1 Complexes.** Providing the oligonucleotide lives long enough in the plasma, it must then penetrate the cell membrane, something particularly difficult for a negatively charged macromolecule. Transfection agents are commonly used *in vitro* to enable cellular uptake of RNA and are simply lipid-based formulations that increase cell membrane affinity and penetration. Similarly, cationic liposomes (or lipoplexes) and complexes with peptides, polymers or antibodies have been used to effect siRNA delivery *in vivo*. For example, liposomal VEGF siRNA was used to reduce neovascularisation in a mouse model of AMD (which laid the ground work for progressing this approach into clinical trials). Liposome-formulated delivery of siRNA has been used to silence multiple targets following systemic administration.[35]

Stable nucleic acid–lipid particles (SNALPs), which contain a variety of lipids and cholesterol, have been used to aid formulation, cellular uptake and endosomal release of siRNA. Alnylam have reported the use of a SNALP ApoB siRNA complex that achieved 90% gene knockdown for 11 days following a single intravenous 2.5 mg kg$^{-1}$ dose.[36]

Complexation with cell-penetrating peptides has also been shown to be effective for siRNA uptake in a variety of cells, although its use may be limited due to as yet uncharacterised effects upon gene expression and a potential for immune activation.[37] Other ligands have been incorporated into the siRNA complex in order to direct the particles to the desired cells.

**6.4.2.2.2 Conjugates.** Generic application of complexes to package and deliver siRNA is currently limited by drawbacks such as toxicity and the need to assess the safety of each component of the complex. As a result, considerable effort in the scientific community has focussed on developing covalent siRNA conjugates, as single chemical entities, to enable (targeted) cell delivery and uptake. A breakthrough was made by Alnylam in 2004 when they reported the use of a cholesterol–siRNA conjugate that silenced an endogenous gene encoding apolipoprotein B after intravenous administration to mice.[38] The cholesterol, which was conjugated to the 3′-end of the sense strand using standard phosphoramidite chemistry, significantly improved cellular import over the naked siRNA and retained similar gene-silencing activity (Figure 6.10).

Follow-up work by Alnylam and the ETH has shown that conjugation to bile acids and long-chain fatty acids, in addition to cholesterol, mediates cellular uptake and gene silencing *in vivo*.[39] Efficient and selective uptake of these conjugates depends on interactions with lipoprotein particles, lipoprotein receptors and transmembrane proteins. High-density lipoprotein (HDL) directs siRNA delivery into liver, gut and kidney, whereas low-density lipoprotein (LDL) targets siRNA primarily to the liver. Cellular uptake also requires the mammalian homologue of the *Caenorhabditis elegans* transmembrane protein Sid1.

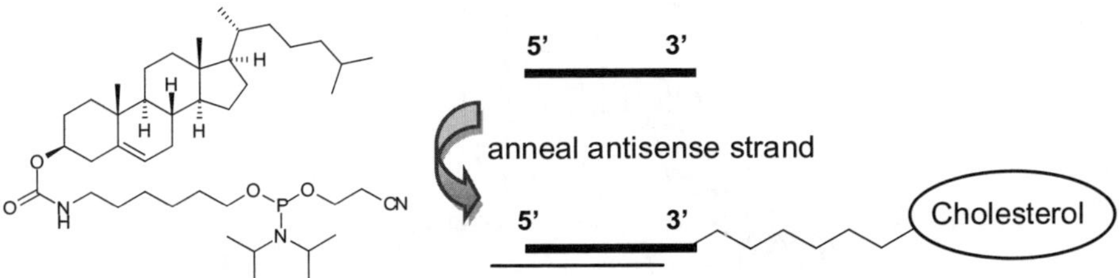


**Figure 6.10** Preparation of a cholesterol-conjugated siRNA.

Another recent approach to targeted cell delivery was made by Mirus Bio where they designed an siRNA polyconjugate that works through the gradual unveiling of its components as it travels to the cytoplasm of the cells.[40] The siRNA is linked by a disulfide bond to a multifunctional polymer backbone, polybutylamino vinyl ether (PBAVE; Figure 6.11). To this backbone, *N*-acetylgalactosamine (NAG), a hepatic targeting ligand, and a shielding agent (PEG), are attached. Following intravenous administration to mice, PEG protects the conjugate from non-specific binding until it reaches the liver, where the NAG ligand binds to the asialoglycoprotein receptor of hepatocytes and receptor-mediated endocytosis ensues. Once in the endosome, the decrease in pH releases the PEG and NAG, revealing the amino groups that cause endo-somolysis (*via* the "proton sponge effect"[41]) and release of the siRNA-PBAVE into the cytoplasm. The disulfide bond is subsequently cleaved in the reducing environment of the cell to reveal the siRNA.

### 6.4.3 Locked Nucleic Acids

Locked-nucleic acids (LNAs) are single-stranded oligonucleotides that target coding and non-coding RNA.[42] The monomeric unit usually contains a methylene linkage between the 2′-oxygen and 4′-carbon of the ribose that locks the sugar into the 3′-endo (N-type) conformation (Figure 6.12). Interestingly, the incorporation of a single LNA into a dodecameric DNA molecule (that can be achieved using standard phosphoramidite chemistry) can propagate this conformational change throughout the entire oligonucleotide.[43] The result of LNA incorporation is a subtle repositioning of the bases that improves the thermodynamics of hybridisation with the target RNA, as measured by an increase in the thermal stability of the duplex. The result is a significant

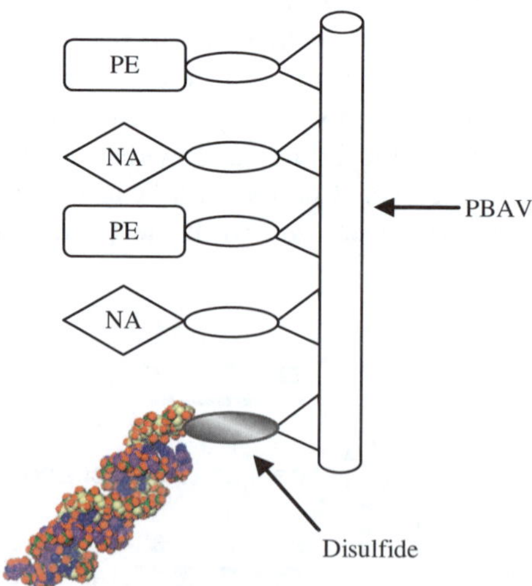


**Figure 6.11**    Mirus Bio siRNA polyconjugate.

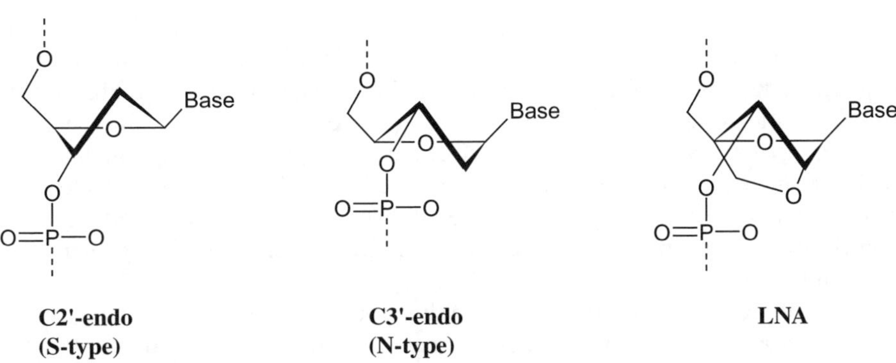

**C2'-endo**
**(S-type)**

**C3'-endo**
**(N-type)**

**LNA**

**Figure 6.12** Ribose conformations and locked nucleic acid.

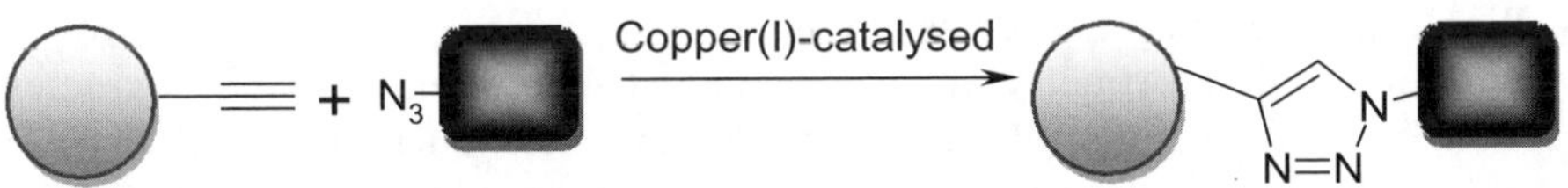



**Figure 6.13** Copper-catalysed triazole formation – a "click" reaction.

improvement in the pharmacological efficacy of the oligonucleotide. Recently, an LNA was developed to silence microRNA-122 in primates with chronic HCV infection which possessed good pharmacokinetic properties, an adequate safety profile and high stability *in vivo*.[44] These results suggest there could be a significant therapeutic potential for therapeutic LNAs.

## 6.5   Conclusions

This chapter does not exhaustively review the area of "chemologics", but rather illustrates the advantages of chemistry-enabled therapeutic modalities using marketed drugs, clinical candidates or cutting-edge novel approaches as relevant examples.

Novel synthetic methods are being developed that broaden the scope of bioconjugation chemistries and that enable the exploration of chemical modifications that alter the properties of the target biomolecule. Probably the best recent example of a synthetic approach that has had a major impact on a wide variety of scientific areas, and specifically bioconjugation, is the "click" reaction. Sharpless at The Scripps Research Institute and Meldal at the Carlsberg Laboratory in Denmark developed an improvement to a reaction called the Huisgen 1,3-dipolar cycloaddition.[45] This reaction, between an azide and an alkyne to form a triazole required high temperatures and gave a number of side products. Therefore, Sharpless and Meldal developed a copper-catalysed variant of this reaction that is able to progress at room temperature (Figure 6.13).

This reaction satisfies the "click" philosophy developed by Sharpless that tailors chemistry to generate complexity in a quick, simple and modular fashion. The power of this technique is that it is resilient to a number of functionalities, including those present in biomolecules, which therefore enables efficient chemistry to be performed on proteins or oligonucleotides, without the need for protecting groups. Additionally, it is compatible with biomedia and useful on the timescale of biological processes, and this has contributed to an explosion in recent years in the applications of click chemistry to chemical biology.

This chemistry has found broad application in materials/polymer science, drug delivery, biological imaging and screening technologies to name but a few. However, improvements to this chemistry are still needed. To illustrate this, the copper used to catalyse the reaction is highly toxic and difficult to remove from conjugated biomolecules, and therefore copper-free click chemistry is being developed by several laboratories to address this issue.[46]

Biophysical techniques such as NMR, mass spectroscopy, circular dichroism, microscopy and calorimetry are being applied to the analysis and characterisation of modified biomolecules and chemologics. These methods are key to the confirmation of structure and the delineation of structure-activity relationships. The importance of this understanding is magnified by the complexity of the activity, or if you like, the biological mechanisms being harnessed or perturbed.

For example, small molecule drug discovery is facilitated by high throughput screening with rapid data turnaround since well-informed modifications to the medicinal chemistry design can be made practically on a weekly basis. In areas such as vaccinology where the "design cycle" is of the order of many months, then there is even greater emphasis on leveraging all available technologies to design, synthesise, purify and analyse candidate molecules. More generally, the synthetic complexity of all chemologics necessitates a long lead time prior to the assessment of the biological activity, metabolic stability, safety and the many features assessed in the early stages of the drug discovery process and therefore our ability to develop greater understanding in the design of these desired properties will be essential to realise the full value of chemistry-enabled biotherapeutics.

# References and Notes


1. J. Peter-Katalinic and F. Hillenkamp, *MALDI MS: A Practical Guide to Instrumentation, Methods and Applications*, Weinheim, Wiley-VCH, 2007.
2. L. Schouls, H. Van der Heide, S. Witteveen, B. Zomer, A. Van der Ende, M. Burger and C. Schot, *BMC Microbiol.*, 2008, **8**, 35.
3. A. Davidson, T. C. Leeper, Z. Athanassiou, K. Patora-Komisarska, J. Karn, J. A. Robinson and G. Varani, *Proc. Nati. Acad. Sci. USA*, 2009, **106**, 11931.

4. M. Cuello, O. Cabrera, I. Martinez, J. M. Del Campo, M. A. Camaraza, F. Sotolongo, O. Pérez and G. Sierra, *Vaccine*, 2007, **25**, 1798.

5. W. Atkinson, C. Wolfe, J. Hamborsky and L. McIntyre, *Epidemiology and Prevention of Vaccine-Preventable Diseases*, 11th edn., Public Health Foundation, Washington DC, 2009.

6. J. D. Warren, X. Geng and S. J. Danishefsky, *Top. Curr. Chem.*, 2007, **267**, 109.

7. T. K. Park, I. J. Kim, S. Hu, M. T. Bilodeau, J. T. Randolph, O. Kwon and S. J. Danishefsky, *J. Am. Chem. Soc.*, 1996, **118**, 11488.

8. G. Ragupathi, F. Koide, P. O. Livingston, Y. S. Cho, E. Atsushi, Q. Wan, M. K. Spassova, S. J. Keding, J. Allen, O. Ouerfelli, R. M. Wilson and S. J. Danishefsky, *J. Am. Chem. Soc.*, 2006, **128**, 2715.

9. For a review of conjugation methods in vaccine development, see: V. Pozsgay and J. Kubler-Kielb, in *Carbohydrate-Based Vaccines*, R. Roy, ed. ACS Symposium Series 989, Oxford University Press, 2008, p. 36.

10. J. Duan, F. Y. Avci and D. L. Kasper, *Proc. Natl. Acad. Sci. U. S. A.*, 2008, **105**, 5183.

11. B. A. Cobb, Q. Wang, A. O. Tzianabos and D. L. Kasper, *Cell*, 2004, **117**, 677.

12. R. A. De Silva, Q. Wang, T. Chidley, D. K. Appulage and P. R. Andreana, *J. Am. Chem. Soc.*, 2009, **131**, 9622.

13. D. C. Jackson, Y. F. Lau, T. Le, A. Suhrbier, G. Deliyannis, C. Cheers, C. Smith, W. Zeng and L. E. Brown, *Proc. Natl. Acad. Sci. U. S. A.*, 2004, **101**, 15440.

14. F. Boato, R. M. Thomas, A. Ghasparian, A. Freund-Renard, K. Moehle and J. A. Robinson, *Angew. Chem. Int. Ed.*, 2007, **46**, 9015.

15. J. Grünewald, G. S. Hunt, L. Dong, F. Niessen, B. G. Wen, M. L. Tsao, R. Perera, M. Kang, B. A. Laffitte, S. Azarian, W. Ruf, M. Nasoff, R. A. Lerner, P. G. Schultz and V. V. Smider, *Proc. Natl. Acad. Sci. U. S. A.*, 2009, **106**, 4337.

16. X. Wu and P. G. Schultz, *J. Am. Chem. Soc.*, 2009, **131**, 12497.

17. www.ambrx.com

18. P. Wirsching, J. A. Ashley, C.-H. L. Lo, K. D. Janda and R. A. Lerner, *Science*, 1995, **270**, 1775.

19. D. Shabat, C. Rader, B. List, R. A. Lerner and C. F. Barbas, *Proc. Natl. Acad. Sci. U. S. A.*, 1999, **96**, 6925.

20. M. Popkov, B. Gonzalez, S. C. Sinha and C. F. Barbas, *Proc. Natl. Acad. Sci. U. S. A.*, 2009, **106**, 4378.

21. P. R. Hamann and M. S. Berger, *Tumor Targeting in Cancer Therapy*, Humana Press, Totowa, New Jersey, 2002, 239.

22. R. G. Bergman, *Acc. Chem. Res.*, 1973, **6**, 25.

23. A mechanism that proceeds via molecular oxygen trapping of 1,4-diradicals to provide a cytotoxic quinone has also been suggested: L. H. Jones, C. W. Harwig, P. Wentworth, A. Simeonov, A. Wentworth, S. Py, J. A. Ashley, R. A. Lerner and K. D. Janda, *J. Am. Chem. Soc.*, 2001, **123**, 3607.

24. For example, *Calicheamicin*: K. C. Nicolaou, C. W. Hummel, M. Nakada, K. Shibayama, E. N. Pitsinos, H. Saimoto, Y. Mizuno, K. U. Baldenius and A. L. Smith, *J. Am. Chem. Soc.*, 1993, **115**, 7625; *Dynemicin*: M. D. Shair, T. Y. Yoon, K. K. Mosny, T. C. Chou and S. J. Danishefsky, *J. Am. Chem. Soc.*, 1996, **118**, 9509; *Uncialamycin*: K. C. Nicolaou, H. Zhang, J. S. Chen, J. J. Crawford and L. Pasunoori, *Angew. Chem. Int. Ed.*, 2007, **46**, 4704.
25. N. K. Damle and P. Frost, *Curr. Opin. Pharmacol.*, 2003, **3**, 386.
26. C. F. Barbas, A. Heine, G. Zhong, T. Hoffmann, S. Gramatikova, R. Bjornestedt, B. List, J. Anderson, E. A. Stura, I. A. Wilson and R. A. Lerner, *Science*, 1997, **278**, 2085.
27. J. I. Gavrilyuk, U. Wuellner and C. F. Barbas, *Bioorg. Med. Chem. Lett.*, 2009, **19**, 1421.
28. C. Tuerk and L. Gold, *Science*, 1990, **249**, 505.
29. E. W. M. Ng, D. T. Shima, P. Calias, E. T. Cunningham, D. R. Guyer and A. P. Adamis, *Nat. Rev. Drug Discovery*, 2006, **5**, 123.
30. P. R. Bouchard, R. M. Hutabarat and K. M. Thompson, *Annu. Rev. Pharmacol. Toxicol.*, 2010, **50**, 237.
31. M. Stevenson, *New Engl J. Med.*, 2004, **351**, 1772.
32. E. Song, S.-K. Lee, J. Wang, N. Ince, N. Ouyang, J. Min, J. Chen, P. Shankar and J. Lieberman, *Nature Med.*, 2003, **9**, 347.
33. A. Birmingham, E. Anderson, K. Sullivan, A. Reynolds, Q. Boese, D. Leake and J. Ka, *Nat. Protocol.*, 2007, **2**, 2068.
34. A. De Fougerolles, H.-P. Vornlocher, J. Maraganore and J. Lieberman, *Nat. Rev. Drug Discovery*, 2007, **6**, 443.
35. M. Khoury, P. Louis-Plence, V. Escriou, D. Noel, C. Largeau, C. Cantos, D. Scherman, C. Jorgensen and F. Apparailly, *Arthritis Rheum.*, 2006, **54**, 1867.
36. T. S. Zimmermann, A. C. H. Lee, A. Akinc, B. Bramlage, D. Bumcrot, M. N. Fedoruk, J. Harborth, J. A. Heyes, L. B. Jeffs, M. John, A. D. Judge, K. Lam, K. McClintock, L. V. Nechev, L. R. Palmer, T. Racie, I. Rohl, S. Seiffert, S. Shanmugam, V. Sood, J. Soutschek, I. Toudjarska, A. J. Wheat, E. Yaworski, W. Zedalis, V. Koteliansky, M. Manoharan, H.-P. Vornlocher and I. MacLachlan, *Nature*, 2006, **441**, 111.
37. S. Moschos, S. W. Jones, M. M. Perry, A. E. Williams, J. S. Erjefalt, J. J. Turner, P. J. Barnes, B. S. Sproat, M. J. Gait and M. A. Lindsay, *Bioconjug. Chem.*, 2007, **18**, 1450.
38. J. Soutchek, A. Akinc, B. Bramlage, K. Charisse, R. Constien, M. Donoghue, S. Elbashir, A. Geick, P. Hadwiger, J. Harborth, M. John, V. Kesavan, G. Lavine, R. K. Pandey, T. Racie, K. G. Rajeev, I. Rohl, I. Toudjarska, G. Wang, S. Wuschko, D. Bumcrot, V. Koteliansky, S. Limmer, M. Manoharan and H. P. Vornlocher, *Nature*, 2004, **432**, 173.
39. C. Wolfrum, S. Shi, K. N. Jayaprakash, M. Jayaraman, G. Wang, R. K. Pandey, K. G. Rajeev, T. Nakayama, K. Charrise, E. M. NDungo, T. Zimmermann, V. Koteliansky, M. Manoharan and M. Stoffel, *Nat. Biotechnol.*, 2007, **25**, 1149.

40. D. B. Rozema, D. L. Lewis, D. H. Wakefield, S. C. Wong, J. J. Klein, P. L. Roesch, S. L. Bertin, T. W. Reppen, Q. Chu, A. V. Blokhin, J. E. Hagstrom and J. A. Wolff, *Proc. Natl. Acad. Sci. U. S. A.*, 2007, **104**, 12982.
41. J.-P. Behr, *Chimica*, 1997, **51**, 34.
42. H. Kaur, B. R. Babu and S. Maiti, *Chem. Rev.*, 2007, **107**, 4672.
43. M. Egli, G. Minasov, M. Teplova, R. Kumar and J. Wengel, *Chem. Commun.*, 2001, **7**, 651.
44. R. E. Lanford, E. S. Hildebrandt-Eriksen, A. Petri, R. Persson, M. Lindow, M. E. Munk, S. Kauppinen and H. Ørum, *Science*, 2010, **327**, 198.
45. (a) H. C. Kolb, M. G. Finn and K. B. Sharpless, *Angew. Chem. Int. Ed.*, 2001, **40**, 2004; (b) C. W. Tornoe, C. Christensen and M. Meldal, *J. Org. Chem.*, 2002, **67**, 3057.
46. J. C. Jewett and C. R. Bertozzi, *Chem. Soc. Rev.*, 2010, **39**, 1273.

# Antibody–Drug Conjugates in Oncology

PHILIP R. HAMANN[a] AND RUSSELL G. DUSHIN[b],*

[a] John Jay College of Criminal Justice, New York, NY 10019, USA;
[b] MS 220-3516, Pfizer, Inc., 445 Eastern Point Road, Groton, CT 06340, USA

## 7.1 Introduction and General Considerations

Antibody conjugates have been the subject of active research for as long or longer than monoclonal antibodies have been available in practical quantities. The species conjugated have varied from fluorescent tags to radioisotopes, from phytotoxins such as ricin to enzymes capable of activating prodrugs, and from traditional chemotherapeutic agents to derivatives of potently cytotoxic natural products. All of these have shown some utility, whether as experimental tools or as clinically approved therapeutic or imaging agents. However, practical considerations for therapeutic applications have led to the pursuit of antibody–drug conjugates (ADCs) of potently cytotoxic natural products as the most active area of research. These ADCs will be the topic of this review, with an emphasis on the evolution of the most common ADCs currently in or approaching clinical trials.

Research into ADCs and their optimization is a multi-variate process and the generic structure in Figure 7.1 belies this complexity. The antibody needs to be producible in quantity, needs to be stable in solution at the necessary pH values for reasonable periods of time, needs to possess acceptable physical properties so as to minimize the tendency to aggregate, needs to be amenable to the desired conjugation chemistry without altering its binding and any desired

RSC Drug Discovery Series No. 5
New Frontiers in Chemical Biology: Enabling Drug Discovery
Edited by Mark E. Bunnage

Published by the Royal Society of Chemistry, www.rsc.org

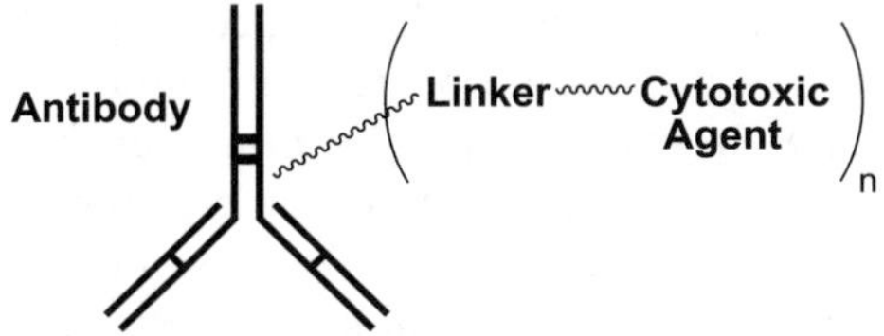


**Figure 7.1**  Generalized structure of an ADC.

Fc functions, and above all needs to target a tumor-associated antigen that is expressed in sufficient quantities on target cells,[1] to be internalized, and to be targeted to cancer cells without any significant toxicities arising from antigen expression on normal cells.

Antibodies used in ADCs today are invariably of the IgG family, but that still leaves a choice of isotype. $IgG_1$ is the most common choice, but this may be due largely to the fact that many antibodies originated in efforts that were looking for activity with an unmodified antibody, where $IgG_1$ is the preferred isotype. $IgG_1$ antibodies are also popular because of the possibility for activity from both the attached cytotoxic agent as well as from effector functions; however, it is not clear that both are obtainable in actual use. For many conjugates the doses are much lower than what is used for the unmodified antibody, and the unmodified antibody is ideally not internalized on its own, while internalization is crucial for ADCs, as all common cytotoxic agents attached to ADCs act on intracellular targets. Some classes of cytotoxic agents have been used primarily with $IgG_4$ antibodies, and conjugates of $IgG_2$ antibodies are also known. $IgG_3$ antibodies have been avoided due to the shorter half-life and the additional complexity from the variations in $IgG_3$s, but recent evidence indicates that $IgG_3$s can be given acceptable half-lives,[2,3] making them an interesting possibility for future.

Attachment of linker or drug-linker to the antibody has been accomplished using a variety of chemistries. The two main approaches are acylation of the lysine ε-amino groups, typically using *N*-hydroxy-succinimide esters, or reduction and alkylation of the inter-chain cysteine disulfides typically using a disulfide reductant (*e.g.* DTT) and a maleimide or α–haloacetyl-based alkylation. Each of these approaches relies upon the inherent nucleophilic character of solvent exposed amino or sulfhydryl moieties present (or generated) on these residues. A plethora of variations in approaches and reaction conditions exist. As examples, lysines can be thiolated using 2-iminothiolane and the introduced sulfhydryl groups alkylated in a manner similar to the reduced cysteines, a linker can be attached to the antibody followed by attachment of the cytotoxic agent (*e.g.* through disulfide exchange) to the linker–Ab intermediate, or the linker and cytotoxic agent can be attached in a single step. Modifications to the chemistry can to some extent control the specific sites of modification and the degree of heterogeneity of the resulting conjugate. Cysteine residues introduced at specific sites through genetic engineering have enabled the production of ADCs that are homogeneous in most practical senses.[4]

It can not be emphasized enough that practically all known ADCs are mixtures at some level, even though modern conjugation strategies are capable of making conjugates with greatly improved homogeneity. The sources of the heterogeneity are due first to the inherent heterogeneity of the antibodies, which includes to various degrees subtle differences in the carbohydrate modifications they naturally have: N-terminal variations, deamidations, iso-aspartates, *etc.* It is not well understood if or how these variations might alter the chemical reactivity of the antibody subfractions, but carbohydrate variations are know to play a significant role in at least one case.[5]

The conjugation chemistry employed provides the main source of variability for most conjugates. The actual sites of modification can vary from one molecule of antibody to another, and the number of sites modified for each antibody can as well. Indeed, for conjugated species with identical loading, there is still the real possibility of a complex mixture being present, as the sites of modification can still vary and this may affect all the physical and even chemical properties in different ways.

The linker used to attach the antibody and cytotoxic agent was greatly under-appreciated in early ADC research, but has become a central focus for more recent conjugate research efforts. The broadest activity for ADCs has been obtained with cleavable linkers that contain disulfides, peptides, or acyl hydrazones that, when properly chosen, all have acceptable stability in circulation but are successfully cleaved under conditions encountered intracellularly. The peptides and hydrazones are most likely cleaved in the lysosomes while most disulfides are believed to be reduced in the cytosol. This approach has the additional advantage that the species released can be predicted, controlled, and modified, albeit within the limits of the linker technology. All of these approaches can be used in a self-immolative mode that cleaves off any residual linker to release a specific desired derivative in un-derivatized form.[6]

In spite of their failure in early ADC research, non-cleavable linkers have also been found to be active with some antibodies and certain payloads, *i.e.* against certain antigens on certain target cells with ADCs containing certain active agents. In these cases the cytotoxic agent is presumably released by gross proteolysis of the antibody, and the cytotoxic agent bearing its linker still contains at least one residue from the antibody still attached, *e.g.* a thiolated lysine or cysteine in the case of maleimide-based conjugations. This extra residue can not be modified without changing the conjugation chemistry, but such released derivatives can be very active if the SAR of the cytotoxic agent is accepting of general substitution at the site of linker attachment. Although conjugates with non-cleavable linkers are not as broadly active, they show comparable activity in some test systems and are often better tolerated. However, caution needs to be exercised in interpreting some of the data because it is incomplete. Publications often show activity at only one or two doses and show only limited maximum tolerated dose (MTD) data. A complete (and clinically relevant) MTD determination and a full dose response in multiple *in vivo* models is really needed for all conjugates being compared to make a valid choice as to the best conjugate. Ideally, conjugate activity is compared in individual xenograft tests at an equal fraction of the MTD for each conjugate being considered.

Additional caution needs to be exercised when comparing cleavable and non-cleavable linkers or when comparing different mechanisms of linker cleavage. This is almost always at least a two-variable comparison, as the mechanism of drug release is different, and the identity of the species being released is almost always different. The most that can be said in the cases where the released species is different is that one conjugate performed better than the other, not that one linker was better than the other. However, the generalization that cleavable linkers tend to give broader activity than non-cleavable linkers appears to be true, and this may be due to insufficient release of payload through incomplete antibody catabolism in some cases, or to differences in the intracellular fate of free drug and released species containing linker and residual residues remaining from the antibody.

The class of cytotoxic agent and choice of which specific derivative to employ as a payload for an ADC provides the last part of this puzzle. This choice is not completely independent of the linker, as some agents lend themselves to certain linker strategies more than others, and a portion of the linker often remains attached to the cytotoxic agent. It is unclear if any particular mechanism of action has advantages, and any experiment that would address this issue would be non-trivial. The four most commonly pursued ADC classes consist of two tubulin agents and two DNA agents. It has become generally accepted in the field that a minimum potency is required for good activity. This is easily understood, as the limited amount of antibody a given antigen can practically internalize must carry a lethal dose to be successful. Antigen levels vary widely, from less than 10 000/cell to more than 1 000 000/cell, and antigens with low expression logically require a more potent conjugate than those with high expression. Of course the potency being discussed here is the *intracellular* potency of the released cytotoxic agent (*versus* a particular cell or cell type), a value that is not the same as determined in cell culture with the unconjugated cytotoxic agent. Because the ideal potency may depend on the antigen being targeted it is useful if the potency of a cytotoxic agent can be varied to optimize it for a specific application.

It could be suggested that if the cytotoxic agent is potent enough, it might well serve as a conjugation partner for all targets. However, when the doses of most substances fall below a certain level, non-linear pharmacokinetics (PK) is observed, and this appears to be true of antibodies and their conjugates as well. At very low doses non-specific sinks presumably take up most of the dose rapidly and leave little remaining ADC in circulation to target the tumor. Also, if these non-specific sites occur in a crucial organ and the cytotoxic agent is extremely potent, unacceptable toxicity may occur. How the minimum acceptable dose for reasonable PK and tumor targeting varies from conjugate to conjugate is unclear and should be explored for each conjugate and the results extrapolated to humans. As one possible example, for conjugates of calicheamicin, the less potent conjugates of *N*-acetyl gamma routinely outperformed the more potent conjugates of gamma made with the same linker and antibody.[7] Although other factors may play a role here, PK differences have been suggested as a possibility.

Another consideration that needs to be made is the stability of the cytotoxic agent. It, and the linker, needs to be systemically stable for many days, and not hours as is the case for typical small molecule drugs. However, as a conjugate that is restricted predominantly to the vascular compartment, an ADC (and the linker-payload it carries) is not subjected to the battery of liver enzymes that metabolize small molecules. It is also commonly suggested that it might be protected by attachment to the antibody, even though there is no published data to support this. In addition, most conjugates go through lysosomal compartments when internalized. Although linker cleavage is desirable at this point, the cytotoxic agent needs to have sufficient stability to the pH and digestive enzymes found in the lysosomes to escape into the cytosol and find its target.

One additional property that is useful in a cytotoxic agent is some degree of water solubility. ADCs have their drug attached by definition to the solvent-accessible surface of the antibody. Their physical properties will, therefore, modify the physical properties of the antibody in solution, and greater loading will likely increase any effects. The most obvious way this change in properties is expressed is by the increased tendency of many conjugates to form aggregates, sometimes to a devastating degree. Even after a practical loading is found and any aggregate is removed by chromatography, the change in physical properties can be important. Many conjugates have circulating half-lives in animal models or in clinical trials that are significantly shorter than that of the unmodified antibody. These effects can be overcome at least in part by modifying the cytotoxic agent to make it more soluble or by modifying the linker portion. Polar, inherently water soluble groups are increasingly used, such as through incorporation of anionic groups and various types of PEG into linker architecture.

Of course all these various parameters need to be either chosen based on literature data (or commonly according to current trends) or worked out with animal models. However, consideration of the interactions in humans needs to be factored in. Even if the antigen is expressed in normal tissues in animal models, the degree and location of expression can vary significantly from species to species. Especially important here are the Fc interactions. Even if effector functions are not desired for a certain conjugate approach, FcRn recycling will be a desirable interaction that increases the half-life of antibodies, and conjugates should ideally be able to take advantage of this as well.

All of the principles described above have been explored with the conjugates described in the chapter. However, many of the conjugates in clinical trials do not contain all of the possible optimizations, *e.g.* they contain earlier versions of linkers that do not appear to be optimal. With the interactions between the various factors that go into the design of ADC, it will take several generations of conjugates to understand and optimize all of them.

# 7.2 Calicheamicin Conjugates

Gemtuzumab ozogamicin (Gem Ozo, Mylotarg®, or CMA-676), an AcBut linked calicheamicin conjugate (Figure 7.2), is the only ADC approved by the FDA for clinical use.[8] The antibody conjugated to calicheamicin, hP67.6,[9] targets the well-internalized myeloid differentiation antigen CD33 (siglec-3[10]) found predominantly on myeloid progenitor cells and on the leukemic blasts from a sizable majority of patients with acute myeloid leukemia (AML). The bifunctional AcBut linker allows for attachment to lysines on the antibody, with unexplained selectivity for eight specific lysines near the hinge region.[11] It also attaches to a hydrazide derivative of calicheamicin to form an acyl hydrazone, which allows for good stability in circulation ($\sim$2% cleavage/day[11]) and good hydrolytic release of the calicheamicin derivative at lysosomal pH.[12] This conjugate was shown to be greatly superior to the amide conjugate for this application.[13]

Gem Ozo is selectively cytotoxic to HL-60 AML cells in tissue culture with an $IC_{50}$ of 0.31 pM.[12] It is active against HL-60 xenografts over a 10-fold dose range (25–300 $\mu$g kg$^{-1}$ of calicheamicin, three doses given every 4 days) with complete cures in all the animals at all but the lowest dose. It also shows significant, selective activity in a colony-forming assay in a significant percentage (1/3 to 1/2, depending on the criteria applied) of bone marrow samples from AML patients. PK studies in rats and cynomologous monkeys gave half-lives of around 100 h, while the half-life in humans was 72 h for the first dose and 94 h for the second dose, when circulating antigen levels were lower.[14] Toxic effects in the two animal species included liver, spleen, kidney, and bone

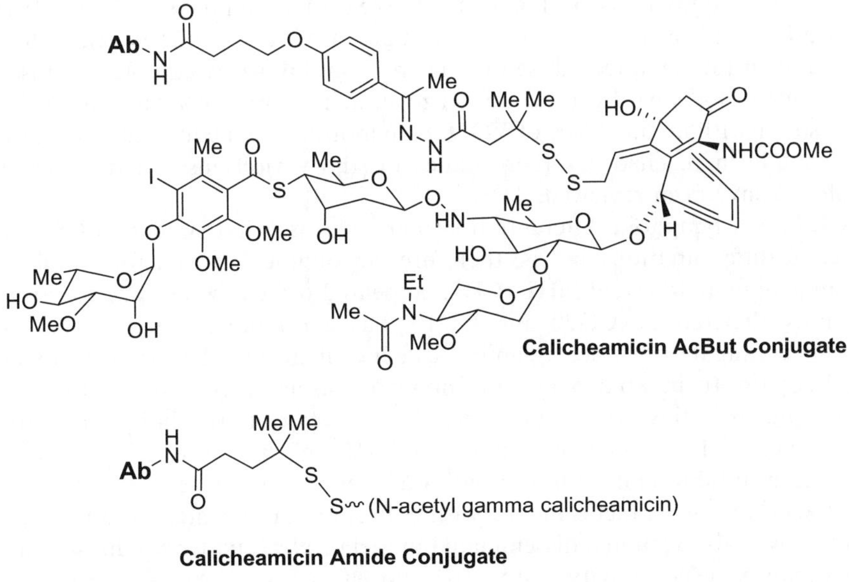

**Figure 7.2** Calicheamicin conjugates.

marrow, with the liver toxicity and myelosupression being reversible, and these were the common side effects seen in humans as well.[8] A tolerance study in chimpanzees at a low dose of $0.5\,mg\,m^{-2}$ resulted in no discernible toxic effects.

A Phase I trial indicated that a dose of $9\,mg\,m^{-2}$ gave antigen saturation, even though formal MTD was not established.[15] However, significant myelosupression was seen consistent with targeting CD33, and liver enzyme elevations indicated that the MTD was not likely to be much higher. Indeed, some subsequent combination trials have been conducted at $3$–$6\,mg\,m^{-2}$ to minimize the toxicity contributed by Gem Ozo.[16,17] The most significant, albeit relatively rare, toxicity seen to date is vascular-occlusive disease (VOD), which occurred in 2% of patients.[18] Prior hematopoietic stem cell transplant was a significant risk factor, and there is some promising indication that it can be prevented prophylactically.[19] Most toxicities decreased with fractionated dosing ($3\times3$ $mg\,m^{-2}$ over 1 week) with no apparent loss in activity.[20]

A complete response in heavily pretreated relapsed or resistant AML patients was seen in 26% of 277 patients treated in a pivotal, open-label, single-arm Phase II trial.[21] About half of the patients failed to recover normal platelet numbers, but were asymptomatic and transfusion free. Recurrence-free survival was 5.5 months. Gem Ozo was approved by the FDA in May 2000 for use in patients with CD33-positive AML in first relapse who are 60 years of age or older and are not candidates for cytotoxic chemotherapy. A significant advantage of Gem Ozo is the reduced toxicity *versus* traditional chemotherapy, which allows many patients to be treated on an out-patient basis, reducing hospitalization costs.[22]

There have been numerous clinical studies of Mylotarg use; the most significant is a comparison of daunorubicin ($45\,mg\,m^{-2}$ on days 1, 2, and 3) and cytarabine ($100\,mg\,m^{-2}$ on days 1 through 7) with or without gemtuzumab ozogamicin (at a reduced dose of $6\,mg\,m^{-2}$ on day 4). A complete remission (CR) rate of 84% with gemtuzumab ozogamicin *versus* 55% without has been seen so far with no incidence of VOD. At 6 months the median duration of CR had not been reached.[11] Other studies on the various uses that have been explored have been reviewed.[23,24]

CD22 is a lymphoid differentiation marker, somewhat similar to CD33 for myeloid differentiation, and is, therefore, an obvious antigen for a calicheamicin conjugate to target NHL. CD22 appeared to be a better target than two obvious alternatives, CD20 and CD19, based on preclinical models.[25] An AcBut conjugate of *N*-acetylgamma calicheamicin and the G544 anti-CD22 antibody (inotuzumab ozogamicin, Ino Ozo) was again chosen over the amide conjugate for this application based on superior preclinical activity.[26] The range of $IC_{50}$ values in a panel of NHL cell lines was $0.006$–$0.6\,nM$.[27] Significant inhibition of tumor growth was seen in various xenograft models at $10$–$160\,\mu g\,kg^{-1}$ of calicheamicin, with 8/8 cures at the highest dose. Good activity was also seen in a disseminated model of NHL under conditions where rituximab gave no activity, with cures (no effect for $>100$ days) seen at the highest dose of $160\,\mu g\,kg^{-1}$.[28]

In an initial Phase I study of Ino Ozo, an MTD of 1.8 mg m$^{-2}$ was seen.[29] Common toxicities were thrombocytopenia, neutropenia, elevated liver enzymes, and the common infusion-related toxicities, with hematopoetic toxicities, especially thrombocytopenia, being the most common dose-limiting toxicity (DLT). Complete responses were observed starting at 0.8 mg m$^{-2}$. At the MTD in a heavily pretreated patient population, the objective response rates were 69% for follicular lymphoma and 33% for DLBCL as monotherapy.[30]

Subsequent studies have focused on combining Ino Ozo with rituximab. This combination has been supported by preclinical pharmacology studies, where an additive effect was seen between the two therapies with no apparent effect on the MTD of Ino Ozo.[31] This was confirmed in clinical studies, where Ino Ozo had the same MTD with or without rituximab with a similar profile of toxicities.[32] The reported interim 6-month progression free survival rate was 100% for FL and 66% DLBCL for all patients in the MTD cohort. Ino Ozo is now in a Phase III clinical trial for folicular NHL.

The two most significant calicheamicin conjugates, the two already described, both contain the AcBut linker, which was shown to be superior to the Amide linker for these conjugates that target hematopoietic malignancies. However, the choice of the optimum conjugate in a preclinical setting is not always the same. One example is the calicheamicin conjugate CMB-401, that was chosen to have the Amide linker due to superior activity in MDR-positive xenograft models.[33] Even though this Amide conjugate showed a strong preclinical profile in both cell culture and numerous xenograft experiments,[34] it failed to show significant clinical activity in two small Phase II trials.[35] The possible reasons that have been suggested for the lack of clinical activity include circulating antigen, undetected resistance mechanisms, and the wrong choice of linker in spite of the preclinical data. The circulating antigen problem was known to be a significant one, and predosing with unconjugated antibody had been used to try to mitigate this problem.[36]

One additional AcBut-linked conjugate has been examined in clinical trials, the anti-Lewis$^{Y}$ conjugate CMD-193. This conjugate of antibody hu3S193 differed from all other calicheamicin conjugates in the choice of an IgG$_1$ isotype instead of a hinge-modified IgG$_4$. This conjugate also had a more modest binding constant of $1.3 \times 10^{-8}$.[37] The IC$_{50}$ seen in preclinical models was <0.01 to 90 ng mL$^{-1}$, depending on the antigen expression and inherent calicheamicin sensitivity of the specific cell line, with antigen selectivity of 2- to over 100-fold. Good activity was seen in several xenograft models with cures seen at the top doses in the N87 gastric cancer model.

In a Phase I clinical trail, half-lives of around 100 h were seen for various doses.[38] Even though this is less than the native antibody,[39] it is not unusual compared with other conjugates. However, when CMD-193 labeled with $^{111}$In was examined in a PK study, prolonged liver uptake was observed with little tumor uptake.[38] The liver toxicity was also more pronounced than with previous calicheamicin conjugates. Although there is no direct data on the cause of these effects, it is possible that the choice of the IgG$_1$ isotype combined with the effect of calicheamicin conjugation on the physical properties of that isotype may have played a role.

One other AcBut-linked calicheamicin conjugate has been published, CME-548, which is a conjugate of the anti-5T4 oncofetal protein antibody huH8. An $IC_{50}$ of $0.04\,ng\,mL^{-1}$ was seen in the MDAMB435 cell line with 5T4 clonally expressed with about 1000-fold selectivity *versus* a non-binding control conjugate.[40] CME-548 and a closely related conjugate with a slightly modified linker gave good activity in MDAMB435/5T4 xenograft experiments and several other *in vivo* models. Continued research will be required to see if 5T4 is an acceptable target antigen in humans.

It should also be mentioned that there have been publications of conjugates of a synthetic calicheamicin analog, calicheamicin θ,[41,42] conjugated through the amino-sugar, although it is questionable whether such a synthetic derivative could be commercialized.

## 7.3  Maytansine Conjugates

The maytansines are a family of very potent (sub-nanomolar) antimitotic agents that inhibit tubulin polymerization. Attempts to use maytansine derivatives themselves in clinical trials have shown them to possess very narrow therapeutic windows.[43] To overcome this limitation, derivatives have been made that allow for conjugation to antibodies.

Ansamitocin P-3 has an isobutyrate ester that can be removed and re-acylated to give very potent maytansine analogs[44] containing thiol derivatives that have been utilized in two basic ways for conjugation. The initial approach used by ImmunoGen formed disulfide linkages with antibodies that had previously been acylated on lysines to introduce multiple reactive thiols.[45] The maytansine thiol derivative chosen, DM1, can be attached to antibodies using a variety of thiolation reagents that attach to lysines in a relatively random process,[46] but they have done so almost exclusively with an activated derivative of 4-mercaptopentanoate (SPP) to give a disulfide that is sterically stabilized on the adjacent methylene groups by a single methyl substituent, as shown in Figure 7.3.[45] This gives rise to conjugates that have, perhaps, less than optimum stability in circulation with a half-life of about 1 day in animal models based on the loss of tritiated maytansine,[47] which is significantly shorter than the half-life of most antibodies. In spite of this, these conjugates have excellent preclinical pharmacology and have been widely examined in the clinic.

The initial conjugate reports on DM1 conjugates were with antibodies A7 (murine, anti-human colon), TA.1 (murine, anti-HER2), 5E9 (murine, anti-human transferin), and B4 (anti-CD19).[45] The TA.1-DM1 disulfide conjugate had an $IC_{50}$ of $1.6\times10^{\times11}$ on SK-Br-3 breast cells, comparable to the methyl disulfide of DM1. Selectivity of 1000-fold was seen by blocking with unconjugated antibody or by determining the $IC_{50}$ on the antigen negative KB cells. The targeted cytotoxicity was shown to be dependent on the loading of the conjugate; increasing the average loading from ~1 to ~4 resulted in a 23-fold

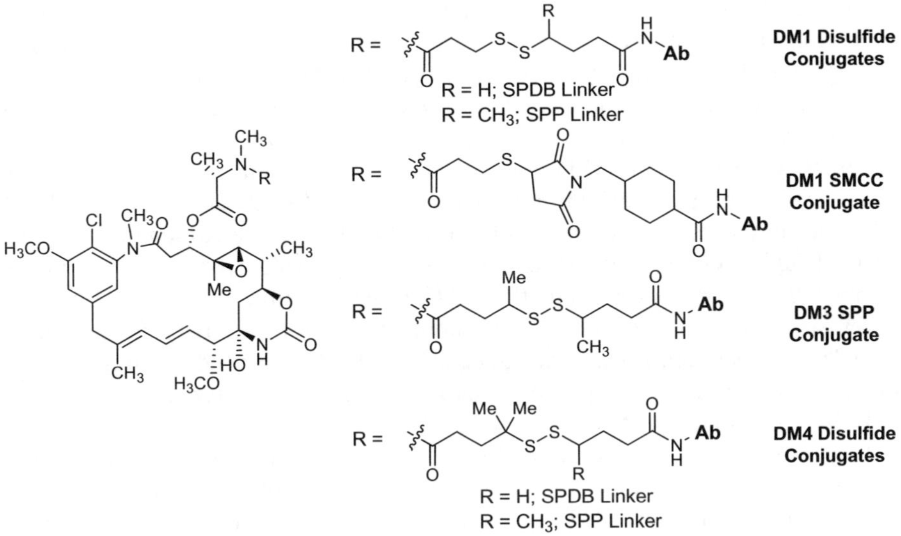

R = H; SPDB Linker
R = CH₃; SPP Linker
**DM1 Disulfide Conjugates**

**DM1 SMCC Conjugate**

**DM3 SPP Conjugate**

**DM4 Disulfide Conjugates**
R = H; SPDB Linker
R = CH₃; SPP Linker

**Figure 7.3** Maytansine conjugates.

lower $IC_{50}$. Similar results were obtained with the A7-DM1 disulfide conjugate. Mice tolerated a dose of $200\,\mu g\,kg^{-1}$ of maytasine as conjugate ($8\,mg\,kg^{-1}$ of antibody) given intravenously.

The first *in vivo* data published for a DM1 disulfide conjugate was with the murine anti-CanAg antibody C242.[48] The CanAg mucin occurs on colon and various other solid tumors. A conjugate with an average loading of $\sim 4$ moles $mole^{-1}$ retained full binding to the antigen on cells, had an $IC_{50}$ on COLO 205 human colon cells of $0.032\,nM$ with 1000-fold selectivity *versus* antigen negative cells, and was curative of COLO 205 xenografts at $300\,\mu g\,kg^{-1}$ of maytansine/ day for 5 days. Almost all the *in vivo* activity was lost at one half that dose. Good activity was also seen in LoVo and HT-29 colon tumors, with tumor re-growth being seen after 40 days. The authors stated this was most likely due to the heterogeneous expression of the target antigen in these tumor models, although bystander killing of antigen negative cells as been shown to play a significant role.[49]

The same DM1 disulfide conjugate, referred to as cantuzumab mertansine or SB-408075, has been made with the humanized version of this antibody, huC242, for use in clinical trials. In Phase I an MTD of $235\,mg\,m^{-2}$ every 3 weeks was found in mostly colon carcinoma patients with observed dose-limiting toxicity being reversible increases in hepatic transaminases.[50] Various other toxicities were observed, including neutropenia and thrombocytopenia, as well as the usual infusion related effects. In spite of a modest half-life of around $40\,h$ seen for intact conjugate, which is consistent with the value found in mice,[47] two minor responses were seen in the 17 patients treated at the MTD.

Due to the marginal activity of huC242-DM1 disulfide in clinical trials, which might be improved by better serum stability of the disulfide, a second conjugate, huC242-DM4 disulfide conjugate, IMGN242, has been investigated. The methyl disulfide of the thiol DM4 (stabilized by two methyl substituents located beside the disulfide, see Figure 7.3) is about 10-fold more potent than the methyl disulfide of DM1 in cell culture.[44] The huC242-DM4 disulfide conjugate (bearing an SPDB linker) had an $IC_{50}$ on COLO 205 cells of 0.04 nM with 100-fold selectivity, comparable to the DM1 disulfide conjugate.[51,52] This DM4 disulfide conjugate showed *in vivo* activity against COLO 205 tumors comparable to or greater than the C242-DM1 disulfide conjugate, but at a lower single dose of $50\,\mu g\,kg^{-1}$ of maytansine. Good activity in xenograft models was also seen for gastric cancers.[53] The DM4 disulfide conjugate gave superior efficacy when compared to the DM1 disulfide in later stage (300 mm$^3$) COLO 205 xenografts.[52]

The improved *in vivo* activity of the DM4 disulfide conjugate is likely due in part to improved stability in circulation. The half-life for the maytansine portion (tritiated) of the DM4 disulfide conjugate was 5–6 days, compared to the 1 day value seen for the DM1 disulfide conjugate.[54] The MTD for IMGN242 in Phase I patients with mostly colon and gastric cancer was determined to be $168\,mg\,m^{-2}$, only a little less than the MTD of the DM1 conjugate.[55] Toxicities were similar to the DM1 disulfide conjugate, with the exception of some eye effects (decreased visual acuity, corneal deposits, and keratitis) that are possibly reversible. Decreases in tumor size of up to 23% were seen in six of 23 evaluable patients. A Phase II study of this conjugate is under way in gastric cancer patients.[56] Although some biological activity has been seen, ocular toxicity that appears to be dependent on circulating shed antigen levels, which in turn appears to predict for drug exposure, has led to modified dosing.

A disulfide is only one way a thiol can be attached to an antibody. A significantly different approach is to form a monosulfide with a Michael acceptor, such as a maleimide. ImmunoGen has done that by pre-attaching the commercially available SMCC (aka MCC) linker to lysines on antibodies and then reacting the attached maleimide with the DM1 thiol, as shown in Figure 7.3.[52] This conjugate contains what is commonly referred to as a non-cleavable linker, one which relies on antibody proteolysis in the lysosomes for the release of a derivative of the cytotoxic agent that still contains the linker and an aminoacid residue from the antibody.[51] Such conjugates are active when the SAR of the cytotoxic agent allows for general substitution at the site of linker attachment, but they tend to be somewhat less potent and activity appears to be more dependent on the target antigen and cell type.

On COLO 205 cells *in vitro*, the huC242-SMCC DM1 conjugate has an $IC_{50}$ similar to the two disulfide conjugates and shows activity on small (100 mm$^3$) COLO 205 xenografts, albeit at higher doses; it is inactive on larger (300 mm$^3$) xenografts.[52] This decreased activity is certainly impacted by the lack of bystander effect seen with this conjugate, which is likely due to the polar nature and lack of membrane penetration of the released derivative.[49] However, the decreased potency of the SMCC-DM1 conjugates is accompanied by an increase in the MTD, making it a viable conjugation partner for favorable antibodies (*vida supra*).

These three conjugate types, DM1 disulfide, DM4 disulfide, and SMCC DM1 (monosulfide) have been applied to various antibodies over the last decade or two. The anti-PSCA (prostate) antibody 8D11-DM1 disulfide conjugate was examined in collaboration with Genentech.[57] *In vitro* cytotoxicity was less than the C242 (anti-CanAg) conjugates with an $IC_{50}$ of about 1 nM and > 70-fold selectivity. *In vivo* the 8D11 antibody itself delayed growth of the PC3.gD.PSCA xenografts, but the DM1 disulfide conjugate had significantly more activity with no tumor re-growth at 60 days.

An anti-PSMA DM1 disulfide conjugate, MLN2704, has also been made for use in prostate cancer, this time in collaboration with Millennium.[58] Similar to the huC242 conjugate, an *in vivo* half-life of about 40 h was seen in mice. In spite of this, prolonged tumor suppression was observed with repetitive bi-weekly dosing, although tumors did eventually reappear when treatment stopped. A Phase I trial has been conducted with this conjugate.[59] Although an MTD was not formally defined, elevated liver enzymes and neuropathy (common for tubulin inhibitors) was seen at the highest dose of 343 mg m$^{-2}$. Activity was seen in two of nine patients treated at the top two doses, as indicated by decreases in PSA levels.

Two DM1 disulfide conjugates have been published for multiple myeloma (MM); the first, an anti-CD56 conjugate known as huN901-DM1 (disulfide), IMGN901, or BB-10901, has been investigated in various preclinical and clinical settings. In an *in vitro* panel of cells $IC_{50}$ values of 10–50 nM were seen.[60] In OPM2 xenografts, doses of 75–150 µg kg$^{-1}$ of conjugate for 5 days resulted in complete suppression of tumor growth for 30 days in an early tumor model, but did not completely suppress the growth of established tumors (200 mm$^2$). Additional preclinical studies indicate the possible benefits of using the huN901-DM1 disulfide conjugate with combination chemotherapy[61] and the potential application to ovarian cancer.[62]

In a Phase I clinical trial for various indications, huN901-DM1 disulfide was given for three consecutive days each week for 3 weeks.[63] There was no remarkable toxicity at the doses reported, and five patients out of 27 had stable disease, with one of those patients achieving a complete remission. A Phase II trial has been reported at a dose of 60 mg m$^{-2}$ given weekly for 4 weeks in patients with CD56-positive tumors.[64] There were two partial responses (one unconfirmed) out of 30 patients and five patients with stable disease lasting up to 18 weeks. The second maytansine conjugate for MM, BT-062, is an DM4 disulfide conjugate of the anti-CD138 antibody[65] that is now in Phase I clinical trials for MM.[66]

Several other maytansine conjugates have been investigated. Bivatuzumab mertansine (BIWI-1), a DM1 disulfide conjugate of the anti-CD44v6 antibody BIWA-4 was discontinued due to skin toxicity that unfortunately resulted in one death.[67] AVE9633, an anti-CD33-DM4 disulfide conjugate, has been examined in a Phase I trial for AML with doses up to 260 mg m$^{-2}$ without any dose-limiting toxicity or activity being observed, in spite of complete antigen saturation.[68] This conjugate replaced an earlier DM1 disulfide conjugate.[69] A DM4 disulfide of the anti-CD19 antibody huB4 is in clinical trials.[70] An anti-Cripto DM4 disulfide conjugate BIIB015 is in clinical trials for solid tumors.[71]

Along with CD19 several other antigens have been examined for targeting maytansine to NHL, including CD20, CD21, CD22, CD72, CD79b and CD180.[72] Two maytansine conjugates, the original SPP linked DM1 disulfide and the SMCC DM1 derivatives with a non-cleavable linker were examined with all these target antigens. Conjugates of the DM1 disulfide were active with all the antibodies, while conjugates of SMCC DM1 were only active with the anti-CD22 and anti-CD79b antibodies. For the disulfide conjugates the data did not allow the choice of a best target, although the anti-CD72 conjugate appeared to be inferior. The best choice between the disulfide and the non-cleavable linker for the CD22 and CD79b antigens was also not clearly determined, however, the conjugates with the non-cleavable linkers were better tolerated. A final choice would require more data, including dose response data (cf. Giles *et al.*[73]).

Maytansinoid conjugates of the anti-tumor fibroblast antibody FAP5 have been profiled.[74] *In vitro* cytotoxicity of 0.22, 0.29 and 0.05 nM was seen for the SMCC DM1, DM4 disulfide and DM1 disulfide conjugates, respectively, with selectivities of 680-, 160- and 540-fold. However, only the two disulfide conjugates showed significant activity in xenograft experiments, perhaps due to the inability of the SMCC conjugate to participate in the bystander killing that has been demonstrated for the disulfide conjugates.[52]

Three maytansine conjugates of the anti-$\alpha_v$ integrin antibody CNTO 95 have been profiled in preclinical models.[75] This antigen is expressed on several solid tumors, as well as new blood vessels. In addition to the DM1 disulfide conjugate (CNTO 364) and the DM4 disulfide conjugate (CNTO 365), they also included a more hindered version of the DM4 disulfide conjugate with the SPP linker (CNTO 366) that has three stabilizing flanking methyl groups (Figure 7.3). The *in vivo* activity of this more hindered DM4 conjugate indicated that the disulfide with the SPP Linker was too stable for efficient drug release, while the SPDB-linked DM4 disulfide conjugate was superior at equivalent doses in xenograft experiments. This conjugate, IMGN388 is now in a Phase I clinical trial.[71]

The most exciting maytansine conjugate to date is that of trastuzumab (Herceptin®), which targets the very highly expressed (commonly 1 000 000 copies/cell) and well-internalized HER2/neu antigen on breast cancer and other solid tumors. Data have been published on five different combinations of maytansine derivatives and linkers (Figures 7.3 and 7.4): the DM1 disulfide SPP linker, a less hindered DM1 disulfide made with the commercially available SPDP linker, the DM1 SMCC (aka MCC) thioether conjugate, a disulfide conjugate of DM3 with the SPP linker (disulfide flanked by a methyl group on each side), and the more hindered disulfide of DM4-SPP.[76]

The $IC_{50}$ values on various cell lines were not dramatically different between the five trastuzumab conjugates examined, with values on high-expressing lines

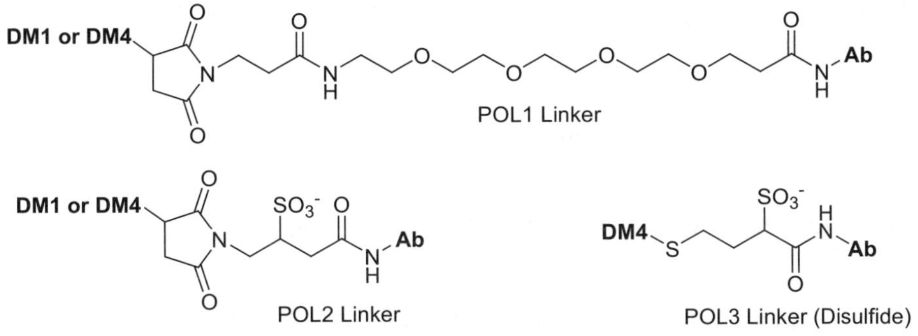

**Figure 7.4**   More soluble maytansine conjugates.

ranging from about 0.2 to 2.7 nM with selectivities of 10- to 100-fold, depending on the specific cell line. However, dramatic differences were seen in PK studies in mice, with the least stable DM1 SPDP disulfide conjugate showing 80% loss of drug within an hour and the original DM1 conjugate (SPP Linker) showing 60% loss in 3 h. The new DM3 disulfide conjugate (SPP Linker) showed 45% loss in 7 h, while the more-hindered DM4 conjugate (SPP linker) and the SMCC-DM1 monosulfide conjugate showed 30% loss in 7 h. The apparent decrease in the amount of maytansine in these later two conjugates may reflect partly or predominantly clearance of more highly loaded species instead of linker instability, a phenomenon that is difficult to distinguish for conjugates that are a mixture of differently loaded species.

In a xenograft experiment on the MMTV-HER2 Fo5 cells, the DM3 disulfide conjugate and the DM1 SMCC (monosulfide) conjugate showed better activity than the most hindered DM4 disulfide conjugate and the original DM1 (SPP) conjugate at equivalent doses. Single dose toxicity studies showed that the SMCC DM1 conjugate was well tolerated at doses significantly higher than required to give therapeutic responses in the xenograft experiment, but did not show data for the DM3 disulfide conjugate. Good activity for the SMCC DM1 conjugate both *in vitro* and *in vivo* was shown for a variety of tumor cell lines, including some complete regressions for up to 125 days in xenografts.

The trastuzumab SMCC DM1 conjugate (PRO132365), commonly referred to as T-DM1, in now making its way through clinical trials. In a Phase I trial in advanced breast cancer with weekly dosing for three weeks, and MTD of 3.6 mg kg$^{-1}$ has been established.[77] DLTs observed were thrombocytopenia and neutropenia; neuropathy, transaminase elevations, and infusion-related symptoms were also seen. Six out of 16 patients had partial responses, and five additional patients had stable disease. In a Phase II trial in metastatic cancer with the same dosing there were 32% confirmed objective responses with thrombocytopenia and hypokalemia being the most commonly observed grade 3–4 toxicity.[78] Two Phase III trials are now under way, making this the first ADC to advance to Phase III in a solid tumor.

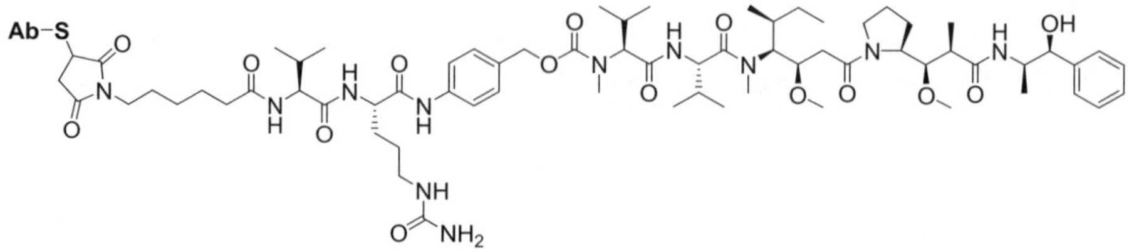

**Figure 7.5**  MalC-ValCit-PABC-MMAE (vcMMAE) conjugate.

Recent poster presentations have profiled three new, more water soluble linkers for maytansine conjugates (Figure 7.4) that allowed for higher levels of conjugation without aggregation.[79] Conjugates of the anti-EpCAM antibody B38.1 and other antibodies showed advantages in PK properties and had improved activity in xenograft experiments and against MDR-positive cell lines.

## 7.4  Auristatin Conjugates

The auristatins are close relatives of dolastatin 10, a very potent inhibitor of tubulin polymerization. Like most very potent natural products, dolastatin 10 showed insufficient activity at tolerable doses in clinical trials.[80] The initially published conjugates of the auristatins included a conjugate with both an ester and acyl hydrazone in the linker (structure not shown) but this conjugate type has not been pursued due to apparent linker instability in plasma and increased toxicity in mice.[81] The more successful conjugate was made with mono-methylauristatin E (MMAE, see Figure 7.5), which was conjugated to antibodies that had the inter-chain disulfides reduced. This is in contrast to the maytansine conjugates, where thiols were introduced by modifying lysine residues. The linker for MMAE contains a maleimide moiety for attachment to free sulfhydryls on the antibody, a protease sensitive valine–citrulline dipeptide as a cleavage site for cathepsin B (confirmed with enzymatic studies), and a self-immolative *p*-aminobenzyl carbamate[82] that releases the MMAE amine after peptide cleavage. A dipeptide linker comprised of Phe-Lys instead of Val-Cit was also explored, but appeared to be less stable in plasma. This conjugation strategy was originally published for doxorubicin conjugates.[83]

Conjugates of mc-ValCit-PABC-MMAE (vcMMAE) were originally made with the anti-CD30 antibody cAC10 and the anti-Lewis[Y] antibody cBR96,[81] which was part of the well-known BR96-DOX conjugate. $IC_{50}$ values of $5$–$90\,ng\,mL^{-1}$ were seen, depending on the specific conjugate and target cell line, with specificities of 50- to >500-fold *versus* a control conjugate. In xenograft experiments with either L2987 (lung, Lewis[Y] positive) or Karpas 299 (ALCL, CD30 positive) tumor lines, multiple doses at $0.5$–$3\,mg\,kg^{-1}$ of total conjugate injected four to six times at 4-day intervals resulted in long-term regression of the tumors with some tumors reappearing at around day 90 in most groups. Single doses of up to $30\,mg\,kg^{-1}$ ($1.1\,mg\,kg^{-1}$ of MMAE) were well tolerated; significant weight loss was seen at 40 mg kg$^{-1}$. *In vivo* activity of the cAC10-vcMMAE conjugate was also demonstrated in a L540cy Hodgkin's lymphoma model at 1 mg kg$^{-1}$, but without complete regressions,[84] while the BR96-vcMMAE conjugate has been shown to be active in an intracerebral xenograft model.[85]

The anti-CD30 cAC110-vcMMAE conjugate (SGN-35, brentuximab vedotin) has gone through complete development and is in a pivotal clinical trial for Hodgkin's lymphoma. The linker has been shown to have good stability in circulation in mice and cynomologous monkeys with half-lives for linker cleavage of 6.0 and 9.6 days, respectively.[86] Cleavage of the linker by lysosomal enzymes, including isolated cathepsin B, has been fully characterized,[87] and bystander killing has been demonstrated.[88] Drug loading has been shown to dramatically affect activity and PK properties.[89] The fully loaded conjugate had a loading of 8 moles mole$^{-1}$, but sub-maximal loading and HIC purification allowed for the separation of conjugate fractions with 2, 4 or 8 moles mole$^{-1}$, referred to as E2, E4 and E8, respectively. All of these conjugate fractions bound antigen equally well, and *in vitro* potency and the MTD were dependent on the MMAE concentration, but xenograft studies gave unexpected results. The E8 conjugate did not give better activity when dosed at twice the amount of MMAE due to a three- to five-fold faster clearance *versus* the E2 fraction, which was the fraction with the greatest therapeutic window. (Such differences have also been seen for anti-CD70 conjugates[90] and conjugates of trastuzumab.[91]) Careful control of the reduction conditions used to generate the free cysteines needed for conjugation lead to partial control of the sites and degree of conjugation, although heterogeneity and modest yield remained as practical issues.[92]

In another attempt to control the sites and degree of conjugation some of the cysteines were replaced with serines.[93] This approach gave largely homogeneous conjugates in high yield. The resultant conjugates with loadings of 2 or 4 moles mole$^{-1}$ have preclinical pharmacology very similar to the previously conjugates of cAC10, including PK and MTD.

An additional approach that has been used with MMAE conjugates is the substitution of unpaired cysteines for other aminoacids in an antibody. This was done previously by finding sites of introduction that leave the cysteine protected from the usual disulfide formation during antibody production and processing,[94] or by finding sites where the blocked cysteine can be selectively reduced.[95] Most recently it has been shown that other sites can be used for cysteine introduction when the antibody is more strongly reduced and the natural cysteine disulfides are allowed to selectively reform, leaving the introduced cysteines free for conjugation.[4] This approach has been successfully applied to the anti-MUC16 (ovarian) antibody hu3A5 and mcMMAE to make what is referred to as a THIOMAB conjugate. The activity of these conjugates is comparable in activity to that of the traditional conjugates made by reduction of the naturally occurring cysteine, but is better tolerated in both rats and cynomologous monkeys.

In a Phase I clinical trial of the cAC10-mc-ValCit-MMAE conjugate made by the original conjugation method (SGN-35) for Hodgkin's lymphoma (HL) and other CD30-positive malignancies with dosing every three weeks, the MTD was $1.8\,\text{mg}\,\text{kg}^{-1}$ with the most common, non-infusion-related toxicity being neutropenia that was dose related.[96] At doses of $1.2\,\text{mg}\,\text{kg}^{-1}$ or higher, 46% of patients obtained an objective response, with 25% CRs. The median duration of response was 22 weeks. A second Phase I trial with weekly dosing is ongoing in HL and systemic anaplastic large cell lymphoma, with seven of eight evaluable patients at the highest doses (0.8 and 1.0) obtaining CRs.[97] Dosing was continued at $1.2\,\text{mg}\,\text{kg}^{-1}$.

Modification of the MMAE structure has lead to additional auristatin conjugates with interesting properties. MMAF is an analog of MMAE where the norephedrine moiety has been replaced with a phenylalanine (hence the "F" in the name) that is negatively charged at physiological pH (Figure 7.6).[98] This charge inhibits membrane penetration, making MMAF itself relatively non-toxic *in vitro* ($IC_{50} > 30\,\text{nM}$), even though the methyl ester, which is presumably cleaved to the acid intracellularly, is extremely cytotoxic ($IC_{50} < 0.01\,\text{nM}$ *versus* MMAE at $0.1$–$2\,\text{nM}$). MMAF is not susceptible to MDR, but conjugates containing MMAF are devoid of bystander killing.

A conjugate of the anti-CD30 antibody cAC10 and MMAF with the mc-ValCit-PABC linkers was examined.[98] $IC_{50}$ values on a panel of targeted cells

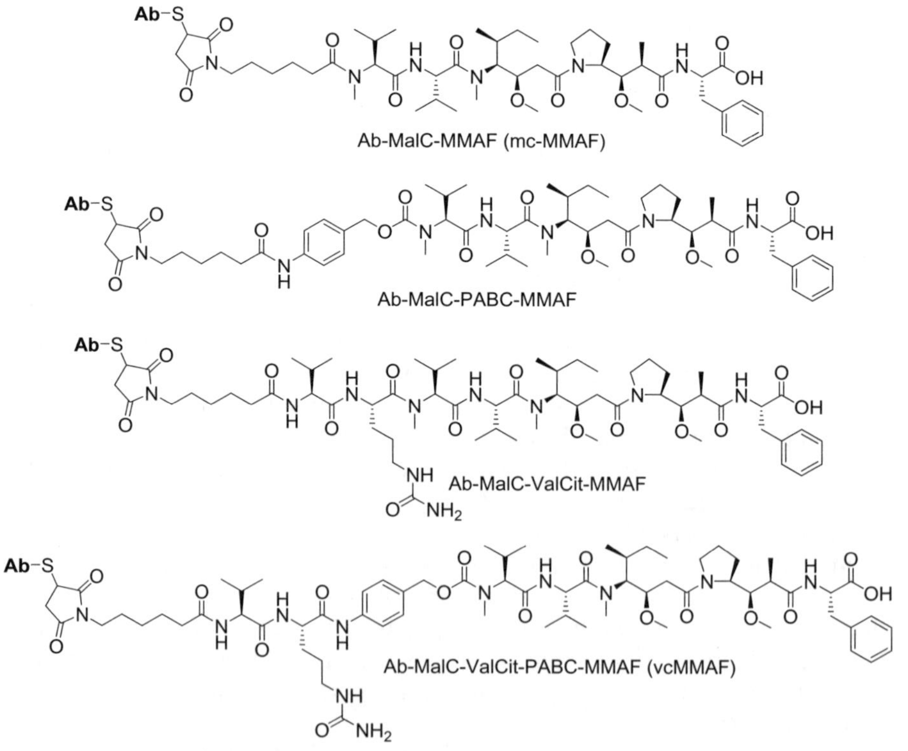

**Figure 7.6**  MMAF conjugates.

averaged 0.034 nM, with >700-fold selectivity *versus* a non-targeted cell line. The other conjugates shown in Figure 7.6 were intended as control conjugates to show the need for the self-immolative linker. The mc-ValCit-MMAF conjugate was indeed inactive, thus demonstrating the need for the PABC spacer, but the presumably non-cleavable mc-PABC-MMAF and the mc-MMAF conjugates had $IC_{50}$ values reduced by only a factor of 2. LCMS analysis of the active species released from the mc-MMAF conjugate incubated with lysomal extracts revealed the presence of cysteine-mc-MMAF that results from complete proteolysis of the antibody. In contrast, active conjugates of MMAE required the proteolytically sensitive mc-ValCit-PABC linker. It therefore appears that either the SAR of MMAE does not tolerate substitution at the amine nitrogen or the complete catabolism of the antibody bearing the non-cleavable linkers attached to MMAE does not proceed at a sufficient rate to permit effective drug release.

The favorable activity of the MMAF conjugates has been confirmed in xenograft experiments, where both the mc-MMAF and the mc-ValCit-PABC-MMAF conjugates showed some activity *versus* Karpas 299 xenografts at $1\,\text{mg}\,\text{kg}^{-1}$ single dose and gave complete regressions up to 100 days at $2\,\text{mg}\,\text{kg}^{-1}$. Activity was also seen with this cell line in a disseminated model, although doses of $8\,\text{mg}\,\text{kg}^{-1}$ were required for good activity. Although good activity was seen in xenografts of hematopoietic cell lines, solid tumors were stated to be less responsive in general to the non-cleavable linker.

The mc-MMAF conjugate has been made with the previously mentioned antibodies with missing disulfides (Cys to Ser substitution).[93,99] The conjugate with a loading of 4 moles mole$^{-1}$ on only the cysteines that normally serve as the disulfides between the heavy and light chains has been profiled in detail. Dual labeling experiments indicate good stability of the conjugate with respect to loss of auristatin, and half-lives of 2.3 and 2.5 days for the auristatin and antibody indicated good overall stability in circulation. Significant accumulation of free auristatin (the cysteine adduct of mc-MMAF) in tumor tissue was demonstrated.

The mc-MMAF conjugates suffer from a small degree of linker instability due to the use of a maleimide and the very slow reversibility of the Michael addition,[100] and this instability would presumably apply to all maleimide conjugates. To overcome this minor instability, MMAF derivatives bearing bromoacetamides instead of a maleimides on the terminal nitrogen of the 6-carbon (caproyl) chain on the N-terminus of MMAF have been made. The maleimide and corresponding bromoacetamide conjugates with either the cleavable (mc- or bac-ValCit-PABC) or non-cleavable (mc- or bac-) linkers showed comparable activity and tolerability in preclinical models.

Choice of antibody isotype has been addressed with MMAF conjugates.[101] Conjugates of the anti-CD70 antibody h1F6 and mc-ValCit-MMAF were made with $IgG_1$, $IgG_1v1$, $IgG_2$, $IgG_4$ and $IgG_4v3$. The $IgG_1$ and $IgG_4$ variants (v1 and v3, respectively) were made with compromised effector functions to see the effects on conjugate activity. The loading for all the conjugates was about 4 moles mole$^{-1}$ and binding constants were comparable. *In vitro* potency tended to be best for the $IgG_1$, $IgG_1v1$ and $IgG_2$ conjugates, while the $IgG_4$ and $IgG_4v3$ conjugates tended to be slightly less potent, with $IC_{50}$ values ranging from about 50 pM to about 500 pM across a panel of six different cell lines. PK data indicated that the two $IgG_4$ antibodies had slightly reduced half-lives (33–50%). The MTD of the $IgG_2$ conjugate appeared to be slightly higher (50%) than the other four isotypes. At equivalent doses in xenograft experiments with four different cell lines, the $IgG_1v1$ conjugate appeared to have marginally better activity, with the $IgG_2$ conjugate being the second most active. Even though the $IgG_1v1$ and the $IgG_2$ conjugates appeared to have advantages over the other conjugates, the differences were not dramatic and it is not certain if these results speak directly to benefits in humans, where the interactions with the various isotypes will be different. MMAF conjugates of h1F6 with the non-cleavable linker (mc-MMAF) have also been shown to have good *in vitro* and *in vivo* activity against a panel of tumor cell lines, including renal cell, glioblastoma, Hodgkin's disease, and multiple myeloma.[90] An anti-CD70 MMAF conjugate is in Phase I clinical trials for NHL.

The favorable activity of the MMAF derivatives has led to the exploration of new derivatives that are modified at the carboxy terminal with potentially cleavable dipeptide linkers attached as amides.[102] Auristatin F with a dimethyl-substituted amine (AF, see Figure 7.7) was mostly used for this work, and two other derivatives with the phenylalanine substituted by either methionine (AM derivatives) or tryptophan (AW derivatives) were also examined (Figure 7.7, AF, AM, and AW differ at R). The dipeptides examined contained mostly aliphatic residues at $AA_1$ and amino acids at $AA_2$ that would discourage cleavage between $AA_1$ and $AA_2$, such as D-amino acids and proline, although valine and lysine were also included due to their success in the earlier dipeptide linkers.

Such carboxy-terminus linked conjugates of the anti-CD70 antibody 1F6 gave $IC_{50}$ values *in vitro* for two renal cell carcinoma lines in the range of 3–56 ng mL$^{-1}$ of antibody component, except when $AA_1$ was beta-alanine, phenylglycine, or proline, which were significantly less potent, as could be anticipated. This is comparable to the mc-ValCit-PABC-MMAF conjugate, which had $IC_{50}$ values of 6 and 9 ng mL$^{-1}$. The MTD in mice determined for these conjugates ranged from $<50$ to $>150$ mg kg$^{-1}$ of antibody component based on weight loss and other overt toxicities, compared to the MMAF

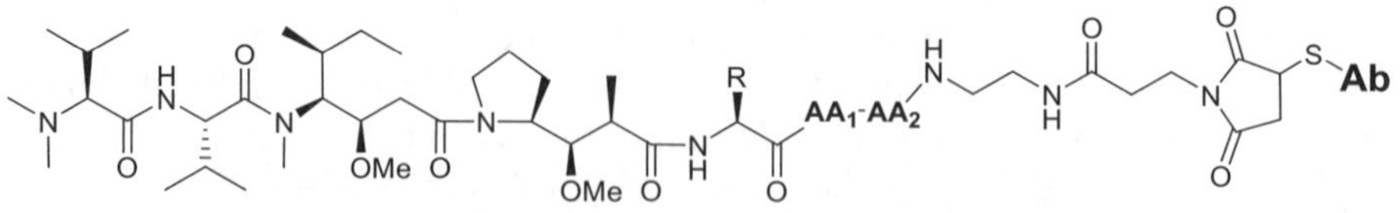


R = $CH_2Ph$: AF Dipeptide Antibody Conjugate
R = $CH_2CH_2SCH_3$: AM Dipeptide Antibody Conjugate
R = $CH_2$-3-inodolyl: AW Dipeptide Antibody Conjugate

**Figure 7.7** Carboxy-linked auristatin conjugates.

conjugate at $50\,\text{mg}\,\text{kg}^{-1}$. Interestingly, the three less potent conjugates were no better tolerated on average than those possessing more potent *in vitro* activity.

Some of the conjugates were examined in xenograft models of RCC and glioblastoma. The most favorable example appeared to be the AF Asn-(D)Lys conjugate; however, a complete evaluation of the series would require dosing at fractions of clinically relevant MTDs. At doses of $3\,\text{mg}\,\text{kg}^{-1}$ IP, AF Asn-(D)Lys conjugate gave better *in vivo* activity than the previously reported mc-ValCit-PABC-MMAF conjugate, but it has double the MTD. Additional improvements in the therapeutic window were claimed with the corresponding AM Asn-(D)-Lys and AW Asn-(D)-Lys conjugates. Further characterization of these conjugate will be of significant interest.

A second approach was briefly examined to conjugate the AF derivative through the carboxy terminus.[103] An amide was made with a *p*-phenylene diamine to allow for acylation with the ValCit dipeptide with the opposite orientation (Figure 7.8). Cleavage of the cathepsin-labile dipeptide would release an aniline that might decompose *in vivo* to release the free AF derivative. This conjugate on the anti-CD70 antibody m1F6 was compared to the vcMMAF conjugate. Both conjugates had $IC_{50}$ values as low as 0.1 nM of auristatin and showed comparable activity in xenograft models of RCC.

These various types of auristatin conjugates, especially the earlier ones, have been applied to numerous antibodies. In an examination of various possible antigens for B-cell lymphoma (non-Hodgkin's) the mc-ValCit-PABC-MMAE and the mc-MMAF conjugates of anti-gp120, anti-CD79b and anti-CD22 antibodies were examined in a Ramos xenograft study.[72] With the non-cleavable linker, the anti-CD22 conjugates were best, the anti-CD79b conjugates were less effective, and the anti-gp120 control conjugates were inactive (cf. results with maytansine conjugates.) *In vitro* activity of conjugates targeting the anti-B-cell maturation antigen (BCMA) has also been demonstrated.[104]

Several auristatin conjugates have been examined for prostate cancer. An anti-PSMA conjugate of vcMMAE showed pM *in vitro* activity and was active in a survival model with 40% survivors at day 500.[105] An anti-E-selectin vcMMAE conjugate also showed preclinical activity *in vitro* and *in vivo*.[106] MMAE and MMAF conjugates of an anti-STEAP1 antibody have also shown

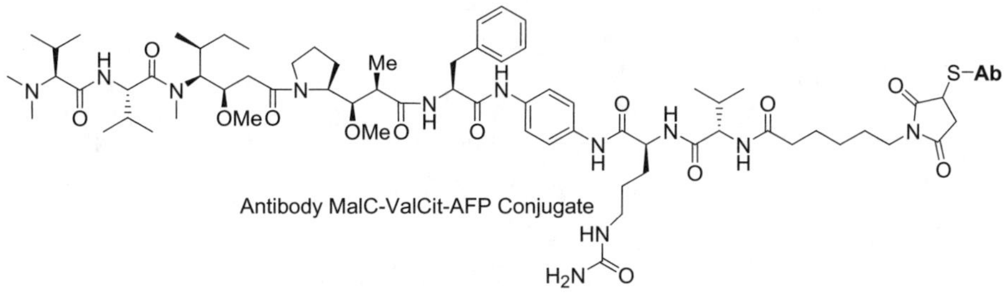

**Figure 7.8**  C-Linked auristatin phenylene diamine conjugate.

favorable activity in preclinical models,[107] as well as the vcMMAE conjugate of an anti-TMEFF2 antibody.[108]

A CR011-vcMMAE conjugate that targets a glycoprotein found on melanoma cells, GPMNB, has shown good activity and is currently in clinical trials. Even though modest $IC_{50}$ values of 200–300 ng mL$^{-1}$ were seen *in vitro*, a good dose response at doses of 0.3–10 mg kg$^{-1}$ was seen in xenografts, with 6/6 animals being tumor-free at 48 days for doses of 2.5 mg kg$^{-1}$ and above.[109] A half-life in mice of over 10 days was observed, which is consistent with the lack of schedule dependency seen for dosing.[110] In a Phase I trial an MTD of 1.88 mg kg$^{-1}$ every 3 weeks was determined.[111] The dose-limiting toxicities were rash, desquamation and neutropenia. There was one PR and seven SD out of nine evaluable patients at the higher doses. The relatively short half-life of 29 h seen for conjugate (41 h for antibody) has prompted additional Phase I studies with more frequent dosing.[112] A vcMMAF conjugate for melanoma targeting melanotransferrin/p97 has demonstrated good *in vitro* cytotoxicity.[113]

Additional conjugates with published preclinical data include an anti-Tim1 vcMMAE conjugate for ovarian and renal cell carcinoma,[114] an anti-EphA2 mc-MMAF conjugate for solid tumors,[115] an anti-EphB2 vcMMAE conjugate for colorectal cancer,[116] an anti-CD133/prominin-1 vcMMAF conjugate for hepatocellular and gastric cancers,[117] and an anti-Her2/neu conjugate of trastuzumab and vcMMAF.[91]


## 7.5 Duocarmycin Conjugates

The class of potent cytotoxic agents that has been investigated as ADCs by the largest number of companies may well be the CC-1065/duocarmycin family of cyclopropylindole DNA alkylating agents.[118] Conjugates of the original member of this family, CC-1065, were first made in collaborations with their discoverer, Upjohn (Figure 7.9), and the structures of the conjugated species were similar to derivatives such as adozelesin that they had investigated in clinical trials. An example of this is the conjugate published by Immunogen, which contains the benzo-indole (CBI) modification discovered by Boger that has become a popular structural variation.[119] Even though conjugates of B4 (anti-CD19) and N901 (anti-CD56) showed good preclinical activity, these

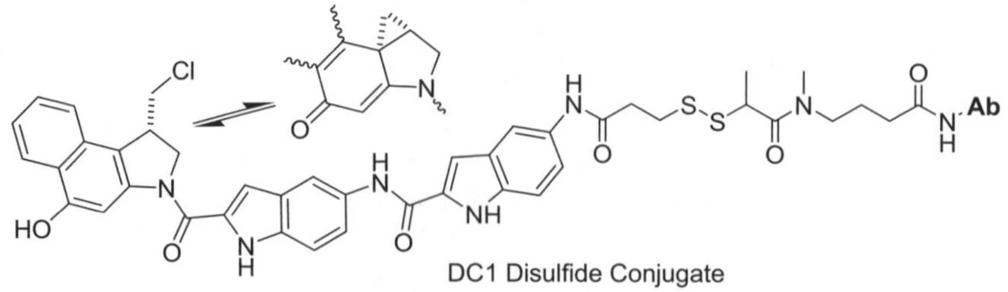

**Figure 7.9** Initial CC-1065 conjugate.

initial conjugates have not been pursued due to practical issues arising from the gross insolubility of this class of cytotoxic agents.

Immunogen has continued their pursuit of these derivatives as evidenced by an ACS poster presentation.[120] This presentation described the increased solubility and stability of the CBI derivatives by making a phosphate prodrug (Figure 7.10). Increased solubility was also achieved by adding PEG to the linker. These conjugates were potently cytotoxic (sub pM) *in vitro* as anti-CD19 or anti-CanAg conjugates, but the phosphates required cleavage with added phosphatase for maximum potency. Conjugates bearing both the disulfide (SPP and SPDB) and monosulfide (SMCC, non-cleavable) linkers had good activity on the COLO 205 cell line.

Seattle Genetics has published *in vitro* data on a variety of conjugates in this class.[121] One example, shown in Figure 7.11, contains the duocarmycin tri-methoxyindole, an aza-CBI that may improved solubility, a cathepsin-clea-vable dipeptide with a *p*-amino benzyl ether linkage that is known to be cleaved less efficiently than a carbamate,[122] and a PEG unit in the linker for increased solubility. The cAC10 conjugate had an IC$_{50}$ around 20 nM *versus* Karpas 299 cells. A company named Syntarga has also published examples of CBI (duocarmycin) conjugates.[123]

Medarex (now part of BMS) has actively pursued a broad variety of conjugates in this class. The most significant of these are the conjugates of

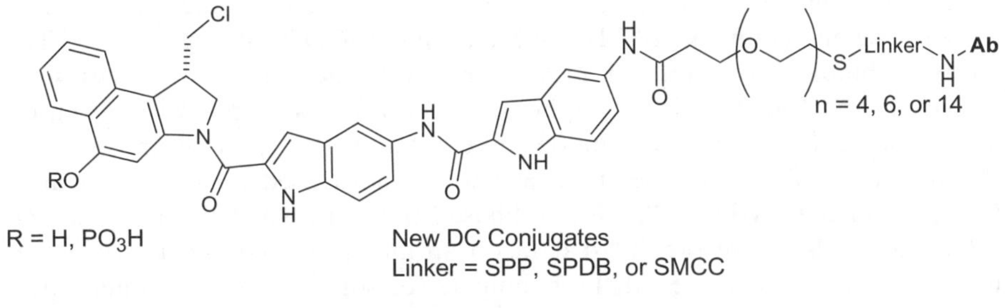

**Figure 7.10**  Solubilized CC-1065 (CBI) conjugates.

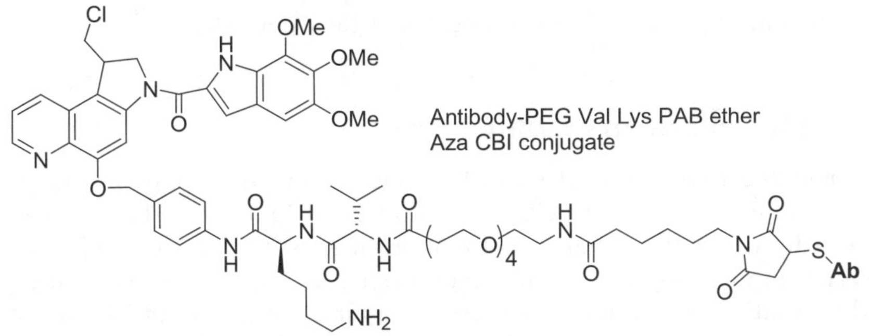

**Figure 7.11**  Duocarmycin conjugate with Aza-CBI.

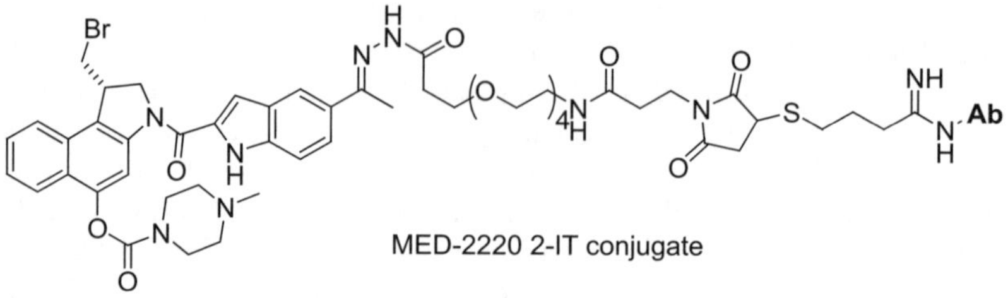

**Figure 7.12**   MED-2220.

MED-2220, which are attached to antibodies with lysines thiolated by 2-IT (2-iminothiolane, Trout's reagent; Figure 7.12). The phenol in the open form of the CBI alkylating unit is protected as a carbamate of *N*-methyl piperazine, which likely increases stability and adds solubility. The cyclopropane precursor contains a bromine instead of chlorine for undisclosed reasons. The linker is attached via an acyl hydrazone similar to the AcBut linker in several calicheamicin conjugates. The linker contains a short PEG unit that would also likely improve solubility.

An anti-PSMA conjugate of MED-2220 has been profiled with an $IC_{50}$ *in vitro* of 0.9 nM *versus* 12 nM for a control conjugate.[124] The antibody has a binding constant of 0.8 nM and is well internalized in a 3 h assay. The conjugate is active in LNCaP xenografts at 0.3 µmol kg$^{-1}$ with some of the tumors being cured. A second conjugate of MED-2220 is an anti-CD70 conjugate.[119,125] The antibody binding constant is 8.2 nM and is well internalized. It shows good activity in the Raji Burkitt's lymphoma xenograft model, but without cures. Two other conjugates with peptide linkers had better activity and a wider therapeutic window, but the structures were undisclosed. One of their anti-CD70 conjugates, MDX-1203, is in a Phase I trial for renal cell carcinoma and NHL. They also have presented data on an anti-CD19 conjugate for leukemia[126] and on an anti-mesothelin conjugate for solid tumors.[127] Other variations in the structure of the duocarmycin and linker have also been presented attached to anti-PSMA and anti-CD70 antibodies, but without the linker structures.[128] An additional presentation shows the importance of having a cleavable linker for anti-CD70 conjugates of their derivatives.[129]

## 7.6   Miscellaneous Conjugates

Although the four major classes of conjugates are the ones that are most actively being pursued in the clinic, additional conjugates have been made that are worthy of mention. Some of the more interesting are the conjugates of doxorubicin (adriamycin), a drug that dates back to the 1950s and that is still used to treat a wide variety of cancers. A hydrazone-linked anti-Lewis$^{Y}$ conjugate of doxorubicin, BR96-DOX, made it into a Phase II clinical trial with some indication of activity, but with no objective responses.[130] Doxorubicin continues to be a common cytotoxic agent to explore new conjugation strategies, as is was in the elucidation of the ValCit and PheLys peptide linkers that Seattle Genetics has used.[131] In addition, the more potent derivatives of doxorubicin 2-pyrrolino-[131] and morpholino-doxorubicin[132] have shown interesting preclinical activity as antibody conjugates.

The taxanes also appear to be of some interest for conjugation. An earlier example with trastuzumab and a novel light-cleavable linker has been published.[133] Even more interesting are the conjugates of new taxanes derivatives that are 10-fold more potent on non-MDR cells, but 100 to 1000-fold more potent on MDR-resistant cells.[134] Conjugates of anti-EGRF antibodies show nanomolar cytotoxicity *in vitro* and good effects in xenograft models of ovarian cancer.

Conjugates of the camptothecins have also been explored by various research groups. An anti-Lewis$^Y$ conjugate of a 10-aminocamptothecin had an *in vitro* $IC_{50}$ of 13 nM, and an anti-CD30 conjugate had a value of 26 nM.[135] *In vivo* activity was seen with an anti-CD70 conjugate in a renal cell carcinoma xenograft model; with an alternate glucuronide-based linker some cures were seen. Alternatively, conjugates of the camptothecin SN-38 and the anti-CEA-CAM5 antibody hMN-14 with a dipeptide self-immolative linker have been explored.[136] Half-lives in serum of up to 66 h were seen with *in vitro* $IC_{50}$ values of 5–10 nM.

Streptonigrin, a DNA-active natural product, has been conjugated to anti-CD30 and anti-CD70 antibodies.[137] Although less potent than unconjugated streptonigrin (0.4–0.8 nM), conjugates with a ValAla-*p*-aminobenzylamine self-immolative linker had $IC_{50}$ values of 2–3 nM (anti-CD30) and 17 nM (anti-CD70). A xenograft experiments against L540ch Hodgkin's lymphoma xenografts (CD70 positive) showed good activity at $10\,mg\,kg^{-1}$.

Conjugates of the Hsp90 inhibitor geldanamycin have also been published.[138] The derivative chosen for conjugation, APA-GA, is much less potent in cell culture that geldanamycin itself (180 nM *versus* 8 nM). However, an anti-HER2 conjugate with a non-cleavable linker showed an $IC_{50}$ of $40\,\mu g\,mL^{-1}$ in cell culture with 40-fold selectivity.

## 7.7 Conclusions

The goal of using antibodies to deliver highly potent but poorly tolerated cytotoxic agents directly to tumor tissue has been sought for over 25 years. Achieving success with an ADC-based therapeutic requires selective targeting by the antibody to tumor tissue, employment of a payload-linker combination that exhibits prolonged systemic stability, tight binding and internalization of the ADC into the targeted tissue, and efficient intracellular drug release at concentrations sufficient to kill the targeted tissue. ADC discovery and development is therefore an extraordinarily complex multivariate problem, and so it comes as no surprise that today there exists but one clinically approved ADC, Mylotarg® (gemtuzumab ozogamicin). Since Mylotarg's approval, interest and activities in ADC research have intensified, and there are presently roughly

20 ADCs undergoing clinical evaluation for treatment of various cancers. Significant advances have been made in payload optimization, linker design and drug release strategies, and in the development of scalable approaches towards achieving site-specific conjugations. Each of these advances has brought incremental improvements in our understanding of the complexities surrounding ADC development, and taken together these advances have helped to launch ADCs into a new era of sophistication.

# References



1. B. A. Teicher, *Curr. Cancer Drug Targets*, 2009, **9**, 982–1004.
2. PCT Intl. Appl. Pat., WO 2009089846, 2009.
3. US Pat. Appl. Publ. Pat., 2009163699, 2009.
4. J. R. Junutula, H. Raab, S. Clark, S. Bhakta, D. D. Leipold, S. Weir, Y. Chen, M. Simpson, S. P. Tsai, M. S. Dennis, Y. Lu, Y. G. Meng, C. Ng, J. Yang, C. C. Lee, E. Duenas, J. Gorrell, V. Katta, A. Kim, K. McDorman, K. Flagella, R. Venook, S. Ross, S. D. Spencer, W. L. Wong, H. B. Lowman, R. Vandlen, M. X. Sliwkowski, R. H. Scheller, P. Polakis and W. Mallet, *Nature Biotechnol.*, 2008, **26**, 925–932.
5. PCT Intl. Appl. Pat., WO 2005089809, 2005.
6. I. Tranoy-Opalinski, A. Fernandes, M. Thomas, J.-P. Gesson and S. Papot, *Anti-Cancer Agents Med. Chem.*, 2008, **8**, 618–637.
7. L. M. Hinman, P. R. Hamann, R. Wallace, A. T. Menendez, F. E. Durr and J. Upeslacis, *Cancer Res.*, 1993, **53**, 3336–3342.
8. P. F. Bross, J. Beitz, G. Chen, X. H. Chen, E. Duffy, L. Kieffer, S. Roy, R. Sridhara, A. Rahman, G. Williams and R. Pazdur, *Clin. Cancer Res.*, 2001, **7**, 1490–1496.
9. R. G. Andrews, J. W. Singer and I. D. Bernstein, *J. Exp. Med.*, 1989, **169**, 1721–1731.
10. P. R. Crocker and P. Redelinghuys, *Biochem. Soc. Trans.*, 2008, **36**, 1467–1471.
11. J. F. DiJoseph, K. Khandke, M. M. Dougher, D. Y. Evans, D. C. Armellino, P. R. Hamann and N. K. Damle, in *Hematology Meeting Reports*, Bologna, Italy, 2008, **5**(6), 74–77.
12. P. R. Hamann, L. M. Hinman, I. Hollander, C. F. Beyer, D. Lindh, R. Holcomb, W. Hallett, H.-R. Tsou, J. Upeslacis, D. Shochat, A. Mountain, D. A. Flowers and I. Bernstein, *Bioconjugate Chem.*, 2002, **13**, 47–58.
13. P. R. Hamann, L. M. Hinman, C. F. Beyer, D. Lindh, J. Upeslacis, D. A. Flowers and I. Bernstein, *Bioconjugate Chem.*, 2002, **13**, 40–46.
14. J. A. Dowell, J. Korth-Bradley, H. Liu, S. P. King and M. S. Berger, *J. Clin. Pharmacol.*, 2001, **41**, 1206–1214.
15. E. L. Sievers, F. R. Appelbaum, R. T. Spielberger, S. J. Forman, D. Flowers, F. O. Smith, K. Shannon-Dorcy, M. S. Berger and I. D. Bernstein, *Blood*, 1999, **93**, 3678–3684.

16. W. J. Kell, A. K. Burnett, R. Chopra, J. A. L. Yin, R. E. Clark, A. Rohatiner, D. Culligan, A. Hunter, A. G. Prentice and D. W. Milligan, *Blood*, 2003, **102**, 4277–4283.
17. L. H. Leopold, M. S. Berger, S.-C. Cheng, J. E. Cortes-Franco, F. J. Giles and E. H. Estey, *Clin. Adv. Hematol. Oncol.*, 2003, **1**, 220–225.
18. L. H. Leopold, M. S. Berger and J. Feingold, *Clin. Lymphoma*, 2002, **2**, S29–S34.
19. B. Versluys, R. Bhattacharaya, C. Steward, J. Comish, A. Oakhill and N. Goulden, *Blood*, 2004, **103**, 1968.
20. A. L. Taksin, O. Legrand, E. Raffoux, T. de Revel, X. Thomas, N. Contentin, R. Bouabdallah, C. Pautas, P. Turlure, O. Reman, C. Gardin, B. Varet, S. de Botton, F. Pousset, H. Farhat, S. Chevret, H. Dombret and S. Castaigne, *Leukemia*, 2007, **21**, 66–71.
21. R. A. Larson, E. L. Sievers, E. A. Stadtmauer, B. Lowenberg, E. H. Estey, H. Dombret, M. Theobald, D. Voliotis, J. M. Bennett, M. Richie, L. H. Leopold, M. S. Berger, M. L. Sherman, M. R. Loken, J. J. M. van Dongen, I. D. Bernstein, F. R. Appelbaum, M. Boogaerts, S. Castaigne, P. Huijgens, R. Spielberger, M. Tallman, C. Bernasconi, J. L. Harousseau, C. Karanes, A. List, D. C. Roy, A. Goldstone, F. Mandelli, M. Schuster, M. Gobbi, P. Mineur, S. Tarantolo, M. Andre, A. Burnett, N. Cambier, P. Cassileth, J. Esteve, M. Gramatzki, G. Heil, G. Juliusson, S. Tura, G. Ehninger, G. A. Granena, J. Karp, J. P. Marie, A. Parreira, C. Paul, K. Rai, G. Schiller, J. Sierra, B. Simonsson, L. Stenke, M. Wernli, R. Willemze, M. Aglietta, D. Clausen, J. Conde, S. Coutre, M. N. Fernandez, D. Fiere, U. Hess, H. A. Horst, W. Linkesch, D. Mediavilla, M. Minden, F. Nobile, D. Schenkein and A. Wahlin, *Cancer (Hoboken, NJ)*, 2005, **104**, 1442–1452.
22. K. Lang, J. Menzin, C. C. Earle and R. Mallick, *Am. J. Health-System Pharm.*, 2002, **59**, 941–948.
23. R. Stasi, *Expert Opin. Biol. Ther.*, 2008, **8**, 527–540.
24. L. Pagano, *Oncogene*, 2007, **26**, 3679–3690.
25. J. F. DiJoseph, D. C. Armellino, M. M. Dougher, A. Kunz, E. R. Boghaert, P. R. Hamann, K. Zinkewich-Peotti and N. K. Damle, *Blood, (ASH Annual Meeting Abstracts)*, 2004, **104**, 2490.
26. J. F. DiJoseph, A. Popplewell, S. Tickle, H. Ladyman, A. Lawson, A. Kunz, K. Khandke, D. C. Armellino, E. R. Boghaert, P. R. Hamann, K. Zinkewich-Peotti, S. Stephens, N. Weir and N. K. Damle, *Cancer Immunol. Immunother.*, 2005, **54**, 11–24.
27. J. F. DiJoseph, D. C. Armellino, E. R. Boghaert, K. Khandke, M. M. Dougher, L. Sridharan, A. Kunz, P. R. Hamann, B. Gorovits, C. Udata, J. K. Moran, A. G. Popplewell, S. Stephens, P. Frost and N. K. Damle, *Blood*, 2004, **103**, 1807–1814.
28. J. F. DiJoseph, M. E. Goad, M. M. Dougher, E. R. Boghaert, A. Kunz, P. R. Hamann and N. K. Damle, *Clin. Cancer Res.*, 2004, **10**, 8620–8629.

29. A. Advani, E. Gine, C. Gisselbrecht, A. Rohatiner, S. Rosen, M. Smith, J. Boni, C. Lejeune and H. Patel, *Blood* (*ASH Annual Meeting Abstracts*), 2005, **106**, 230.

30. L. Fayad, H. Patel, G. Verhoef, M. Czuczman, J. Foran, E. Gine, A. Rohatiner, M. R. Smith, M. Shapiro and A. Advani, *Blood* (*ASH Annual Meeting Abstracts*), 2006, **108**, 2711.

31. J. F. DiJoseph, M. M. Dougher, L. B. Kalyandrug, D. C. Armellino, E. R. Boghaert, P. R. Hamann, J. K. Moran and N. K. Damle, *Clin. Cancer Res.*, 2006, **12**, 242–249.

32. L. Fayad, H. Patel, G. Verhoef, M. R. Smith, P. W. M. Johnson, M. S. Czuczman, B. Coiffier, G. Hess, E. Gine, A. Advani, F. Offner, E. R. Vandendries, M. Shapiro and N. H. Dang, *Blood* (*ASH Annual Meeting Abstracts*), 2008, **112**, 266.

33. P. R. Hamann, L. M. Hinman, C. F. Beyer, L. M. Greenberger, C. Lin, D. Lindh, A. T. Menendez, R. Wallace, F. E. Durr and J. Upeslacis, *Bioconjugate Chem.*, 2005, **16**, 346–353.

34. P. R. Hamann, L. M. Hinman, C. F. Beyer, D. Lindh, J. Upeslacis, D. Shochat and A. Mountain, *Bioconjugate Chem.*, 2005, **16**, 354–360.

35. S. Y. Chan, A. N. Gordon, R. E. Coleman, J. B. Hall, M. S. Berger, M. L. Sherman, C. B. Eten and N. J. Finkler, *Cancer Immunol. Immunother.*, 2003, **52**, 243–248.

36. H. M. Prinssen, C. F. M. Molthoff, R. H. M. Verheijen, T. J. Broadhead, P. Kenemans, J. C. Roos, Q. Davies, A. C. Van Hof, M. Frier, W. Den Hollander, A. J. Wilhelm, T. S. Baker, M. Sopwith, E. M. Symonds and A. C. Perkins, *Cancer Immunol. Immunother.*, 1998, **47**, 39–46.

37. E. R. Boghaert, L. Sridharan, D. C. Armellino, K. M. Khandke, J. F. DiJoseph, A. Kunz, M. M. Dougher, F. Jiang, L. B. Kalyandrug, P. R. Hamann, P. Frost and N. K. Damle, *Clin. Cancer Res.*, 2004, **10**, 4538–4549.

38. R. A. Herbertson, N. C. Tebbutt, F.-T. Lee, D. J. MacFarlane, B. Chappell, N. Micallef, S.-T. Lee, T. Saunder, W. Hopkins, F. E. Smyth, D. K. Wyld, J. Bellen, D. S. Sonnichsen, M. W. Brechbiel, C. Murone and A. M. Scott, *Clin. Cancer Res.*, 2009, **15**, 6709–6715.

39. A. M. Scott, N. Tebbutt, F.-T. Lee, T. Cavicchiolo, Z. Liu, S. Gill, A. M. T. Poon, W. Hopkins, F. E. Smyth, C. Murone, D. MacGregor, A. T. Papenfuss, B. Chappell, T. H. Saunder, M. W. Brechbiel, I. D. Davis, R. Murphy, G. Chong, E. W. Hoffman and L. J. Old, *Clin. Cancer Res.*, 2007, **13**, 3286–3292.

40. E. R. Boghaert, L. Sridharan, K. M. Khandke, D. Armellino, M. G. Ryan, K. Myers, R. Harrop, A. Kunz, P. R. Hamann, K. Marquette, M. Dougher, J. F. Di Joseph and N. K. Damle, *Int. J. Oncol.*, 2008, **32**, 221–234.

41. H. N. Lode, R. A. Reisfeld, R. Handgretinger, K. C. Nicolaou, G. Gaedicke and W. Wrasidlo, *Cancer Res.*, 1998, **58**, 2925–2928.

42. K. M. Bernt, A. Prokop, N. Huebener, G. Gaedicke, W. Wrasidlo and H. N. Lode, *Bioconjugate Chem.*, 2009, **20**, 1587–1594.

43. S. V. Smith, *Curr. Opin. Mol. Ther.*, 2004, **6**, 666–674.

44. W. C. Widdison, S. D. Wilhelm, E. E. Cavanagh, K. R. Whiteman, B. A. Leece, Y. Kovtun, V. S. Goldmacher, H. Xie, R. M. Steeves, R. J. Lutz, R. Zhao, L. Wang, W. A. Blaettler and R. V. J. Chari, *J. Med. Chem.*, 2006, **49**, 4392–4408.

45. R. V. J. Chari, B. A. Martell, J. L. Gross, S. B. Cook, S. A. Shah, W. A. Blattler, S. J. McKenzie and V. S. Goldmacher, *Cancer Res.*, 1992, **52**, 127–131.

46. L. Wang, G. Amphlett, W. A. Blattler, J. M. Lambert and W. Zhang, *Protein Sci.*, 2005, **14**, 2436–2446.

47. H. Xie, C. Audette, M. Hoffee, J. M. Lambert and W. A. Blaettler, *J. Pharmacol. Exp. Ther.*, 2004, **308**, 1073–1082.

48. C. Liu, B. M. Tadayoni, L. A. Bourret, K. M. Mattocks, S. M. Derr, W. C. Widdison, N. L. Kedersha, P. D. Ariniello, V. S. Goldmacher, J. M. Lambert, W. A. Blattler and R. V. Chari, *Proc. Natl Acad. Sci. U. S. A.*, 1996, **93**, 8618–8623.

49. Y. V. Kovtun, C. A. Audette, Y. Ye, H. Xie, M. F. Ruberti, S. J. Phinney, B. A. Leece, T. Chittenden, W. A. Blaettler and V. S. Goldmacher, *Cancer Res.*, 2006, **66**, 3214–3221.

50. A. W. Tolcher, L. Ochoa, L. A. Hammond, A. Patnaik, T. Edwards, C. Takimoto, L. Smith, J. de Bono, G. Schwartz, T. Mays, Z. L. Jonak, R. Johnson, M. DeWitte, H. Martino, C. Audette, K. Maes, R. V. J. Chari, J. M. Lambert and E. K. Rowinsky, *J. Clin. Oncol.*, 2003, **21**, 211–222.

51. H. K. Erickson, P. U. Park, W. C. Widdison, Y. V. Kovtun, L. M. Garrett, K. Hoffman, R. J. Lutz, V. S. Goldmacher and W. A. Blaettler, *Cancer Res.*, 2006, **66**, 4426–4433.

52. H. K. Erickson, W. C. Widdison, M. F. Mayo, K. Whiteman, C. Audette, S. D. Wilhelm and R. Singh, *Bioconjugate Chem.*, 2010, **21**, 84–92.

53. C. N. Carrigan, M. F. Mayo, H. Xie, M. Murphy, J. M. Lambert, R. V. J. Chari, G. Payne and R. J. Lutz, *AACR-EORTC-NCI Meeting*, San Francisco, CA, 2007, B53.

54. M. F. Mayo, H. Xie, H. K. Erickson, P. S. Wunderli, L. M. Garrett, K. R. Whiteman, B. A. Leece and R. J. Lutz, in *Annual Proceedings of the AACR*, Philadelphia, PA, 2005.

55. K. Sankhala, A. Mita, J. Sarantopoulos, A. Richart, M. M. Mita, C. Lin, L. Wood, J. Watermill, S. Zildjian and A. Qin, *AACR-EORTC-NCI Meeting*, San Francisco, CA, 2007, B70.

56. L. W. Goff, K. Papadopoulos, J. A. Posey, A. T. Phan, A. Patnaik, J. G. Miller, S. Zildjian, J. J. O'Leary, A. Qin and A. Tolcher, *J. Clin. Oncol. (Meeting Abstracts).*, 2009, **27**, e15625.

57. S. Ross, S. D. Spencer, I. Holcomb, C. Tan, J. Hongo, B. Devaux, L. Rangell, G. A. Keller, P. Schow, R. M. Steeves, R. J. Lutz, G. Frantz, K. Hillan, F. Peale, P. Tobin, D. Eberhard, M. A. Rubin, L. A. Lasky and H. Koeppen, *Cancer Res.*, 2002, **62**, 2546–2553.

58. M. D. Henry, S. Wen, M. D. Silva, S. Chandra, M. Milton and P. J. Worland, *Cancer Res.*, 2004, **64**, 7995–8001.

59. M. D. Galsky, M. Eisenberger, S. Moore-Cooper, W. K. Kelly, S. F. Slovin, A. DeLaCruz, Y. Lee, I. J. Webb and H. I. Scher, *J. Clin. Oncol.*, 2008, **26**, 2147–2154.

60. P. Tassone, A. Gozzini, V. Goldmacher, M. A. Shammas, K. R. Whiteman, D. R. Carrasco, C. Li, C. K. Allam, S. Venuta, K. C. Anderson and N. C. Munshi, *Cancer Res.*, 2004, **64**, 4629–4636.

61. K. R. Whiteman, O. Ab, L. M. Bartle, K. Foley, V. S. Goldmacher and R. J. Lutz, *Annu. Proc. AACR,* 2008, 2146.

62. K. R. Whiteman, M. F. Murphy, K. Prince Kohan, W. Sun, C. N. Carrigan, M. F. Mayo, Y. Li and R. J. Lutz, *Annu. Proc. AACR*, 2008, 2135.

63. P. Lorigan, P. Woll, M. O'Brien, Y. Clinch, K. Donaldson, H. Xie, P. Wunderli, J. Hutchison and R. J. Fram, *AARC-NCI-EORTC*, Philadelphia, PA, 2005, B97.

64. J. McCann, F. V. Fossella, M. A. Villalona-Calero, A. W. Tolcher, P. Fidias, R. Raju, S. Zildjian, R. Guild and R. Fram, *ASCO Annual Meeting Proceedings*, 2007, 18084.

65. H. Ikeda, T. Hideshima, M. Fulciniti, R. J. Lutz, H. Yasui, Y. Okawa, T. Kiziltepe, S. Vallet, S. Pozzi, L. Santo, G. Perrone, Y.-T. Tai, D. Cirstea, N. S. Raje, C. Uherek, B. Daelken, S. Aigner, F. Osterroth, N. Munshi, P. Richardson and K. C. Anderson, *Clin. Cancer Res.*, 2009, **15**, 4028–4037.

66. A. Chanan-Khan, S. Jagannath, T. Heffner, D. Avigan, K. Lee, R. J. Lutz, T. Haeder, M. Ruehle, C. Uherek, A. Wartenberg-Demand, N. Munshi and K. Anderson, *Blood* (*ASH Annual Meeting Abstracts*), 2009, **114**, 1862.

67. U. Rupp, E. Schoendorf-Holland, M. Eichbaum, F. Schuetz, I. Lauschner, P. Schmidt, A. Staab, G. Hanft, J. Huober, H.-P. Sinn, C. Sohn and A. Schneeweiss, *Anti-Cancer Drugs*, 2007, **18**, 477–485.

68. F. Giles, R. Morariu-Zamfir, J. Lambert, S. Verstovsek, D. Thomas, F. Ravandi and D. Deangelo, *Blood*, 2006, **108**, 4548.

69. R. J. Lutz, H. Xie, C. Dionne, R. Steeves, V. S. Goldmacher, B. Leece, L. Bartle and R. Chari, Annual Proceedings of the AACR, 2002.

70. A. Aboukameel, A.-S. Goustin, R. Mohammad, C. Zuany-Amorim, M.-C. Bissery and A. M. Al-Katib, *Blood* (*ASH Annual Meeting Abstracts*), 2007, **110**, 2339.

71. http://clinicaltrials.gov/ct2/show/NCT00721669.

72. A. G. Polson, J. Calemine-Fenaux, P. Chan, W. Chang, E. Christensen, S. Clark, F. J. de Sauvage, D. Eaton, K. Elkins, J. M. Elliott, G. Frantz, R. N. Fuji, A. Gray, K. Harden, G. S. Ingle, N. M. Kljavin, H. Koeppen, C. Nelson, S. Prabhu, H. Raab, S. Ross, J.-P. Stephan, S. J. Scales, S. D. Spencer, R. Vandlen, B. Wranik, S.-F. Yu, B. Zheng and A. Ebens, *Cancer Res.*, 2009, **69**, 2358–2364.

73. F. Giles, R. Morariu-Zamfir, J. Lambert, S. Verstovsek, D. Thomas, F. Ravandi and D. Deangelo, *Blood* (*ASH Annual Meeting Abstracts*), 2006, **108**, 4548.

74. E. Ostermann, P. Garin-Chesa, K. H. Heider, M. Kalat, H. Lamche, C. Puri, D. Kerjaschki, W. J. Rettig and G. R. Adolf, *Clin. Cancer Res.*, 2008, **14**, 4584–4592.

75. Q. Chen, H. J. Millar, F. L. McCabe, C. D. Manning, R. Steeves, K. Lai, B. Kellogg, R. J. Lutz, M. Trikha, M. T. Nakada and G. M. Anderson, *Clin. Cancer Res.*, 2007, **13**, 3689–3695.

76. G. D. Lewis Phillips, G. Li, D. L. Dugger, L. M. Crocker, K. L. Parsons, E. Mai, W. A. Blaettler, J. M. Lambert, R. V. J. Chari, R. J. Lutz, W. L. T. Wong, F. S. Jacobson, H. Koeppen, R. H. Schwall, S. R. Kenkare-Mitra, S. D. Spencer and M. X. Sliwkowski, *Cancer Res.*, 2008, **68** 9280–9290.

77. M. Beeram, I. H. A. Burris, S. Modi, M. Birkner, S. Girish, J. Tibbitts, S. N. Holden, S. G. Lutzker and I. E. Krop, *J. Clin. Oncol.*, 2008 **26**(suppl), ASCO Mtg abst 1028.

78. C. L. Vogel, H. A. Burris, S. Limentani, R. Borson, J. O'shaughnessy, S. Vukelja, S. Agresta, B. Klencke, M. Birkner and H. Rugo, *J. Clin. Oncol.*, 2009, **27**, 1017.

79. B. Kellogg, C. Audette, L. Clancy, S. Emrich, H. Erickson, N. Fishkin, G. Jones, Y. Kovtun, E. Maloney, R. Mastice, M. Mayo, M. Okamoto, H. Pinkas, R. Singh, X. Sun, S. Wilhelm and R. Zhao, *Annu. Proc. AACR*, 2009, 5480.

80. U. Vaishampayan, M. Glode, W. Du, A. Kraft, G. Hudes, J. Wright and M. Hussain, *Clin. Cancer Res.*, 2000, **6**, 4205–4208.

81. S. O. Doronina, B. E. Toki, M. Y. Torgov, B. A. Mendelsohn, C. G. Cerveny, D. F. Chace, R. L. DeBlanc, R. P. Gearing, T. D. Bovee, C. B. Siegall, J. A. Francisco, A. F. Wahl, D. L. Meyer and P. D. Senter, *Nature Biotechnol.*, 2003, **21**, 778–784.

82. P. L. Carl, P. K. Chakravarty and J. A. Katzenellenbogen, *J. Med. Chem.*, 1981, **24**, 479–480.

83. G. M. Dubowchik, R. A. Firestone, L. Padilla, D. Willner, S. J. Hofstead, K. Mosure, J. O. Knipe, S. J. Lasch and P. A. Trail, *Bioconjugate Chem.*, 2002, **13**, 855–869.

84. J. A. Francisco, C. G. Cerveny, D. L. Meyer, B. J. Mixan, K. Klussman, D. F. Chace, S. X. Rejniak, K. A. Gordon, R. DeBlanc, B. E. Toki, C.-L. Law, S. O. Doronina, C. B. Siegall, P. D. Senter and A. F. Wahl, *Blood*, 2003, **102**, 1458–1465.

85. L. Muldoon, M. Pagel, P. Senter and E. Neuwelt, *Annu. Proc. AACR*, 2007, 3332.

86. R. J. Sanderson, M. A. Hering, S. F. James, M. M. C. Sun, S. O. Doronina, A. W. Siadak, P. D. Senter and A. F. Wahl, *Clin. Cancer Res.*, 2005, **11**, 843–852.

87. M. S. K. Sutherland, R. J. Sanderson, K. A. Gordon, J. Andreyka, C. G. Cerveny, C. Yu, T. S. Lewis, D. L. Meyer, R. F. Zabinski, S. O. Doronina, P. D. Senter, C.-L. Law and A. F. Wahl, *J. Biol. Chem.*, 2006, **281**, 10540–10547.

88. N. M. Okeley, J. B. Miyamoto, X. Zhang, R. J. Sanderson, D. R. Benjamin, E. L. Sievers, P. D. Senter and S. C. Alley, *Clin. Cancer Res.*, 2010, **16**, 888–898.

89. K. J. Hamblett, P. D. Senter, D. F. Chace, M. M. C. Sun, J. Lenox, C. G. Cerveny, K. M. Kissler, S. X. Bernhardt, A. K. Kopcha, R. F. Zabinski, D. L. Meyer and J. A. Francisco, *Clin. Cancer Res.*, 2004, **10**, 7063–7070.

90. E. Oflazoglu, I. J. Stone, K. Gordon, C. G. Wood, E. A. Repasky, I. S. Grewal, C.-L. Law and H.-P. Gerber, *Clin. Cancer Res.*, 2008, **14**, 6171–6180.

91. D. Leipold, N. Jumbe, D. Dugger, L. Crocker, W. Leach, M. Sliwkowski, D. Meyer, P. Senter and J. Tibbitts, *Annu. Proc. AACR*, 2007, 1551.

92. M. M. C. Sun, K. S. Beam, C. G. Cerveny, K. J. Hamblett, R. S. Blackmore, M. Y. Torgov, F. G. M. Handley, N. C. Ihle, P. D. Senter and S. C. Alley, *Bioconjugate Chem.*, 2005, **16**, 1282–1290.

93. C. F. McDonagh, E. Turcott, L. Westendorf, J. B. Webster, S. C. Alley, K. Kim, J. Andreyka, I. Stone, K. J. Hamblett, J. A. Francisco and P. Carter, *Protein Eng. Design Select.*, 2006, **19**, 299–307.

94. A. Lyons, D. J. King, R. J. Owens, G. R. Yarranton, A. Millican, N. R. Whittle and J. R. Adair, *Protein Eng.*, 1990, **3**, 703–708.

95. J. B. Stimmel, B. M. Merrill, L. F. Kuyper, C. P. Moxham, J. T. Hutchins, M. E. Fling and F. C. Kull Jr, *J. Biol. Chem.*, 2000, **275**, 30445–30450.

96. A. Younes, A. Forero-Torres, N. L. Bartlett, J. P. Leonard, C. Lynch, D. A. Kennedy and E. Sievers, *Blood (ASH Annual Meeting Abstracts)*, 2008, **112**, 1006.

97. N. Bartlett, A. Forero-Torres, J. Rosenblatt, M. Fanale, S. J. Horning, S. Thompson, E. L. Sievers and D. A. Kennedy, ASCO Annual Meeting Proceedings, 2009.

98. S. O. Doronina, B. A. Mendelsohn, T. D. Bovee, C. G. Cerveny, S. C. Alley, D. L. Meyer, E. Oflazoglu, B. E. Toki, R. J. Sanderson, R. F. Zabinski, A. F. Wahl and P. D. Senter, *Bioconjugate Chem.*, 2006, **17**, 114–124.

99. S. C. Alley, X. Zhang, N. M. Okeley, M. Anderson, C.-L. Law, P. D. Senter and D. R. Benjamin, *J. Pharmacol. Exp. Ther.*, 2009, **330**, 932–938.

100. S. C. Alley, D. R. Benjamin, S. C. Jeffrey, N. M. Okeley, D. L. Meyer, R. J. Sanderson and P. D. Senter, *Bioconjugate Chem.*, 2008, **19**, 759–765.

101. C. F. McDonagh, K. M. Kim, E. Turcott, L. L. Brown, L. Westendorf, T. Feist, D. Sussman, I. Stone, M. Anderson, J. Miyamoto, R. Lyon, S. C. Alley, H.-P. Gerber and P. J. Carter, *Mol. Cancer Ther.*, 2008, **7**, 2913–2923.

102. S. O. Doronina, T. D. Bovee, D. W. Meyer, J. B. Miyamoto, M. E. Anderson, C. A. Morris-Tilden and P. D. Senter, *Bioconjugate Chem.*, 2008, **19**, 1960–1963.

103. C.-L. Law, K. A. Gordon, B. E. Toki, A. K. Yamane, M. A. Hering, C. G. Cerveny, J. M. Petroziello, M. C. Ryan, L. Smith, R. Simon, G. Sauter, E. Oflazoglu, S. O. Doronina, D. L. Meyer, J. A. Francisco,

P. Carter, P. D. Senter, J. A. Copland, C. G. Wood and A. F. Wahl, *Cancer Res.*, 2006, **66**, 2328–2337.

104. M. C. Ryan, M. Hering, D. Peckham, C. F. McDonagh, L. Brown, K. M. Kim, D. L. Meyer, R. F. Zabinski, I. S. Grewal and P. J. Carter, *Mol. Cancer Ther.*, 2007, **6**, 3009–3018.

105. D. Ma, C. E. Hopf, A. D. Malewicz, G. P. Donovan, P. D. Senter, W. F. Goeckeler, P. J. Maddon and W. C. Olson, *Clin. Cancer Res.*, 2006, **12**, 2591–2596.

106. V. Bhaskar, D. A. Law, E. Ibsen, D. Breinberg, K. M. Cass, R. B. DuBridge, F. Evangelista, S. M. Henshall, P. Hevezi, J. C. Miller, M. Pong, R. Powers, P. Senter, D. Stockett, R. L. Sutherland, U. von Freeden-Jeffry, D. Willhite, R. Murray, D. E. H. Afar and V. Ramakrishnan, *Cancer Res.*, 2003, **63**, 6387–6394.

107. T. McKenna, J. Batson, S. Ross, P. Polakis and B. Rubinfeld, *AACR Meeting Abstracts*, 2007, **2007**, 4468.

108. D. E. H. Afar, V. Bhaskar, E. Ibsen, D. Breinberg, S. M. Henshall, J. G. Kench, M. Drobnjak, R. Powers, M. Wong, F. Evangelista, C. O'Hara, D. Powers, R. B. DuBridge, I. Caras, R. Winter, T. Anderson, N. Solvason, P. D. Stricker, C. Cordon-Cardo, H. I. Scher, J. J. Grygiel, R. L. Sutherland, R. Murray, V. Ramakrishnan and D. A. Law, *Mol. Cancer Ther.*, 2004, **3**, 921–932.

109. K. F. Tse, M. Jeffers, V. A. Pollack, D. A. McCabe, M. L. Shadish, N. V. Khramtsov, C. S. Hackett, S. G. Shenoy, B. Kuang, F. L. Boldog, J. R. MacDougall, L. Rastelli, J. Herrmann, M. Gallo, G. Gazit-Bornstein, P. D. Senter, D. L. Meyer, H. S. Lichenstein and W. J. LaRochelle, *Clin. Cancer Res.*, 2006, **12**, 1373–1382.

110. V. A. Pollack, E. Alvarez, K. F. Tse, M. Y. Torgov, S. Xie, S. G. Shenoy, J. R. MacDougall, S. Arrol, H. Zhong, R. W. Gerwien, W. F. Hahne, P. D. Senter, M. E. Jeffers, H. S. Lichenstein and W. J. LaRochelle, *Cancer Chemother. Pharmacol.*, 2007, **60**, 423–435.

111. P. Hwu, M. Snozl, H. Kluger, L. Rink, K. B. Kim, N. E. Papadopoulis, D. Sanders, P. Boasberg, C. E. Ooi and O. Hamid, *ASCO Annual Meeting Proceedings*, 2008, Abstract 9029.

112. M. Sznol, O. Hamid, P. Hwu, H. Kluger, T. Hawthorne, E. Crowley, R. Simantov and A. Pavlick, *ASCO Annual Meeting Proceedings*, 2009, Abstract 9061.

113. L. M. Smith, A. Nesterova, S. C. Alley, M. Y. Torgov and P. J. Carter, *Mol. Cancer Ther.*, 2006, **5**, 1474–1482.

114. M. E. Jeffers, V. A. Pollack, K. F. Tse, M. Y. Torgov, M. Shadish, J. R. MacDougall, D. A. McCabe, N. V. Khramtsov, C. S. Hackett, S. G. Shenoy, M. Gallo, G. Gazit-Bornstein, P. D. Senter, D. L. Meyers, H. S. Lichenstein and W. J. LaRochelle, *AACR*, 2006, Abstract 1163.

115. D. Jackson, J. Gooya, S. Mao, K. Kinneer, L. Xu, M. Camara, C. Fazenbaker, R. Fleming, S. Swamynathan, D. Meyer, P. D. Senter, C. Gao, H. Wu, M. Kinch, S. Coats, P. A. Kiener and D. A. Tice, *Cancer Res.*, 2008, **68**, 9367–9374.

116. W. Mao, E. Luis, S. Ross, J. Silva, C. Tan, C. Crowley, C. Chui, G. Franz, P. Senter, H. Koeppen and P. Polakis, *Cancer Res.*, 2004, **64**, 781–788.
117. L. M. Smith, A. Nesterova, M. C. Ryan, S. Duniho, M. Jonas, M. Anderson, R. F. Zabinski, M. K. Sutherland, H. P. Gerber, K. L. Van Orden, P. A. Moore, S. M. Ruben and P. J. Carter, *Br. J. Cancer*, 2008, **99**, 100–109.
118. K. S. MacMillan and D. L. Boger, *J. Med. Chem.*, 2009, **52**, 5771–5780.
119. D. J. King, J. Terrett, C. Pan, C. Rao, P. M. Cardarelli, S. Deshpande, V. Rangan, D. Passmore, J. Sung, L. Green, C. Chong, E. Kwok, O. Cortez, V. Guerlavais, A. Zhang, L. Chen, B. Sufi, C. Cong, M. Huber, P. Sattari, K. Vemuri, J. F. Mirjolet, F. Bichat and S. Gangwar, *AACR Meeting*, 2008.
120. R. Y. Zhao, D. Sun, E. Cavanagh, M. Miller, B. Leece, H. Erickson, R. Singh, Y. Kovtun, V. Goldmacher and R. Chari, *Abstracts of Papers, 233rd ACS National Meeting*, Chicago, IL, United States, March 25–29, 2007, MEDI-138.
121. S. C. Jeffrey, M. Y. Torgov, J. B. Andreyka, L. Boddington, C. G. Cerveny, W. A. Denny, K. A. Gordon, D. Gustin, J. Haugen, T. Kline, M. T. Nguyen and P. D. Senter, *J. Med. Chem.*, 2005, **48**, 1344–1358.
122. P. J. Carter, *Nature Rev. Immunol.*, 2006, **6**, 343–357.
123. F. de Groot, J. Joosten, H. Spijker, L. Tietze, F. Major and P. Beusker, *AACR Meeting Abstracts*, 2007, **2007**, 1501.
124. L. Chen, S. Gangwar, C. Pan, C. Rao, B. Sufi, S. Boyd, M. Huber, P. Sattari, M. Do, R. Dai, C. Chong, C. Soderberg, H. Li, H. Huang, H. Chen, L. Green, J. Sung, D. Passmore, V. Rangan, V. Guerlavais, K. Horgan, N. Sharkov, K. Heaton, T. Kempe, M. Srinivasan, S. Deshpande, P. Cardarelli and D. J. King, *Abstracts of Papers, 232nd ACS National Meeting, San Francisco, CA, United States, 10–14 September*, 2006, MEDI-090.
125. J. A. Terrett, S. Gangwar, C. Rao-Naik, C. Pan, V. Guerlavais, M. Huber, C. Chong, L. Green, P. Cardarelli, D. King, S. Deshpande, V. Rangan, M. Coccia, L. Lu, D. Passmore, D. Blansett, R. Dai, B. Sufi, Q. Zhang, L. Chen, C. Soderberg, E. Kwok, K. Horgan, O. Cortez, P. Sattari, M. Srinivisan, F. Bichat and J.-F. Mirjolet, *AACR National Meeting*, 2007, Abstract 4112.
126. C. Rao, C. Pan, M. Huber, P. Sattari, C. Chong, C. Dai, C. Sonderberg, L. Chen, V. Gueravais, K. Horgan, A. Zhang, B. Sufi, H. Huang, H. Chen, S. Gangwar, P. Cardarelli and D. King, *AACR*, 2007, Abstract 4104.
127. C. Rao, M. Huber, K. Vemuri, Q. Zhang, B. Chen, J. Phillips, M. Greenbaum, J. Sung, D. Derwin, D. Passmore, R. Vangipuram, J. Terrett, S. Deshpande, P. M. Cardarelli, D. Blanset and S. Gangwar, *AACR National Meeting*, 2009.
128. V. Guerlavais, Q. Zhang, K. Horgan, S. Boyd, B. Sufi, L. Chen, L. Green, D. Passmore, J. Sung, R. Vangipuram, L. Thevanayagam, M. Srinivasan,

M. Do, R. Dai, E. Kwok, C. Chong, C. Soderberg, C. Pan, M. Huber, P. Sattari, C. Rao, S. Deshpande, P. Cardarelli, D. J. King and S. Gangwar, *Abstracts of Papers, 234th ACS National Meeting, Boston, MA, United States, 19–23 August*, 2007, MEDI-318.

129. S. Gangwar, C. Pan, Q. Zhang, C. Rao, B. Sufi, L. Green, D. Passmore, M. Huber, C. Chong, E. Kwok, O. Cortez, P. Sattari, L. Chen, V. Guerlavais, V. Rangan, S. Deshpande, P. Cardarelli, D. J. King and J. A. Terrett, *AACR National Meeting*, 2009, Abstract 1722.

130. J. A. Ajani, D. P. Kelsen, D. Haller, K. Hargraves and D. Healey, *Cancer J.*, 2000, **6**, 78–81.

131. S. C. Jeffrey, M. T. Nguyen, J. B. Andreyka, D. L. Meyer, S. O. Doronina and P. D. Senter, *Bioorg. Med. Chem. Lett.*, 2006, **16**, 358–362.

132. B. M. Mueller, R. A. Reisfeld, M. H. Silveira, J. D. Duncan and W. A. Wrasidlo, *Antibody, Immunoconjugates, Radiopharm.*, 1991, **4**, 99–106.

133. C. W. Gilbert, E. B. McGowan, G. B. Seery, K. S. Black and M. D. Pegram, *J. Exp. Ther. Oncol.*, 2003, **3**, 27–35.

134. I. Ojima, *Accounts Chem. Res.*, 2008, **41**, 108–119.

135. P. J. Burke, P. D. Senter, D. W. Meyer, J. B. Miyamoto, M. Anderson, B. E. Toki, G. Manikumar, M. C. Wani, D. J. Kroll and S. C. Jeffrey, *Bioconjugate Chem.*, 2009, **20**, 1242–1250.

136. S.-J. Moon, V. Govindan Serengulam, M. Cardillo Thomas, A. D'souza Christopher, J. Hansen Hans and M. Goldenberg David, *J. Med. Chem.*, 2008, **51**, 6916–6926.

137. P. J. Burke, B. E. Toki, D. W. Meyer, J. B. Miyamoto, K. M. Kissler, M. Anderson, P. D. Senter and S. C. Jeffrey, *Bioorg. Med. Chem. Lett.*, 2009, **19**, 2650–2653.

138. R. Mandler, C. Wu, E. A. Sausville, A. J. Roettinger, D. J. Newman, D. K. Ho, C. R. King, D. Yang, M. E. Lippman, N. F. Landolfi, E. Dadachova, M. W. Brechbiel and T. A. Waldmann, *J. Natl Cancer Inst.*, 2000, **92**, 1573–1581.

# Drug Discovery by DNA-encoded Libraries

YIZHOU LI, ZHENG ZHU AND XIAOYU LI*

Beijing National Laboratory for Molecular Sciences (BNLMS), Key Laboratory of Bioorganic Chemistry and Molecular Engineering of Ministry of Education, Institute of Organic Chemistry, College of Chemistry and Chemical Engineering, Peking University, Beijing, China, 100871

## 8.1 Introduction

### 8.1.1 The Lack of Discovery Tools Interrogating Undrugged Targets

In the era of molecular medicine, the campaign against human diseases comprises two fronts: understanding the diseases at the molecular level and discovering therapeutic agents to treat them. With the explosion of technological developments in genomics, transcriptomics and proteomics, and the emerging metabomics, scientists are rapidly gaining insights into the molecular mechanisms of diseases on an unprecedented high level of resolution. However, the pace of the second front, discovering therapeutic agents to interrogate a variety of validated drug targets, is significantly lagging behind, resulting in "The gap between scientists' aspirations and society's expectations", as precisely put by Professor Stuart Schreiber at the Broad Institute.[1] It has been conceived that the drug targets currently being pursued by major pharmaceutical companies are highly limited. For example, for nearly 30 000 genes in the human genome, there are expected to be around 3000 disease-modifying genes;

RSC Drug Discovery Series No. 5
New Frontiers in Chemical Biology: Enabling Drug Discovery
Edited by Mark E. Bunnage
© Royal Society of Chemistry 2011
Published by the Royal Society of Chemistry, www.rsc.org

while there are about 300 molecular targets of FDA-approved drugs, only $\sim 10\%$ of all possible target genes. If limited to human genome targets of approved small molecule drugs, the number is reduced to only $\sim 200$.[2] This dramatic gap represents the lack of enough drug molecules to tackle available drug targets; however, somewhat ironically, scientists do *not* lack the tools to synthesize any particular chemical structures, but *do* lack the knowledge to know which ones to synthesize. This reality in drug discovery imposes urgent needs in developing new strategies and novel technologies that are able to rapidly and efficiently access the vast chemical space in a holistic fashion, to interrogate those currently undrugged targets, and to obtain active molecules from a large population, providing starting points for further development and optimization.

## 8.1.2 Applying Nature's Strategy in Discovering Functional Molecules

On the molecular level, the goal of drug discovery is to obtain the optimal chemical structure able to physically bind to the target and impose perturbation ultimately leading to the desired therapeutic effects in disease treatment. On the other hand, this chemical structure is "optimal" only when it can work within the complexity of the biological milieu around the target; and also, can tolerate the complicated process to reach there after administration. These demanding requirements on drug molecules make drug discovery solely based on rational design and predictive computation, at least with current technologies, extremely difficult.[3] Currently, the mainstream approach in drug discovery is more or less a stochastic one: brute-force screening of a large collection of compounds to look for promising starting points for further optimizations. Among the various technology platforms developed to accomplish this task, high-throughput screening (HTS) has emerged as the preferred choice by pharmaceutical companies in generating hit compounds for targets lacking starting structures.[4,5] Currently, screening of a couple of million compounds against drug targets is a routine practice in major pharmaceutical companies. Screening at such a scale requires tremendous investments in compound management, screening robotics, consumable logistics and informatics. It is not a viable platform for any but the few pharmaceutical companies with financial and technological means.

In addition, as the name indicates, screening subjects each compound candidate to the target individually, then medicinal chemists attempt to interpret data to deduce structure–activity relationships (SARs) for further development and optimization. This process is highly empirical and can be time-consuming and difficult. Nature, on the contrary, uses a fundamentally different approach in discovering molecules for desired functions: *selection*. As shown in Figure 8.1, an example of Nature's approach is compared with the chemist's approach. Instead of individually accessing and improving specific chemical structure's functions, Nature imposes the selection pressure on a large population of candidate molecules, such as proteins and natural products *in the same*

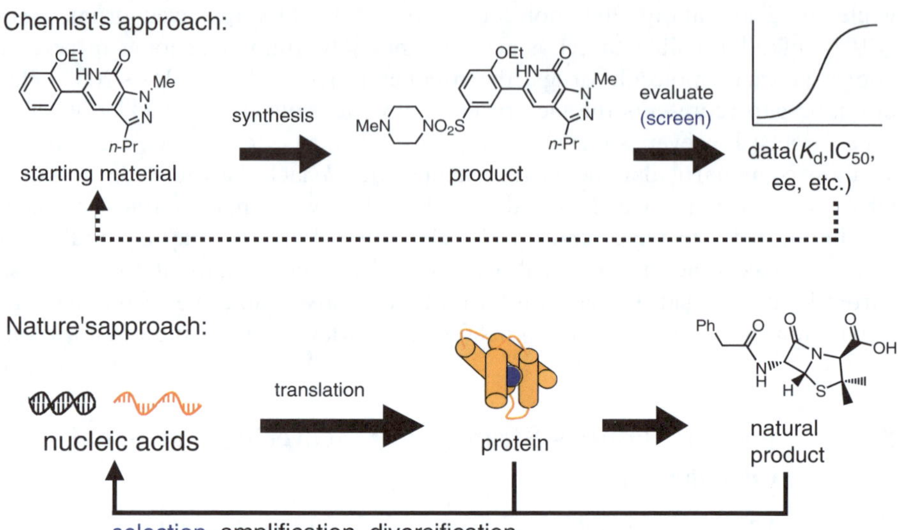



**Figure 8.1** Comparison of Nature's approach to discovery with the chemist's approach.[31] Reprinted, with permission, from Li and Liu.[31] © 2004 Wiley-VCH Verlag GmbH & Co. KGaA.

*system, not individually*. Cells that survive the selection will therefore be enriched; their genetic information will be amplified, diversified and translated into the next population of functional molecules for further selection. Nature's evolutional approach offers benefits on various aspects;[6,7] and has actually been taken advantage of by scientists in laboratories into a number of display technologies to evolving functional biopolymers using native cellular machinery, including phage display,[8–10] ribosomal display,[11,12] mRNA display,[13–16] *etc.*[17] Indeed, these technologies have proved to be tremendously useful in discovering molecules with desired biological functions.[18] The highlights of these display systems include:

- *Unprecedented library size.* For example, display libraries with Nature's approach can reach the size of $10^{15}$;[6,11] while with the sophisticated "one-bead, one-compound" approach, a $\sim$2 000 000-member library has been created by chemists but usually a typical combinatorial library ranges from a few hundred to a few thousand different compounds.[20]
- *Enrichment.* Active members of the library are *enriched* in each round of selection. With each iteration, the library population is evolved into a distribution with higher abundance of active molecules; and eventually the most active ones will emerge from the selection regardless their initial abundances in the library.
- *Amplifiability.* All the library members are displayed on the DNA or RNA which encodes the structural information. Therefore, the library can be

manipulated with molecular biology techniques, such as PCR, to amplify the encoding DNA/RNA tags. Even though the amplification itself will not change the composition of the library, it allows the selected sub-population to be subjected for further iteration. Therefore, libraries can be prepared on a miniature scale, in picomoles to nanomoles rather than in milligrams to grams, as in conventional libraries.

However, one key issue has prevented Nature's approach being applied to small molecules: the information coded in DNA or RNA base sequences cannot be directly translated into small molecule structures with Nature's machinery. New technologies and platforms need to be designed and implemented. The DNA-encoded library has emerged as one of the most intriguing approaches towards this goal.

### 8.1.3  Topics in this Chapter

In this chapter, we will first discuss the basic principles of DNA-templated organic synthesis, as it is the mechanistic foundations for DNA-encoded libraries; then we will discuss the progressive development of DNA-templated libraries and the evolution into a novel drug discovery tool. We will also discuss the DNA-recorded library, which also encodes library molecules with DNA but is conceptually different. These discussions will naturally involve specific drug discovery programs in which these libraries were applied; and finally, we will discuss the outlook of DNA-encoded libraries in the future of drug discovery.

Our focus in this chapter is the DNA-encoded library as a tool in exploring chemical space for drug discovery. For the applications of DNA-templated synthesis in other fields, such as the origin of life, biodetection, nano-technology, *etc.*, we refer readers to various excellent reviews.[21–25]

## 8.2  DNA-templated library

### 8.2.1  Basic Principles of DNA-templated Organic Synthesis

Macromolecule-templated synthesis is one of the most fundamental biological processes in life and is the mechanism underlying genetic information replication, transcription and translation into biological functions in nearly every life form. In one impressive example, Nature uses mRNA typically ranging in nanomolar to micromolar cellular concentration as templates to direct programmed polypeptide syntheses from dozens of amino acid-esterified tRNAs. While the base pairing specificity between tRNAs and mRNAs controls the translation fidelity, the whole ribosomal machinery and the mRNA templates provide precise regulation of the syntheses. Indeed, since the cell is the single vessel for tens of thousands of materials but only at picomolar to nanomolar concentrations, it is necessary for Nature to use macromolecules as templates to virtually compartmentalize and coordinate all the complex biological processes simultaneously.

Nature's ability to control a reaction caught scientists' attention many years ago. It has been speculated that the earliest biological functions may arise from an autocatalytic templated reaction between primitive building blocks in the pre-biotic period.[21,26] Besides the involvement in the study of origin of life, chemists started to use DNA, RNA or other nucleic acid analogs, to perform templated reactions to facilitate chemical reactions in various forms, including structural mimics of native DNA/RNA's phosphodiester bond formations,[27–33] completely unrelated synthetic organic reactions,[34] polymerization reactions,[29,35–37] functional group transformations,[38] base-filing reactions,[39] and double-stranded DNA-templated reactions.[40,41] Chemists have also investigated new modes of DNA-templated synthesis, such as directing otherwise incompatible reactions in the same vessel,[42,43] DNA-templated synthesis in organic solvents,[44,45] stereoselective reactions templated by DNA,[46] and discovering new chemical reactions.[44,47,48] Chemists have also accomplished DNA-templated multistep reactions to synthesize complex structures[49,50] and DNA-templated combinatorial library syntheses,[51] either with distance-independent reactions[49] or creative manipulations of DNA template architectures.[52] Progressively, DNA-templated organic synthesis has become a novel approach to exploring chemical space and, recently embraced by the pharmaceutical companies, discovering drug candidates.

### 8.2.1.1 Components and Architectural Designs of DNA-templated Synthesis

Figure 8.2b shows the components of a typical reactant and several architectural designs of DNA-templated reactions. The reactant in DNA-templated reactions is a chemically reactive group conjugated to a DNA strand, which is either used as a template ("the template") or a hybridizing strand ("the reagent"). When "the template" and "the reagent" are annealed together, the two reactive groups are confined in a limited space and the effective molarity between the two reactive groups increases. Figure 8.2b shows several commonly used architectures. The $A + A'$ architecture, also called "end-of-helix" architecture, is the most widely used due to its ease of synthesis, flexibility in incorporating reactive groups, the compatibility to multistep synthesis, downstream DNA amplification and sequencing, and more importantly, the selection against biological targets.

Regardless of which one of the above architectures is used, a templated reaction becomes pseudo-intramolecular and the effective molarity created by this effect could exceed $1000\,M$.[53] In principle, if the reagent DNA is placed at a relatively distal position from the template's reactive group, the effective molarity is expected to decrease. However, interestingly, Liu and co-workers have observed that the rates of some DNA-templated reactions, *e.g.* amide formation, remain nearly constant when the two reactants are separated by even up to 30 bases.[54] They hypothesized, for these so-called "distance-independent reactions", the reaction rates are faster than the hybridization;

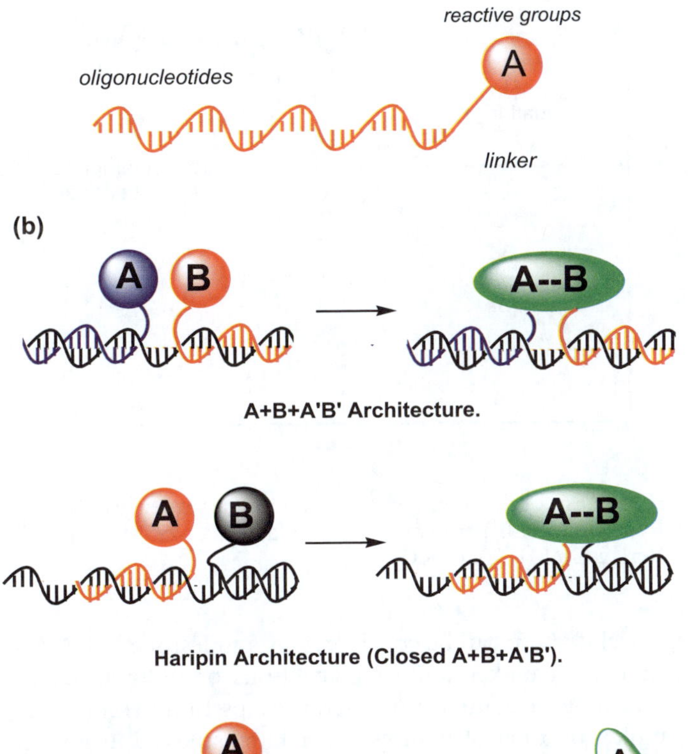



**Figure 8.2** (a) Components of reactants for DNA-templated reactions. (b) Reactions architectures. X and X' refer to reactants containing complementary oligonucleotides, and + symbols indicate separate molecules.[31] Reprinted, with permission, from Li and Liu.[31] © 2004 Wiley-VCH Verlag GmbH & Co. KGaA.

therefore the hybridization is rate-limiting so that the templated reaction rate remains unchanged until the separation is extreme enough to drive the rate below the hybridization rate (Figure 8.3).

Furthermore, in order to be able to perform distal reactions efficiently for all types, Liu and co-workers designed an Omega architecture in which the reagent DNA anneal to two regions on the template: one distal region and one region proximal to the reactive group (Figure 8.4a)[52] and the intervening region loops out forming an $\Omega$ shape. This architecture enables a proximal reaction, even though the reagent DNA is hybridized to a distal position on the template.

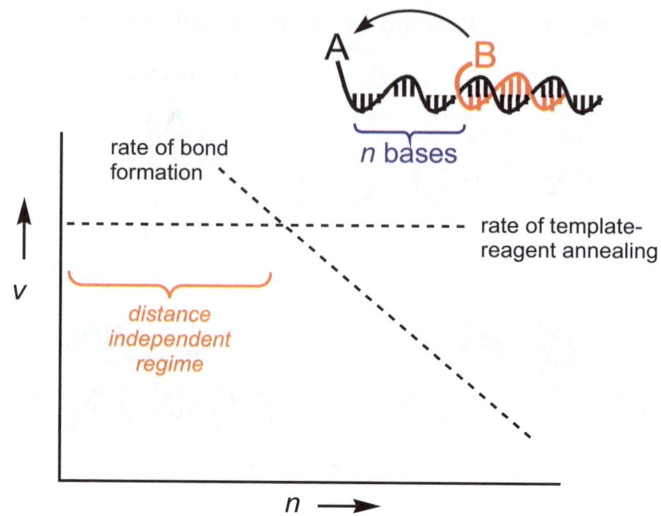


**Figure 8.3**    Hypothesis of the origin of the distance independence of some reactions.[34] Reprinted, with permission, from Gartner *et al.*[34] © 2002 Wiley-VCH Verlag GmbH & Co. KGaA.

The Omega architecture can be considered as a milestone in DNA-templated organic synthesis as it makes multistep reactions on a single DNA template possible by annealing reagents from different steps to different coding regions to create complex product structures. In a later study, Liu and co-workers found that the fine structure of the $\Omega$ region affects the reactivity in an interesting way. They observed that a DNA-templated reaction with a completely unstructured $\Omega$ region proceeds poorly, while a partially structured $\Omega$ region is necessary for an efficient reaction.[55] It has been concluded that the partial structure in the $\Omega$ region offsets some of the entropic penalty of looping out bases, while a highly structured $\Omega$ region may compete with reagent DNA hybridization. This set of principles in DNA template architecture design has been applied in the design of actual library synthesis in their later studies.

In addition, Liu and co-workers have designed the T architecture by annealing two reagent DNA on both sides of the reactive group in the middle of a template; and von Kiedrowski and co-workers used a Y-shaped DNA template to mediate the DNA-templated formation of three hydrazones simultaneously (Figure 8.4b and c).[52,56] However, these two architectures are not as widely used, except Liu and co-workers have used a combination of the $\Omega$ and T architecture in the synthesis of two *N*-acyloxazolidine molecules.[50]

## 8.2.1.2    *The Reaction Scope for DNA-templated Synthesis*

Initially, the studies of DNA-templated reactions were geared towards the syntheses of analogs of native DNA's phosphodiester linkage,[34] until recently

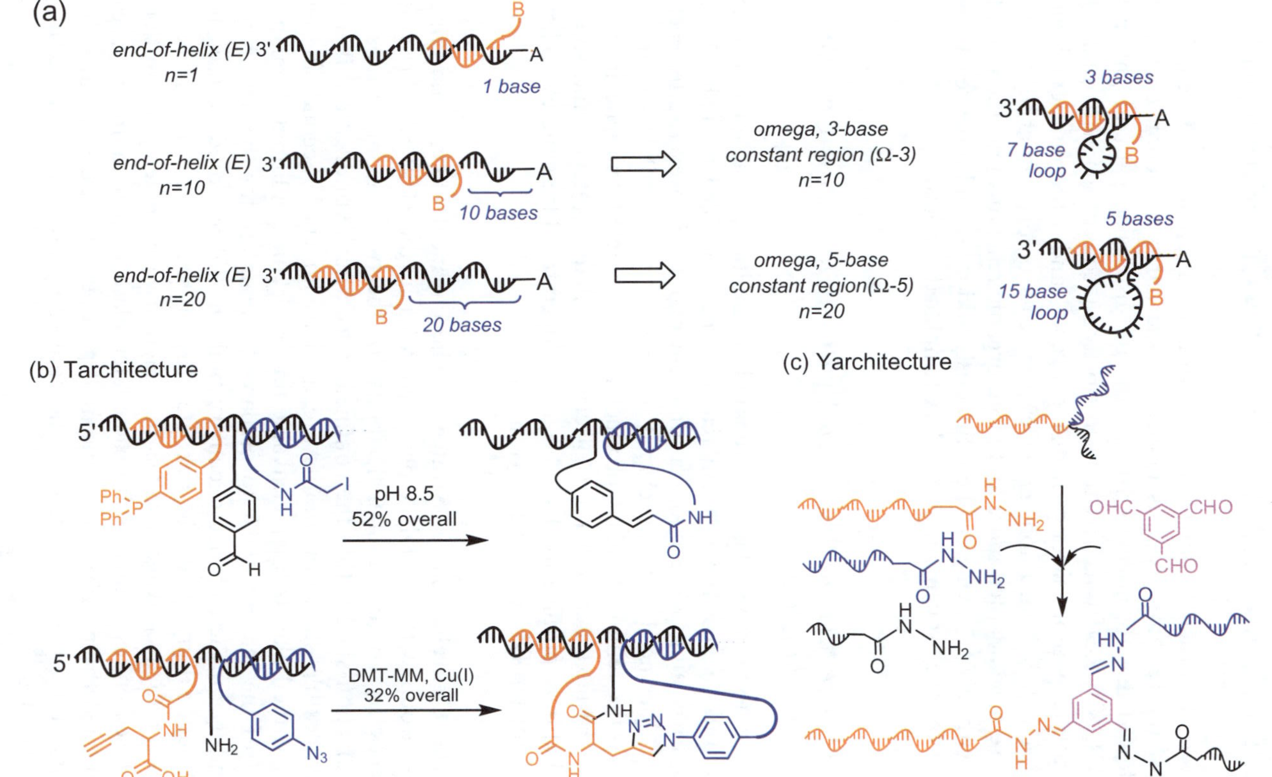



**Figure 8.4** Other architectures for DNA-templated synthesis. (a) Use of Ω architecture to bring the reactive groups together for distal reactions. (b) T architecture. (c) A Y-shaped template mediates tris-hydrazone formation.[52,56] Reprinted in part, with permission, from Gartner *et al.*[52] © 2003 Wiley-VCH Verlag GmbH & Co. KGaA.

Liu and co-workers adapted a wide range of chemical reactions to DNA-templated format, a comprehensive toolkit for the syntheses of complex synthetic structures is created.[34] These studies have proved that DNA-templated reactions are not limited to the structural analogs of native DNA. Provided with sufficient conformational flexibility for the reactants to access each other, there should be no limitation on structures that can be synthesized on a DNA template. Nevertheless, the reaction scope is limited by the narrow reaction condition window allowed due to, not surprisingly, the intrinsic chemical and physical properties of DNA. Reactions have to be in mostly aqueous media, with mild pH and temperature, without strong oxidizing agents, and with sufficient ionic strength for DNA hybridization, *etc.* Developments of organic chemistry in aqueous phase have adapted more and more organic reactions into the format compatible with DNA,[45] which will continuously add more and more reactions to the toolkit for the growing need of DNA-templated synthesis.

### 8.2.1.3  DNA-templated Multistep Synthesis to Access Complex Chemical Structures

Structurally complex molecules are prepared by multistep reactions. The $\Omega$ architecture and distance-independent reactions have enabled a single DNA template to direct multiple reactions at different regions. However, after each reaction, strategies need to be developed to remove the reagent DNA strand and to purify the template-linked product from unreacted template in preparation for the next step. A variety of cleavable linker strategies and solid-phase-based affinity purification have been developed to address these two key aspects of multistep DNA-template synthesis.

**8.2.1.3.1  The Cleavable Linker Strategy.**  The chemical linker connecting the reacting group with the DNA play two roles in the synthesis. First, it provides necessary conformational flexibility for the two reacting groups to form the transition state within the confined space after hybridization; therefore usually flexible structural motifs are used as linkers.[54] Second, in the reagent DNA, the linker is often cleavable under certain conditions to enable the removal of the reagent DNA to prepare the template for the next reaction. Depending on the cleavage chemistry built in, linkers could be used to furnish a functional group for the next step (useful scar linker); or for auto-cleaving linkers, their cleavage involves the reaction with the template so that the product self-elutes from the unreacted template, which in many cases greatly facilitated the purification of the final products. Figure 8.5 shows the structure and cleavage conditions of some commonly used linkers in DNA-templated reactions.[49,50,57,58]

**8.2.1.3.2  Solid Phase-based Affinity Purification.**  Cleavage of the linker does not automatically remove the reagent DNA from the template. In order

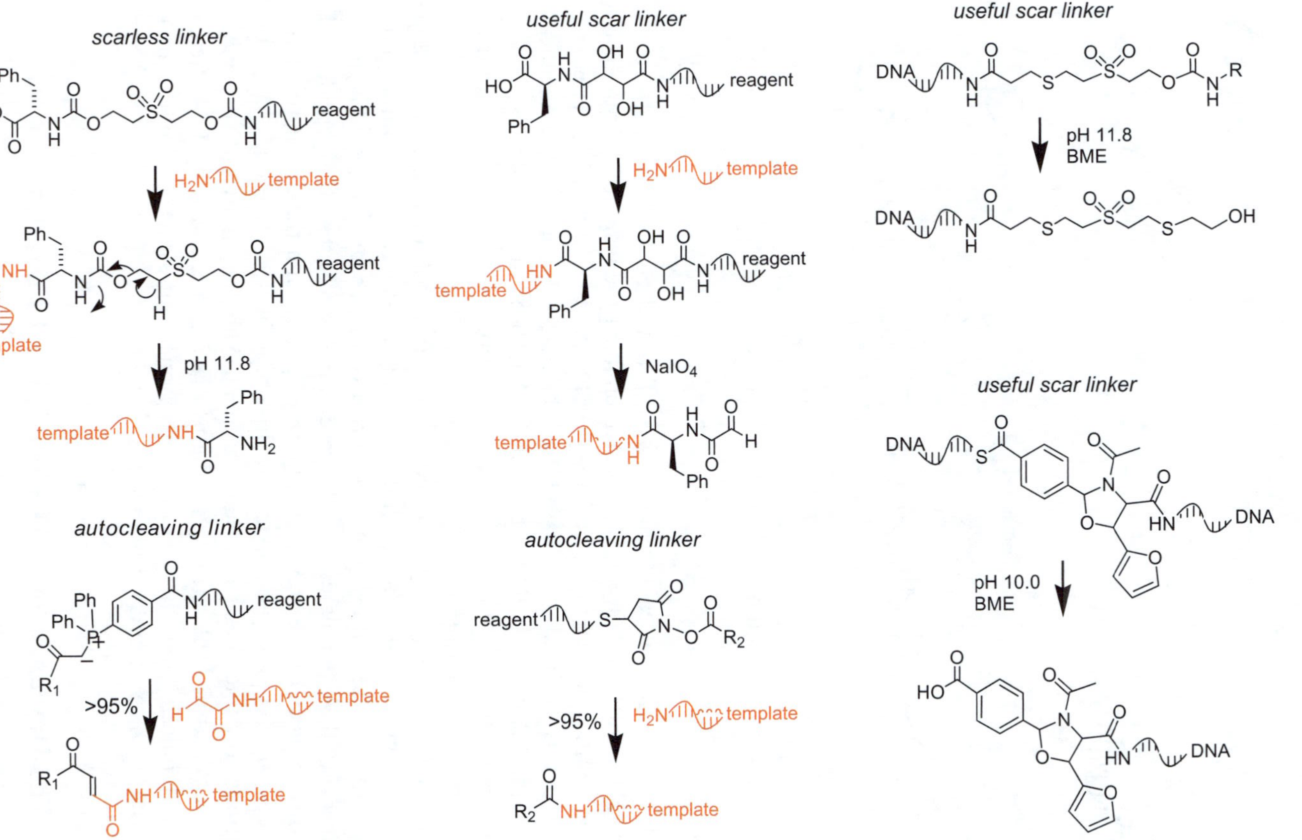


**Figure 8.5** Commonly used linkers in DNA-templated reactions: scarless linker; autocleaving linker; useful scar linkers.[49,50,57,58] Reprinted in part, with permission, from Gartner *et al.*[49] © 2002 American Chemical Society.

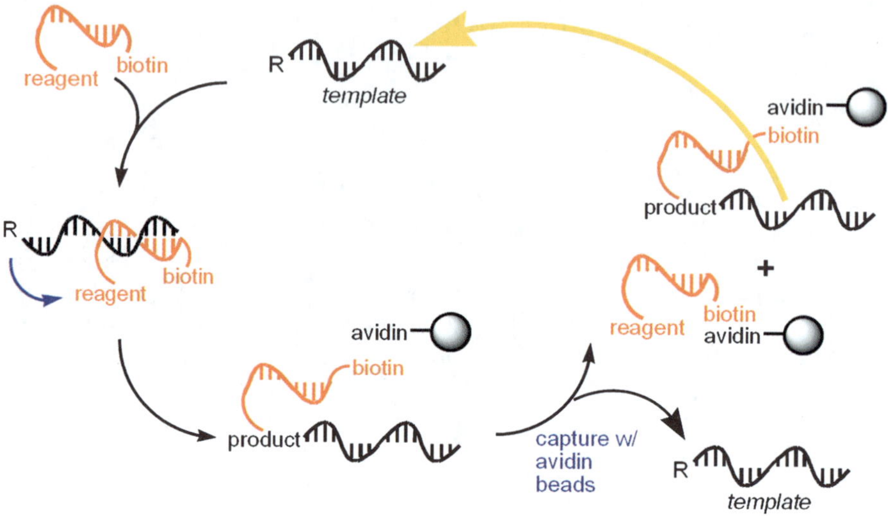


**Figure 8.6**   Solid-phase-based purification strategy for DNA-templated synthesis. Reprinted with permission from Professor David R. Liu.

to purify the product in each reaction step, Liu and co-workers tagged each reagent DNA with a biotinyl group, which enabled the isolation of pure product from unreacted templates and reagent DNA (Figure 8.6).[49] It is worth noting that this purification is based on DNA strands, not on the chemical motifs displayed on the DNA, therefore it has been successfully used as a library-wide purification method in the preparation of combinatorial libraries by DNA-templated synthesis.[51]

Liu and co-workers have performed a series of DNA-templated multistep syntheses integrating template architecture, linker design, and affinity purification into a streamlined synthetic process to create complex chemical structures on DNA templates.[49,50] Representative syntheses of tripeptide and oxazolidine molecules are shown in Figure 8.7. Although these structures are still modest in complexity compared with compounds routinely prepared by conventional organic chemistry, they are great examples that creative manipulations can lead to sophisticated synthetic structures encoded by DNA templates; and indeed later they became the pioneering structures of the first generation of DNA-templated libraries.

## 8.2.2   The Development of DNA-templated Libraries

### 8.2.2.1   The Concept

Besides the control of effective molarity, another fundamental principle of DNA-template synthesis is its sequence specificity, which controls the

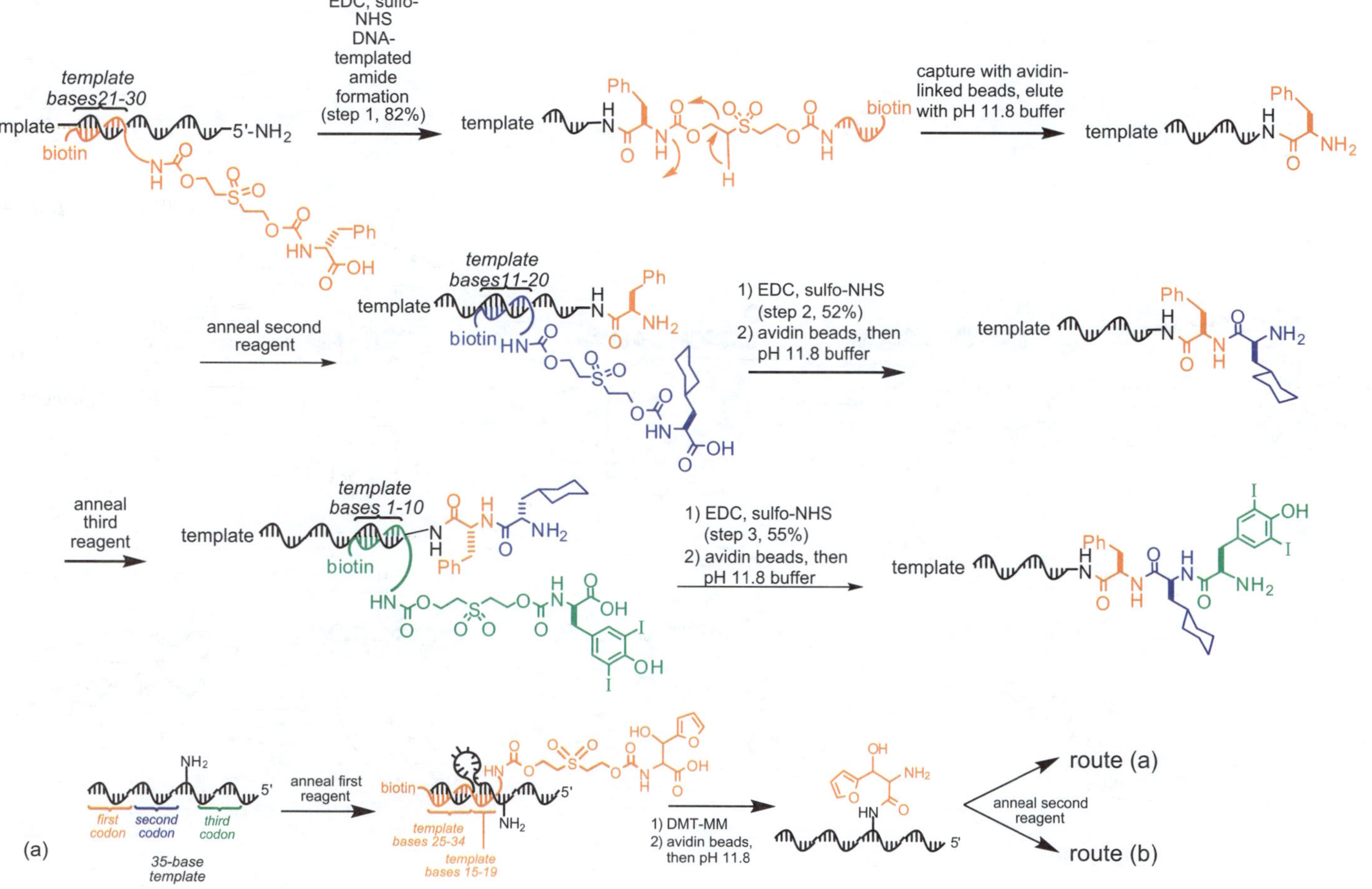



**Figure 8.7** DNA-templated multistep synthesis of tripeptide and *N*-acyloxazolidines.[49,50] Reprinted, with permission, from Gartner *et al.*[49] and Li *et al.*[50] © 2002 and 2004 American Chemical Society.

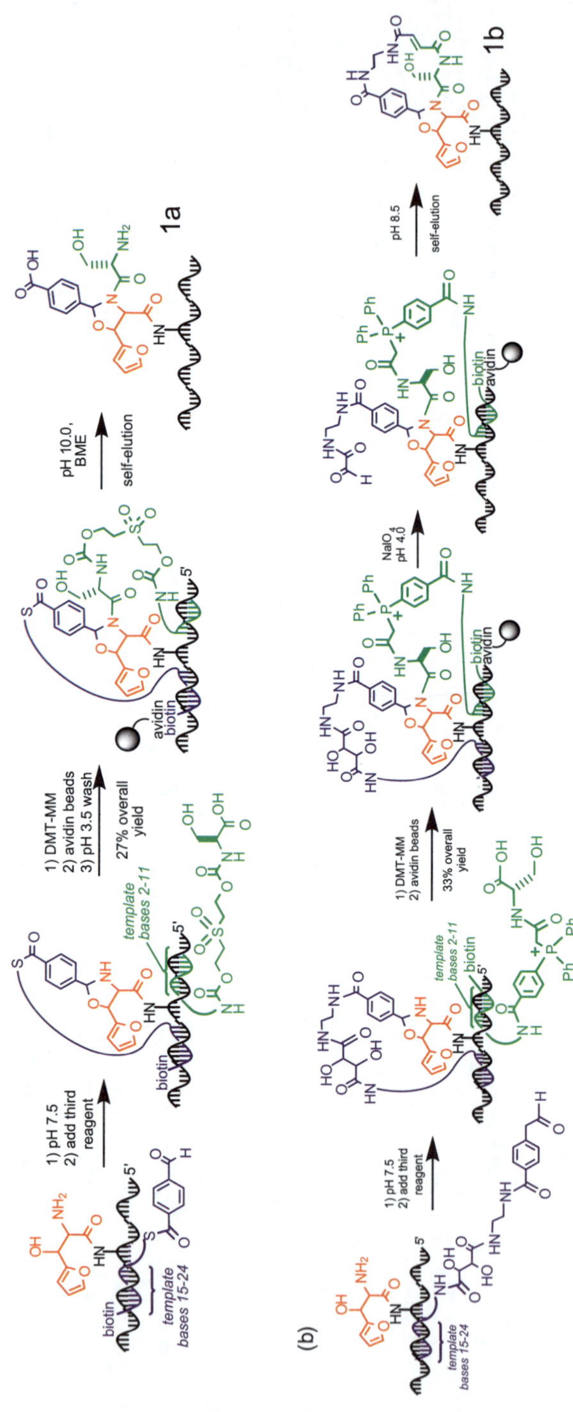



**Figure 8.7** Continued.

hybridization between DNA strands based on the rules of Watson–Crick base pairing (in the case of triplex DNA formation, Hoogsteen base-paring is also involved, which is outside the topic of this chapter). When the base sequences of two DNA strands are mismatched, at low concentrations (nanomolar to micromolar), intermolecular reactions between these mismatched strands are negligible.

This sequence fidelity lays the foundation for constructing a DNA-templated library of synthetic small molecules. At low concentrations, each DNA template can be considered as an individual "nano-reactor". These reactors will not interfere with each other and are capable of directing the hybridization only with DNA reagents with complementary base sequences. For a DNA reagent with a mismatched base sequence, the DNA template is "invisible" and will not hybridize. Therefore, parallel reactions can proceed in the same vessel without spatial separation and libraries with many different compounds can be prepared as a whole. A scheme for DNA-templated library synthesis is shown in Figure 8.8. A pool of templates with different "codons", regions of base sequences varying from template to template but complementary to one of the reagent DNAs carrying chemical building blocks, are prepared in one pot. After incubation the templates pool with *all* reagents, the reagent DNA will hybridize onto the correct codon region on the templates as directed by sequence specificity. The templated reactions take place and the building blocks are delivered. After affinity purification/ linker cleavage (Figure 8.6), the intermediates are purified library-wide and ready for the next round of reaction. Usually this process is repeated to incorporate three to four dimensions of structural diversities to create the full library.

In a DNA-templated library, the DNA template serves two purposes. First, it is a nano-reactor virtually separated from each other spatially so that parallel reactions can proceed simultaneously. Second, it encodes the information of the specific chemical structure to be synthesized on the template in form of DNA sequences. In other words, the sequence of each template pre-determines the specific structures to be synthesized on it. This is analogous to the genetic information translation from mRNA to proteins: the sequence of a particular mRNA determines the specific protein synthesized on it. This feature differentiates a DNA-templated library from other types of DNA-encoded libraries, which will be discussed in later sections.

The DNA-templated libraries are subjected to selections against interested targets as a whole. The selection, as discussed above, will enrich a sub-population of compounds with higher affinities. Iterated selections will further enhance the enrichment so that compounds with higher affinity against the target can emerge from the selection, even though they may have low abundance in the original library. The DNA tags of selected population can be amplified by PCR and the structures of the active molecules "hits" are identified by DNA sequencing.

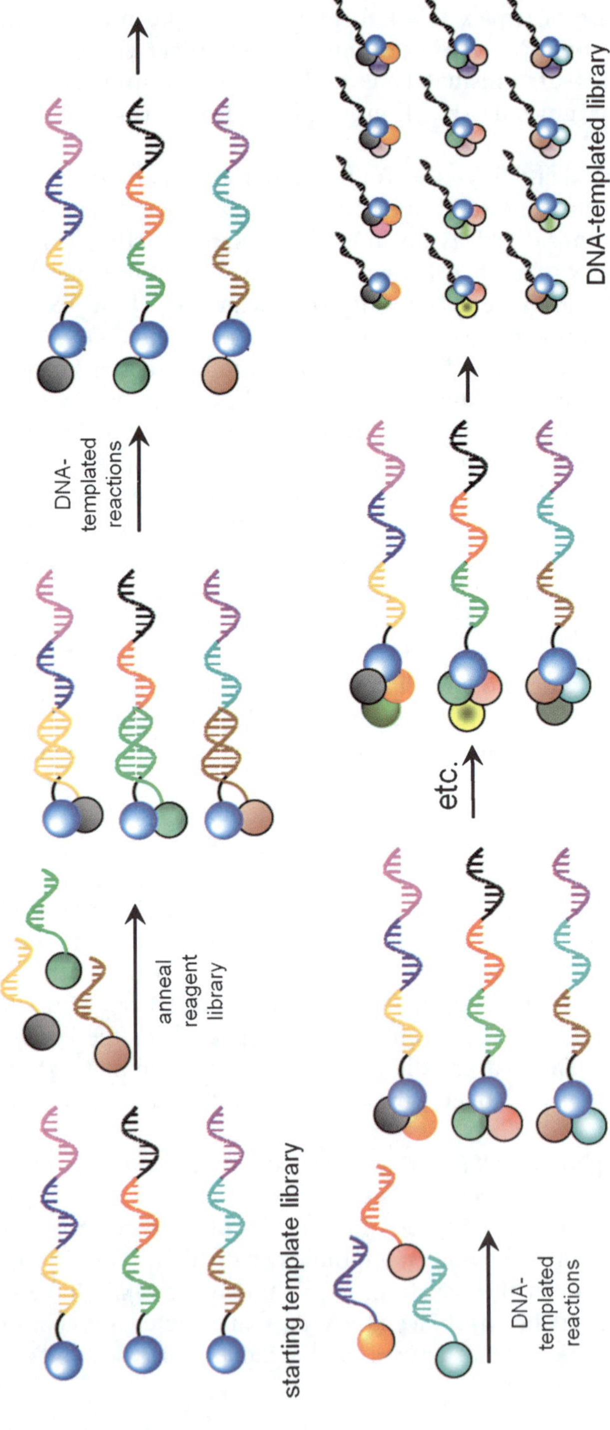



**Figure 8.8** Scheme for the construction of DNA-templated libraries and identification of active molecules after selection. Reprinted with permission from Ensemble Discovery.

## 8.2.2.2 The Syntheses of DNA-templated Macrocyclic Fumaramide Libraries

As an example, Liu and co-workers have constructed two macrocyclic fumaramide libraries.[51,59] The first library is 65-membered (4×4×4, plus a built-in sulfonamide as positive control for proof-of-principle selection), and the second one is much larger with, theoretically, 13 824 different structures. As an example, Figure 8.9 shows the scheme of the construction of the second larger library. This library was constructed on a pool of templates with three coding regions for templated synthesis. Each coding region has four variable codons responsible for templating reactions with reagent DNAs with complementary anti-codons. Flanking the coding regions are two PCR primer binding sites for PCR amplification. At the 5′-terminal of the DNA template, a diamine motif with the distal amino group capped by a tartarimide which plays an important role in creating the macrocyclic structure; and the α-amino group in the diamine serves as the starting point of the synthesis.

In both libraries, Liu and co-workers employed amidation reactions for installing each building blocks, mediated by a combination of sulfo-NHS and

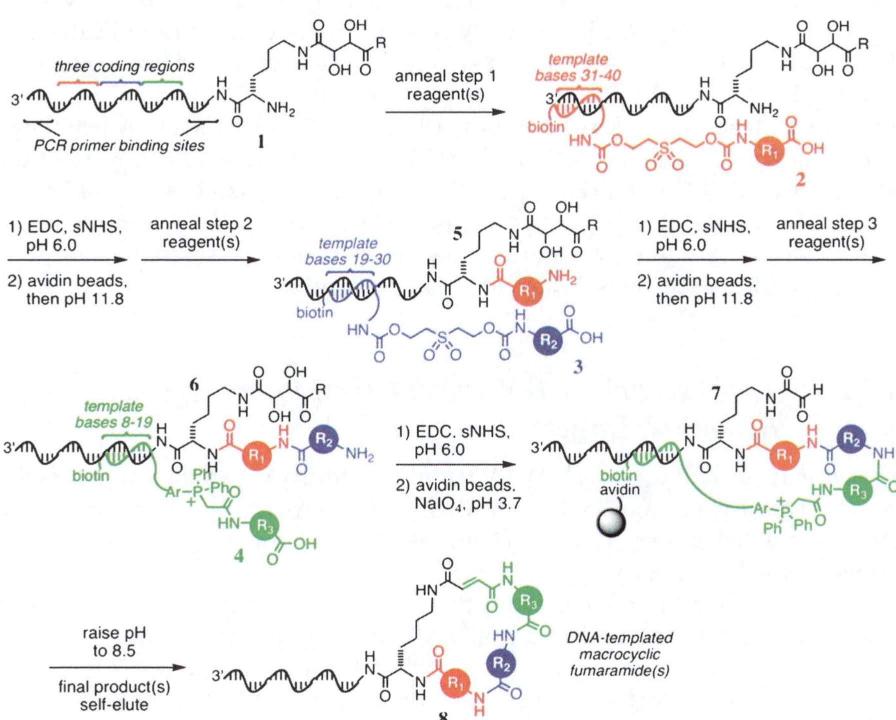

**Figure 8.9** Scheme for DNA-templated synthesis of a 13 824-membered macrocyclic fumaramide library.[59] Reprinted, with permission, from Tse *et al.*[59] © 2008 American Chemical Society.

EDC. As discussed above, this reaction is one of the "distance-independence" reactions and therefore no $\Omega$ architecture was intentionally designed in the template. However, Liu and co-workers utilized the results from one of their studies, showing that a partially structured intervening region can actually help the distal reagents to interact with each other. After three rounds of templated reactions and affinity purifications, an on-beads cleavage of the diol group exposed the aldehyde under mildly acidic condition. After raising the pH to 8.5 to activate the built-in Wittig linker, the structure cyclizes and the final macrocyclic product self-elutes from avidin beads. Any uncyclized precursors stay on the solid phase and will not contaminate the final library.

It is worth noting that the diversity of the library comes not only from the building blocks delivered by the three templated reactions, but that the pre-installed diamine also can provide an additional dimension of scaffold diversity. Liu and co-workers have shown that, with the same set of building blocks but different diamines, highly different macrocycles (ring size and orientations of building blocks on the ring) are generated based on three-dimensional molecular modeling after energy minimization (Figure 8.10).

With all the library members in the same solution at such a miniature scale, the choice of analytical methods to characterize the library is critical. In the first library, Liu and co-workers used MALDI-TOF to analyze the products from each step respectively. An increasingly complex MALDI spectra indicates the diversification of the library as each step takes place. However, the sensitivity and resolution of MALDI-TOF do not meet the needs to analyze large libraries. Instead, S1 nuclease digestion was able to remove the DNA template, so that the small molecule portion can be directly analyzed with nanospray LC-MS (Figure 8.11).[59] Remarkably, for the sampled analysis of one sub-library, the correct masses of 94% of all 1728 possible structures have been observed with clean background, proving the quality of the library synthesis and providing confidence for the subsequent *in vitro* selection.


### 8.2.2.3  *The Selection of DNA-templated Library Against Biological Targets*

The ultimate goal of creating a DNA-templated library is to interrogate protein targets. Due to the poor cell permeability of DNA-conjugated molecules, a DNA-templated library is mostly suitable for *in vitro* selection of purified proteins based on affinity.

Liu and co-workers piloted the study by using a collection of small molecules' DNA conjugates and performed selection against their known protein targets.[60] The conjugates were PCR-amplified before and after selection; specific restriction enzyme cleavage sites are built-it to identify binders from background non-binders. Two remarkable results were observed. First, the sensitivity of the mock selection can reach as low as $10^{-20}$ mol, because of the extremely high sensitivity of PCR amplification; and second, a net enrichment factor of $5\times10^6$ was achieved with just three rounds of selection, which is able

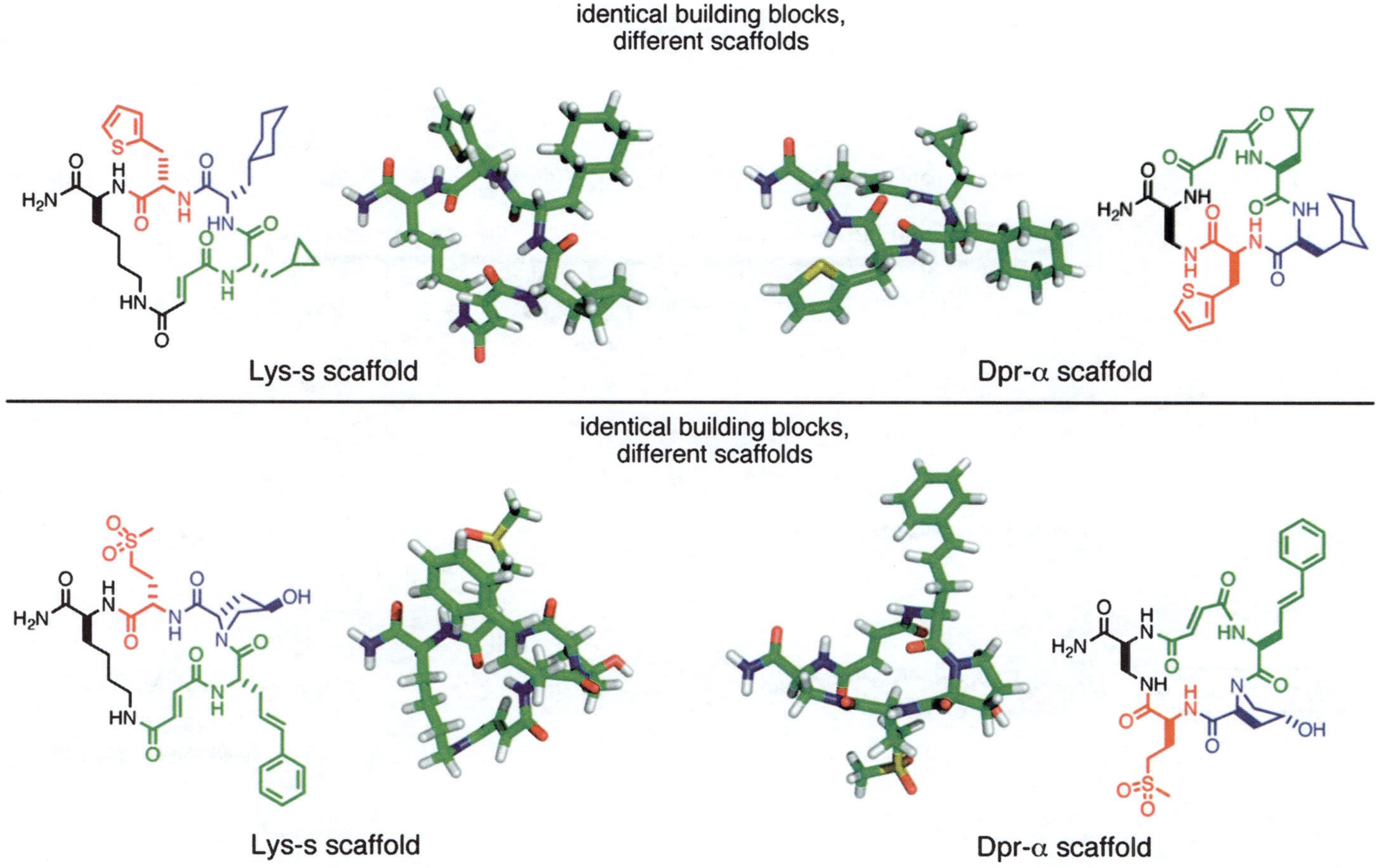

**Figure 8.10** Changes of the diamine linker dramatically alter the conformation of macrocycle scaffolds.[59] Reprinted, with permission, from Tse *et al.*[59] © 2008 American Chemical Society.

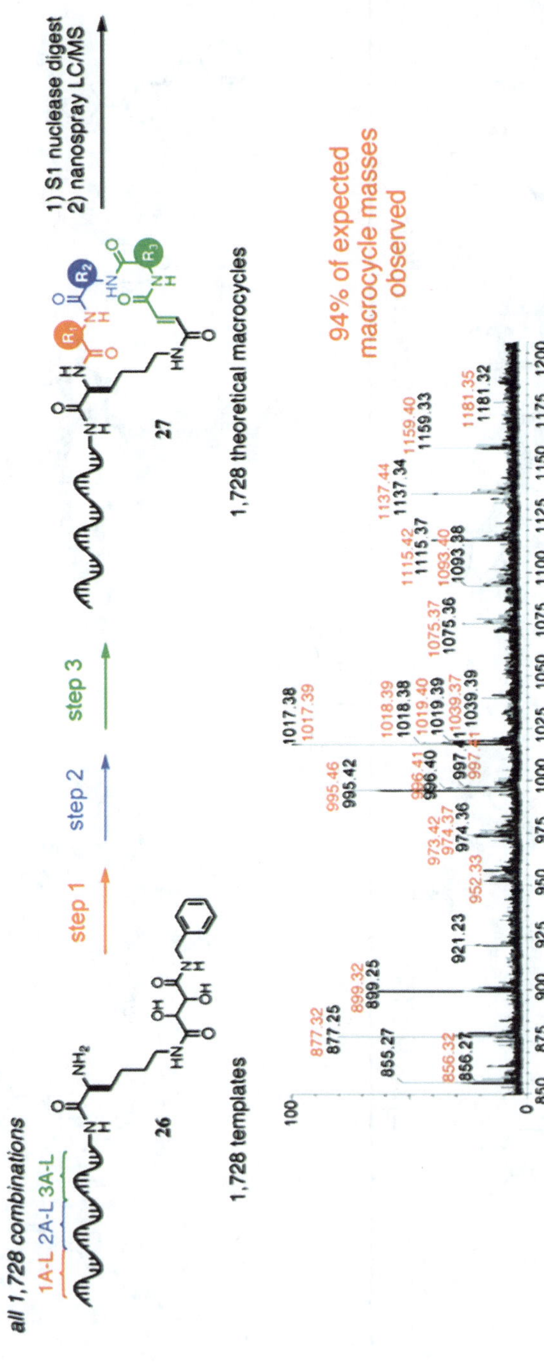



**Figure 8.11**   MS-based characterization of DNA-templated macrocycle library.[59] Reprinted, with permission, from Tse *et al.*[59] © 2008 American Chemical Society.

to make high affinity/low abundance binders, undetectable before the selection, be dominant after the selection. This transformation of the population and the enrichment of a high-affinity binder proved that selection-based screening of a DNA-templated small molecule library is a viable approach.

Liu and co-workers have performed a true selection with their 65-membered macrocycle library against immobilized carbonic anhydrase.[51] With only 100 fmol of the library, they have successfully enriched the positive control, a library member with a sulfonamide group, from the library pool. Identified by a built-in *Nla*III cleavage site, the positive control was transformed from undetectable to dominant after two rounds of selections.

Even though Liu and co-workers did not report the selection of the large macrocycle library and the hits generated, the two studies above have provided important groundwork regarding the process by which a DNA-templated library could become a practical tool for drug discovery and be applied to various drug discovery programs by pharmaceutical companies.

## 8.2.3 The Application of DNA-templated Libraries in Drug Discovery

In 2004, a company named Ensemble Discovery Corporation was founded on the groundwork provided by Liu and co-workers on DNA-templated organic synthesis and library construction. A new name for this technology, called "DNA-Programmed Chemistry (DPC™)" was also coined. Indeed, during the next a few years, DPC™ was developed into an integrated platform composing DNA-programmed organic synthesis, construction of diversity libraries, affinity-based high throughput selections, and, for each target, establishment of comprehensive SAR data set (Figure 8.12).

At Ensemble, the basic framework of the technology has not changed, but the company has implemented more sophisticated computational, experimental and instrumental tools to expand the applicability of DPC into previously difficult to access chemical structures, to increase the scale of a DPC library to hundreds of thousands of compounds, and to enhance the selection and hit identification throughputs.

### 8.2.3.1 *Ensemble's Macrocycle Libraries by DPC™ (Ensemblins™)*

Macrocycles are a class of small molecules with a ring structure with 12 or more atoms. Macrocyclic scaffolds have several unique features in terms of interacting with protein targets. First, the pre-organized cyclic structure can reduce the entropic penalty upon binding to the protein surface; second, it can display appendages (diversity groups) around the cyclic scaffold to explore the protein surface for optimal binding; finally, conformations of macrocycles are very sensitive to structural changes within its scaffold or appendages,[61,62] therefore display of different appendages on different cyclic scaffolds can result in

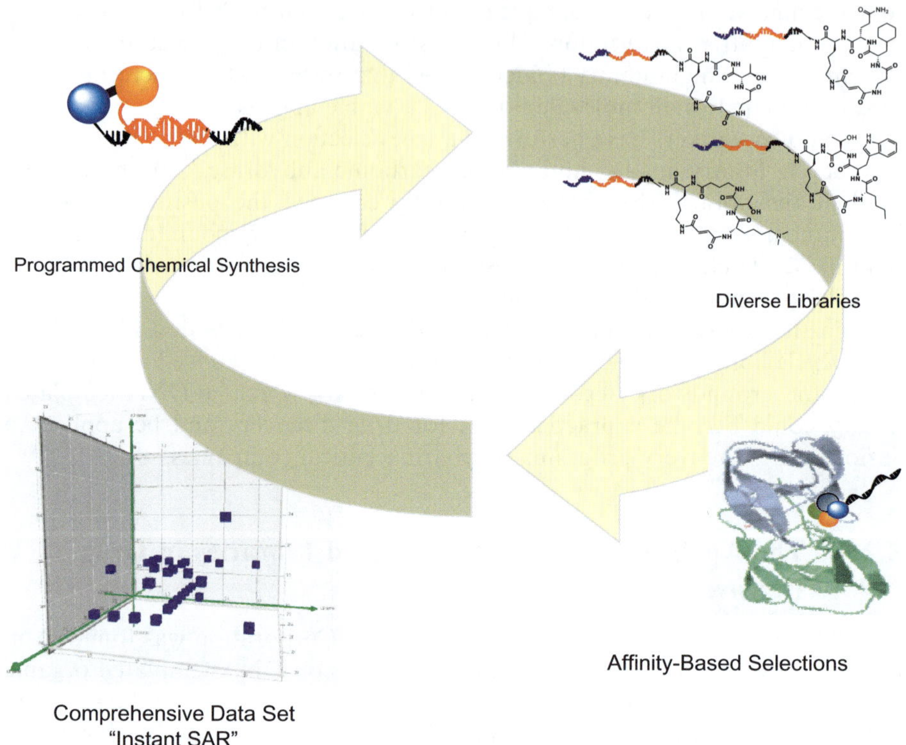



**Figure 8.12**  Ensemble's integrated drug discovery platform based on DPC®. Reprinted with permission from Ensemble Discovery.

dramatic variations in the global conformations of these molecules; and conformational flexibility allows macrocycles to mold the surface of the protein target, further enhancing the complexity and diversity presented by these macrocycles. Macrocycles represent one of the most intriguing structural features of drug molecules. Their higher molecular weights place these molecules outside of the typical Lipinski 'Rule of 5' compliant structural space, but there is growing interest in compounds in the 500–2000 molecular weight range, recently named 'millamolecular' space.

Ensemble Discovery placed their focus on macrocycles because (i) DNA-templating is a unique tool to creating macrocyclic structures, and (ii) macrocycles represent a poorly explored class of compound suitable for interaction with difficult drug targets such as protein–protein interactions. In addition, the implications and advantages of natural and synthetic macrocycles as a class of drug molecules have been thoroughly reviewed[63] and we will only briefly discuss the benefits of using DNA-templated approach for the synthesis of macrocycle libraries.

Even though individual macrocycle molecules can be feasibly synthesized from their linear precursors via macrocyclization reactions,[64] large collections of macrocycle libraries, especially ones with rich diversities in both appendages and cyclic scaffolds, are not easily accessible. Partially this is due to the high dilution condition, used in macrocyclization to suppress intermolecular reactions, is not practically applicable to the construction of a large collection of macrocycles, regardless of whether these molecules are spatially separated or in one pot. Solid-phase-based approaches have been used to address this problem by immobilizing the linear precursor to the macrocycle to the beads and then perform on-beads cyclization and cleavage.[65,66] However, the size of the library is up to ~10 000 compounds as reported and a further increase in scale will proportionally increase the time and efforts required with a solid-phase-based approach.[66]

At Ensemble, scientists took advantage of the nano-reactor function of the DNA template and the self-eluting feature in the synthesis of peptidic macrocycles depicted in Figure 8.13. First, since the synthesis of each molecule is controlled by a single DNA-template, *i.e.* a single nano-reactor, it creates a virtual dilution and precludes the possibility of intermolecular reactions; second, the purification strategy warrants that only cyclized products can be eluted from the beads, excluding contamination from an undesired linear precursor; finally, unlike other approaches, all macrocycles are synthesized in one pot and the effort needed for library construction remain mostly the same regardless the size of the library. Ensemble has prepared a series of macrocycle libraries as shown in Figure 8.13. Besides building block variations, different cyclization linkers are used to provide scaffold diversity in each library. The overall diversity of the libraries is in range of the hundreds of thousands of compounds. Ensemble has given the name "Ensemblins™" to these macrocycles to represent the unique chemical space accessed by these libraries.

After library construction, Ensemble focused the selection on drug targets previously considered to be difficult, or even "undruggable". Ensemble's intended drug targets, based on publicly available information, include three classes: phosphatases, proteases and protein–protein interaction (PPI) targets. These targets are mostly hard-to-drug proteins by current means, and the diseases in which they are implicated all have unmet medical needs with significant commercial values.

As an example, ~20 000 unique Ensemblins were selected against cancer target Bcl-xL (Figure 8.14).[67] The reported amount of library used in each selection is 30 pmol total per selection, which corresponds to only ~1.5 fmol per compound. The low quantity of each compound is compensated by relatively high effective concentration of protein target on the solid phase. In this case, a recombinant human Bcl-xL is immobilized via an His-tag onto nickel resin; and about 300 pmol (<10 μg) of the protein is used. The excess amount of target protein enables the selection can capture even the least abundant or weakest binding compounds in the library. After the selection, active compounds are eluted from the target, PCR-amplified and sequenced to decode the identity of selected hits.

In the first selection campaign, a small number of compounds (~0.02%) were identified as active hits against Bcl-xL. The top 10 hits have an enrichment

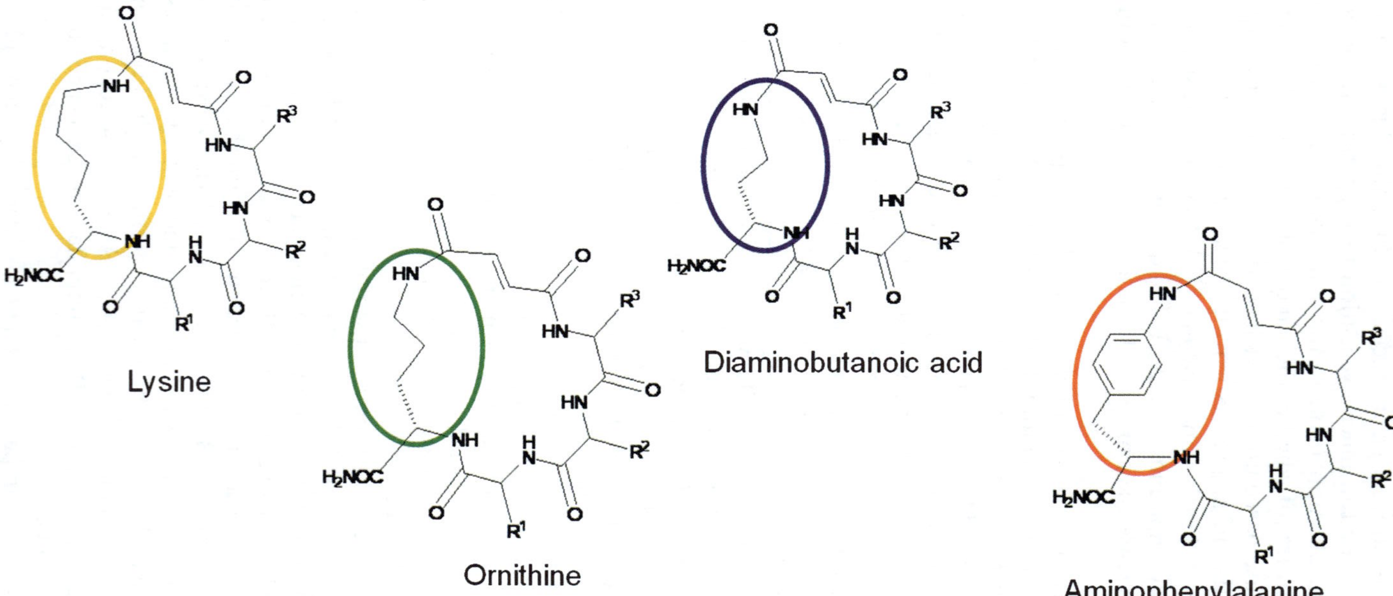



**Figure 8.13** Structures of the four macrocycle libraries prepared by Ensemble. Each library contains ∼5000 different structures. The total number of macrocycles in this series is about 20 000. Reprinted with permission from Ensemble Discovery.

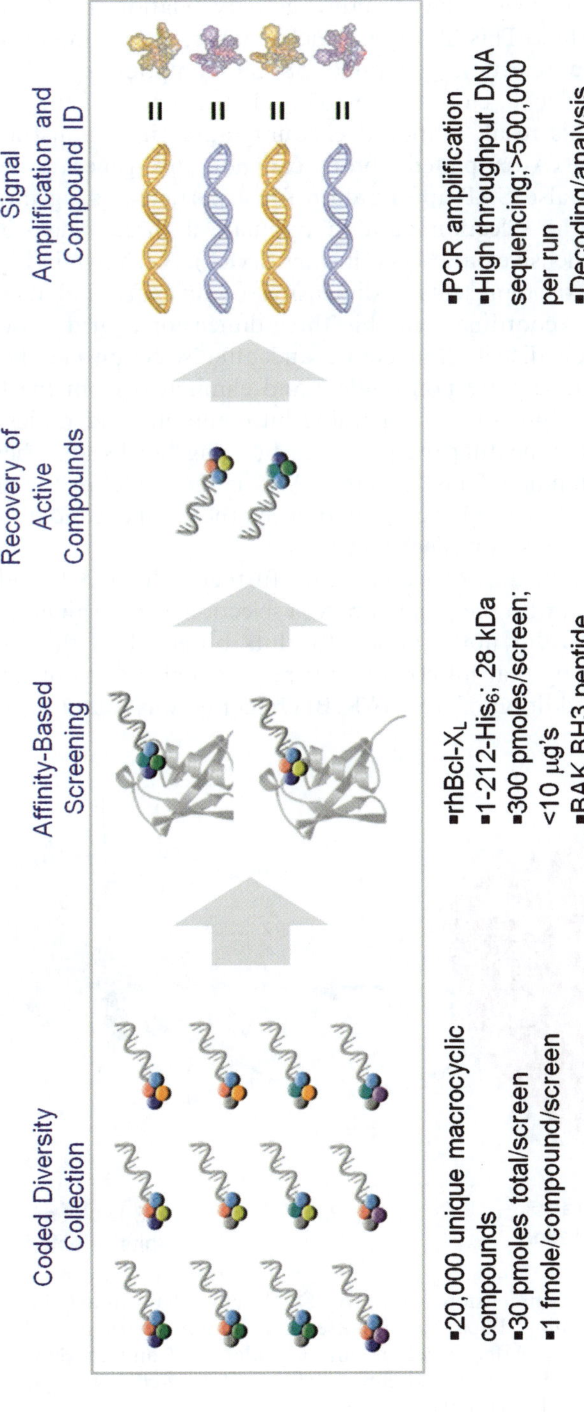



**Figure 8.14** Selection of a DNA-templated macrocycle library against Bcl-xL and hit identification. Reprinted with permission from Ensemble Discovery.

fold from 11 to 171. Next, Ensemble designed a second generation of more focused library based on the structure–activity relationship (SAR) obtained from the top 10 hits. This library is much smaller, containing only 1728 compounds, but the selection generated 58 hit compounds with the highest enrichment fold at 650, close to the 800-fold enrichment of the natural BAK-BH3 peptide. This result is indeed encouraging as it demonstrated that the selection of a DNA-templated library can not only generate individual hit compounds, but also SAR information for a particular drug target. Indeed, Ensemble used the selection data to populate a three-dimensional plot to generate comprehensive data sets ("instant SAR"). As shown in Figure 8.15, in the plot, each axis represents a dimension of diversity and therefore every compound has a coordinate in this three-dimensional grid. Comparison of library population distribution before and after selection clearly shows that most library members are non-binders and eliminated from the library after selection. The analysis of the remaining hit compounds identifies patterns of planes and lines, indicating the preferred building blocks and their combinations for target binding. This "instant SAR" is the guide for the design of the next generation library, hit compound re-syntheses and validation, and conclusion of the target's "privileged structures".

A few important control experiments further validated the identified hit compounds. First, with the same library, a selection against an immobilized but denatured Bcl-xL did not generate the hits observed in the native Bcl-xL selection, indicating that hit compounds require active form of Bcl-xL. In the second control, addition of free BAK BH3 peptide was found to suppress hits,

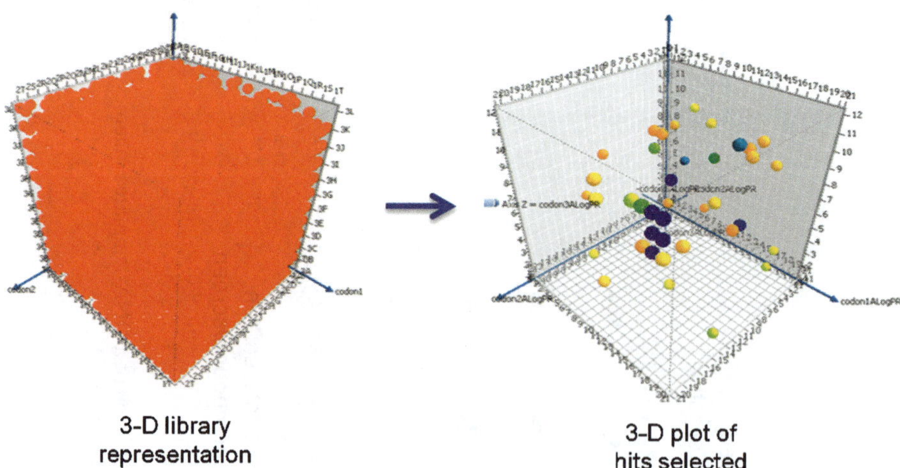



**Figure 8.15**  Three-dimensional plot of DNA-templated library before and after selection. Each ball represents a unique compound structure in the library. After selection, hits are identified and the enrichment fold is represented by different colors. Reprinted with permission from Ensemble Discovery.

confirming that hit compounds bind to the same location on the Bcl-xL and is competitive to the native BAK BH3 binding. Finally, a selection of the "mock library", a control library without the small molecule portion but with the exact same DNA template pool, excluded any false positives arising from DNA–protein interactions.

In the selection of DNA-templated libraries, since the active hits are identified *indirectly*, it is extremely important to resynthesize hit compounds as discretes without the DNA template and then confirm their binding activity against the target. Among the hits identified by Ensemble against Bcl-xL, the confirmed $IC_{50}$ values are in the low micromolar range, roughly consistent with the level of their enrichment factor in primary selections, while the $IC_{50}$ values of the non-binders are $>200\,\mu M$ in the assay. Even though the potencies of these hits are still modest, the results validated the fidelity of the affinity-based selection system with DPC-based libraries. In addition, considering the difficulty of Bcl-xL as a drug target, the low micromolar range is a good starting point for either further library design or direct medicinal chemistry optimization. Indeed, the three-dimensional plot of the selection gives a very clear SAR, providing solid data for these further campaigns.

Besides Bcl-xL, Ensemble has also reported preliminary selection results of their DPC libraries against diabetes targets PTP1B. An inhibitor of PTP1B with an $IC_{50} \sim 2.8\,\mu M$ was reported.[68] Currently, with their Ensemblin™ macrocyclic and other libraries, Ensemble is carrying an extensive array of selections against a variety of high-valued drug targets. Even though, so far, there is no report from the company regarding any clinical trial candidates, recently Ensemble has established strategic alliances and collaborations with Pfizer (on protein–protein interaction targets, 2010) and Bristol–Myers Squibb (target class not disclosed, 2009). In addition, Ensemble's biodetection technology, which is also based on DPC, has been the subject of a collaboration with Roche on the development of diagnostic technology to optimize cancer therapy (see http://www.ensemblediscovery.com/news/index.html). These collaborations with major pharmaceutical companies are solid endorsements of the strong potential and versatile applicability of this technology in drug discovery.


### 8.2.3.2 DNA-templated Library by DNA Routing

The DNA-templated library approach used by Liu and Ensemble Discovery has a pre-requisite that for any building blocks to be delivered to the templates they must have at least two functional groups: one conjugated with the reactive group on the template, the other conjugated with the reagent DNA strand. This requirement limits the number of usable building blocks to multifunctional motifs, such as the commonly used amino acids. Also, the preparation of the reagent DNA with building blocks via cleavable linker conjugation requires considerable synthetic efforts.

As a solution to this problem, Harbury and co-workers developed a method to physically partition DNA template pool into sub-population on individual codon bases.[69–72] As shown in Figure 8.16, a mixture of templates with several codon regions, similar to the template design by Liu and Ensemble, was subjected to flowing through several columns with individually immobilized anti-codon DNA strands. The templates therefore are separated based on sequence specificity and then can be processed for discreet chemical reactions, *e.g.* reaction with free, non-DNA-conjugated building blocks. The sub-pools after the reactions can be combined and flown through a second set of partitioning columns based on the codons in the second region for further discrete reactions. This iterative process is repeated until all codon regions are split-pooled and all building blocks have been delivered to the templates.

Harbury and co-workers initially prepared a library of 1 million templates and then chemically translated them into acylated pentapeptides. The peptide library included ten different monomers at each position.[72] At each position, structural elements (specific amino acids) of a known opioid peptide, Leu-enkephalin, were intentionally included so that the library will have a built-in positive control. Through two rounds of selections against 3-E7 antibody, Harbury and co-workers observed significant enrichment of codons corresponding to the Leu-enkephalin. In addition, another similar peptide library with different coding assignments was subjected to the selection, again enrichment of Leu-enkephalin was observed on a similar level.

In a more recent study, Harbury and co-workers extended their work on DNA-routing-based library synthesis and constructed an octamer peptide library with complexity ∼ 100 million compounds.[69] The size of the library was unparalleled at that time. Via six rounds of selection against drug target N-CrkSH3 (N-terminal Src homology 3 domain of the c-Crk protein), a few peptoid hit compounds previously unknown were identified. Hit resyntheses and assay confirmed their activities in binding the N-CrkSH3 target with $K_d = 16$ to $97\,\mu M$ (Figure 8.17).

Actually, Harbury's approach exploits DNA templating in a different way. Strictly speaking, the library is not synthesized by templated reactions. DNA templates are not used to bring reactants in proximity to boost the effective molarity; instead DNA templating is the mean to spatially partition the templates into sub-pools. The advantage of this approach is mostly that chemical building blocks can be directly used without the need to couple to a DNA strand and no cleavable linkers are needed; second, the chemical reactions take place on solid support (DEAE Sepharose column), therefore potentially more chemical reactions can be applied to create libraries with more diverse structures. However, so far there has been no report of library synthesis other than peptidic structures.

There are two additional noteworthy features of this approach. First, the construction of the initial template pool is not accomplished by standard split-pool, but a two-stage enzymatic assembly/subcloning method to diversify limited number of codon fragments into a full pool of templates. Second, as the actual chemical reactions are not templated by hybridizing DNA strands, there is no issue with reactions encoded by distal codons. Therefore the template DNA could be much longer with larger numbers of coding regions, *i.e.* higher

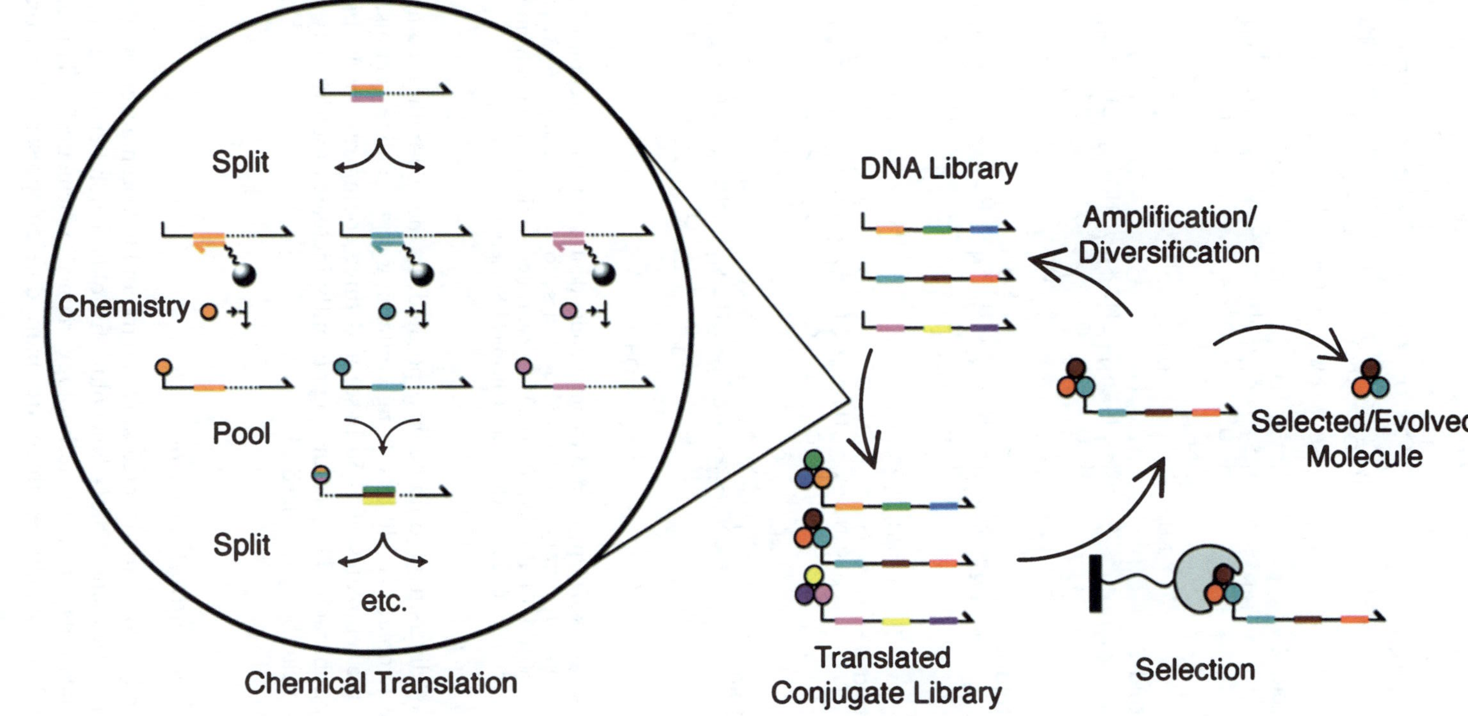



**Figure 8.16** Harbury's DNA routing method in synthesizing DNA-template libraries.[69] Reprinted, with permission, from Wrenn *et al.*[69] © 2007 American Chemical Society.

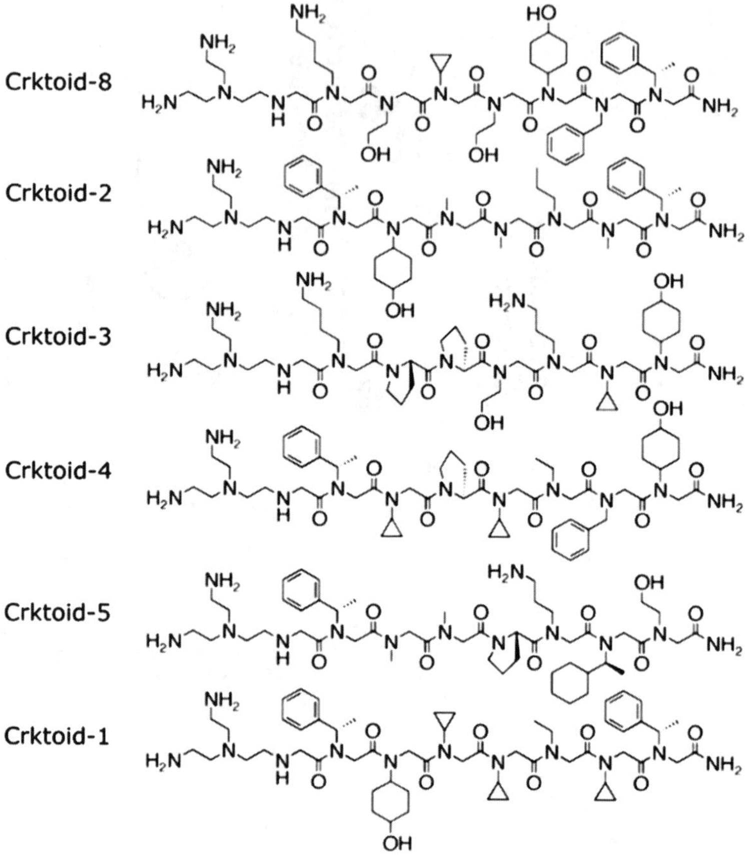

**Figure 8.17** N-CrkSH3 peptoid hit compounds identified from selection of a 100-million library. Re-assayed activities (top to bottom): 16 µM, 59 µM, 60 µM, 61 µM, 75 µM, 97 µM.[69] Reprinted, with permission, from Wrenn *et al.*[69] © 2007 American Chemical Society.

dimension of diversities, could be incorporated. In the 100-million-compound library, the template is 340-base long containing eight coding regions, compared with the templates of ~50 bases and three coding regions in Liu and Ensemble's approach. This feature significantly increases the diversity of the libraries that can be synthesized.

## 8.2.3.3 Nuevolution and Chemetics™

Nuevolution is a company located in Denmark and has been developing DNA-templated libraries since 2001. Nuevolution holds a number of patents and patent applications on their technology named Chemetics™. Throughout Nuevolution's patent filings, there are many different presentations on how

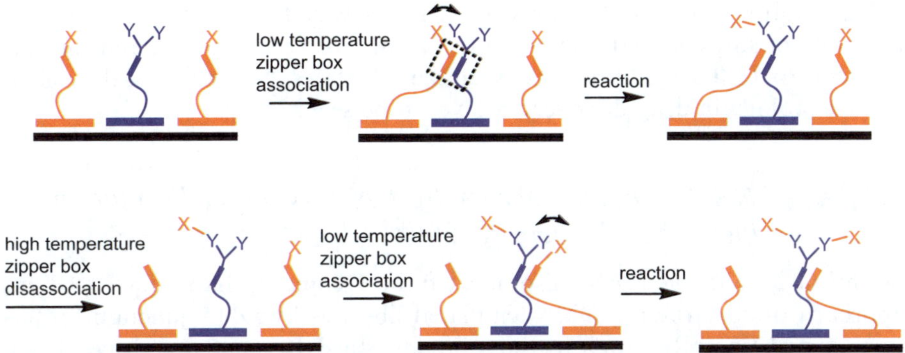


**Figure 8.18**  Nuevolution's "zipper box" approach in synthesizing DNA-templated libraries.[73]

their DNA-templated libraries are constructed and encoded. Here we only give some representative examples. In a more recent patent filed in 2008, Nuevolution presented a "Zipper-Box" approach.[73] As shown in Figure 8.18, a DNA template can direct the reaction between two chemically reactive groups conjugated with DNA strands. However, the two DNA strands have a 5–10 nt region with complementary sequences therefore they form a "zipper-box". By performing thermal cycling, similar to PCR, they are able to control the association/disassociation of the zipper-box. If one DNA reagent is conjugated to a multi-functional reaction group (with two identical Y groups), thermal cycling is able to react with two adjacent reagents with reactive X groups. The information of the product XY–YX is then encoded by the template DNA associated.

In another example as presented in their patent filed in 2004,[74] Nuevolution used a biotinylated DNA template with random coding sequences to direct the polymerization of multiple DNA nucleotides. Each DNA nucleotide carries a reactive group and a cleavable linker. Under the templating effect, the reactive groups polymerize and then linker cleavage yield the product; the library formed with the DNA duplex is subjected to selection and the templates of selected library compounds are amplified and iterated into the next round of synthesis and selection, mimicking a true small molecule evolution directed by DNA. In a later patent, this strategy is extended to using hairpin DNA template to "prime" the annealing of DNA strands and the reactions between reactive groups are triggered by DNA denaturation.[75] Even though there is no publication of the specific libraries constructed with the Chemetics™ technology, from Nuevolution's website, they built a 100-million library in 2–3 days; and, in the summer of 2009, their library has passed the 1-billion compound mark (http://www.nuevolution.com/Technology/Chemetics.htm). In addition, as proof of its applicability in drug discovery, Nuevolution announced collaborations with major pharmaceutical companies including Lexingcon, Merck, GSK, and Novartis (http://www.nuevolution.com/News/Press_releases.htm).

Nuevolution's approach represents a novel way taking advantage of the flexibility of single strand DNA and the rigidity of DNA duplex, and the feasible control of the equilibrium between the two states. Nevertheless, the role of this technology in drug discovery has yet to be seen.

### 8.2.3.4 DNA-templated Library by DNA Junction Formation: Vipergen and YoctoReactor®

Scientists at Vipergen AS, also located in Denmark, have developed an approach of constructing DNA-templated libraries by DNA junction formation, named YoctoReactor® by the company, since the templated reactions are confined in a space at yoctoliter level ($10^{-24}$ L). In a recent publication, Vipergen described a three-step templated synthesis of a library with 100 peptidic compounds.[57] Their approach is indeed unique. There are no distinctly different "templates" and "reagents". Each of the DNA-conjugated reagents is the template for another reagent. As shown in Figure 8.19, three hairpin-shaped DNA with reactive groups form a three-way junction step by step. Each step's reaction delivers a building block to the first reagent's reaction site. After enzymatically ligating the nicks among these DNA reagents, linker cleavage and primer extension, the library is transformed to an architecture with small molecules displayed on duplex DNA templates, compatible to subsequent selections and PCR amplification. In a model selection, Vipergen's library enriched a known binder to the immobilized mAb 3-E7 antibody.

One unique feature of Vipergen's approach is its "rolling translation" process they used to synthesize the next generation library directly out of the selected and amplified library templates, ready for the next round of selection. It is an ingenious design as it seamlessly connects all the components necessary for a small molecule evolution process, albeit the process itself is rather

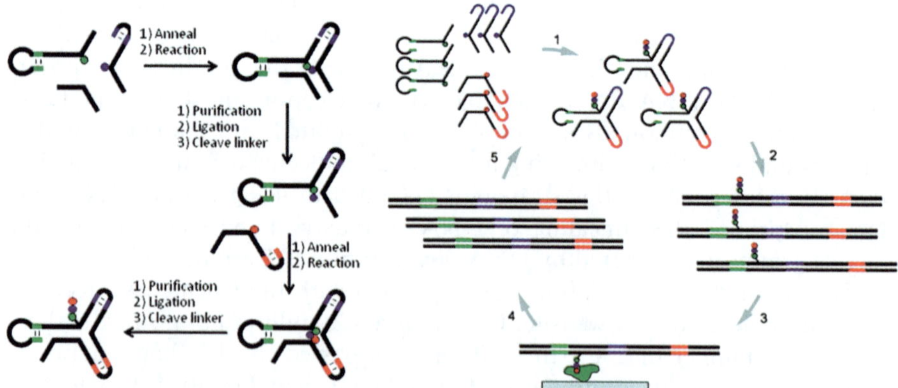


**Figure 8.19** Vipergen's YoctoReactor® approach in synthesizing a DNA-templated peptide library.[57] Reprinted, with permission, from Hansen *et al.*[57] © 2009 American Chemical Society.

complex, involving many chemical and enzymatic reactions and purification steps, which might prevents its practical implementations as a general drug discovery tool.

In principle, Vipergen's DNA junction-enabled templated reactions can also take place with higher dimensions of junctions. Indeed, Vipergen proposed using a four-way or even a five-way junction to template library synthesis with higher level of diversities (http://www.vipergen.com/library/documents/Vipergen-2-pager-I.pdf).[76] However, to date, there has been no report of previously unknown small molecule binders discovered by this technology against biological targets.

## 8.3 DNA-recorded Library

DNA-recorded library is another approach to encode small molecule libraries on DNA templates. Instead of preparing a template pool in which every library compound has already been encoded, in a DNA-recorded library initially all templates share an identical sequence, or even are simply one template. Usually, the chemical reactions on each template are encoded spatially such as the "split–pool–split" method; *i.e.* the initial template pool is separated into several batches for discrete reactions with various building blocks. Before all the sub-pools are combined, an enzymatic tag is added to the template to record the synthesis. For every step, an additional tag is added. As a matter of fact, DNA-recorded library, similar to Harbury's DNA routing strategy discussed above, does not use DNA to template the reactants into proximity and to increase the effective molarity.

DNA-recorded libraries exist in various formats and in the following sections we will discuss them based on the companies/research groups who invented them.

### 8.3.1 GSK/Praecis

GSK (formerly Praecis Pharmaceuticals) employed a duplex tagging technology to record the history of reactions on individual DNA template.[77] As shown in Figure 8.20, a starting reaction site is conjugated with a DNA duplex with same sequence throughout the library. Four steps of splitting, chemical reaction, tagging, and pooling rapidly generated the library with 802 160 640 theoretical members. A three-round selection of DEL-B against p38 MAPK enriched a very distinct SAR pattern in a three-dimensional data analysis (similar to Ensemble's data analysis method). As shown in Figure 8.21, a line of compounds all sharing a single structure scaffold is shown below. These compounds are only different at the variable position on the triazine ring. These "hit" compounds were re-synthesized as discretes and confirmed their binding potency against p38 MAPK. The $EC_{50}$ values of compounds with highest potency are in the single nanomolar range. More remarkably, they obtained a crystal structure of one hit compound bound to p38 MAPK; and indeed, the

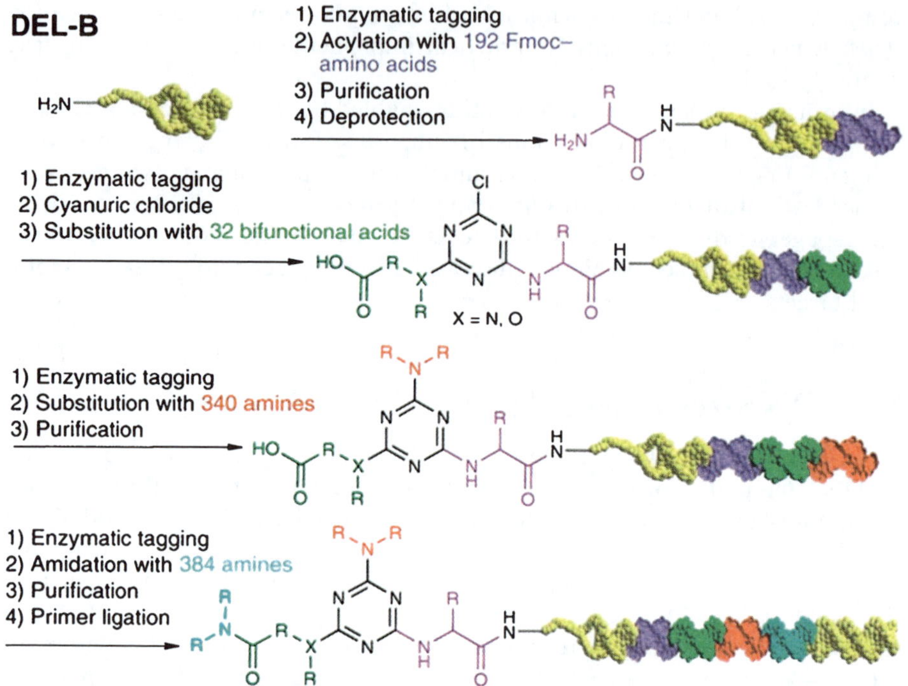


**Figure 8.20** GSK's synthesis of a 800-million member DEL-B library of substituted triazines.[77] Reprinted, with permission, from Clark *et al.*[77] © 2009 Macmillan Publishers Ltd.

DNA-linking carboxamide is found being exposed to solvent. Along with assayed binding potency of re-synthesized hit compounds, these data solidly validated the fidelity of the selection system enabled by the DNA tags used in a DNA-recorded library.

The DNA-recorded library from GSK has a few advantages. First, there is no need to prepare a pool of DNA templates with all specific sequences for each of the library member; there is no need to use sophisticated algorithm to design DNA sequences to avoid undesired hybridizations. Second, in principle, the DNA tagging can continue to many steps of reactions to incorporate higher dimensions of diversities. Third, duplex DNA are rigid structure motif, it is unlikely to form secondary structures to incur undesired interaction with the protein target. And finally, the DNA duplex is maintained throughout the library construction, and the bases are protected from being modified during chemical reactions.

The discovery of the novel p38 MAPK inhibitor is the first at that time, at least within the realm of published journals. It is worth noting that the spatial encoding method is conceptually similar to Harbury's DNA routing, despite in which the DNA encoding by DNA happens *before* the spatial separation and chemical reaction.

## 8.3.2 Neri Group/Philochem

Professor Neri's group at ETH Zurich developed a series of programs of several different types of DNA-encoded libraries. In one approach,[78–80] Neri and co-workers encoded the first dimension of building blocks by a DNA template with a coding region and an additional common sequence. After mixing, splitting into sub-pools, and chemical reactions with second batch of building blocks, a second set of DNA strands is hybridized with the common region on the template. The hybridization forms a staggered DNA structure and a Klenow extension fills in the bases to form the full duplex DNA and therefore encode the chemical synthesis. This process can continue to encode multiple steps of reactions in the library construction. Figure 8.22 shows an example of DNA-recorded library synthesis with two dimensions of diversities encoded by this approach. Neri's group prepared a 4000-member library by a Diels–Alder

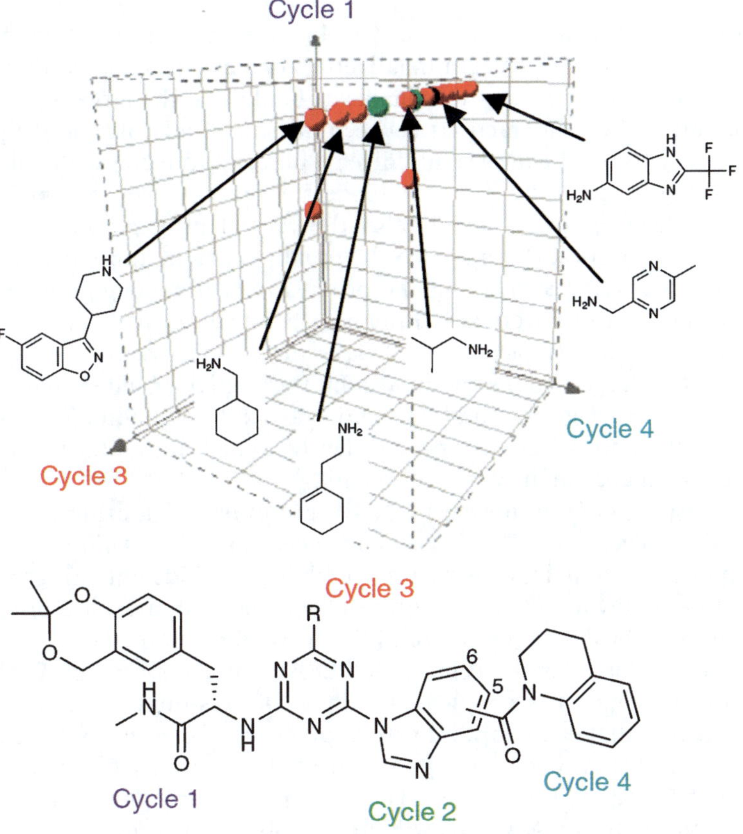


**Figure 8.21** Selection results of DEL-B against p38 MAPK enriched a series of compounds sharing a common structure scaffold.[77] Reprinted, with permission, from Clark *et al.*[77] © 2009 Macmillan Publishers Ltd.

reaction between a set of dienes and maleimide dienophiles[78,79] and a 4000-member dipeptide library by amide formation.[80]

Selections of these libraries against a variety of targets include human IgG, streptavidin , MMP-3, TNF, Bcl-xL, trypsin and albumin. One notable aspect of Neri's approach is its single round of selection to identify selective binders by thoroughly sampling the selection results by high throughput sequencing. Novel binders discovered by Neri's DNA-recorded libraries include a Bcl-xL binder and a TNF binder with $K_d \sim 10\,\mu M$; and the TNF binder has been shown to be active in inhibiting a TNF-mediated killing of fibroblasts *in vitro*.[79] The company Philochem is currently performing drug discovery programs under the license of Neri's technology.

### 8.3.3 A DNA-recorded Library by DNA Assembly: ESAC Library

All the DNA-encoded libraries discussed involve chemical reactions between building blocks to construct a relatively complex structure with multiple dimensions of diversities. It is reasonable as binding to the target is a complex combination of many interactions between the small molecule and the protein. It is necessary to build complex and diverse libraries to increase the likelihood of discovering potent and selective binders. This has been the mainstream strategy in drug discovery and is also the central principle in constructing combinatorial libraries. Recently, a conceptually different approach, fragment-based drug discovery, is emerging and progressively becoming more and more important in drug discovery.[81,82] Fragment-based drug discovery starts with low-molecular-weight (typically $<250\,Da$) structural fragments to screen/select against protein targets to discover low affinity/selectivity binders. The selected fragments are either grown to other regions around the binding site to obtain higher affinity and selectivity, or covalently linked with other low-affinity fragment to enhance affinity and selectivity.

This concept has been utilized by Neri's group in synthesizing DNA-recorded libraries called encoded self-assembling chemical (ESAC) libraries,[83–87] and by Hamilton's group in constructing a library of bidentate fragments for binding to streptavidin.[88] As a matter of fact, these libraries are not synthesized, but assembled by combinatorial DNA hybridization.

Figure 8.23 illustrates the principle of ESAC library assembly. Each pharmacophore (fragment) is conjugated with a DNA strand, which includes a common region that can hybridize with another set of pharmacophores. At the tail end of the DNA a coding region is located. When mixed together, DNA strands hybridization brings the pharmacophores in proximity, forming the library displayed by a DNA duplex. Figure 8.23b–d shows several variations of the ESAC library. For example, if one of the pharmacophores is a known binder, then the library can be used to find additional fragment to improve the binding properties to the target, called affinity maturation (Figure 8.23b). In

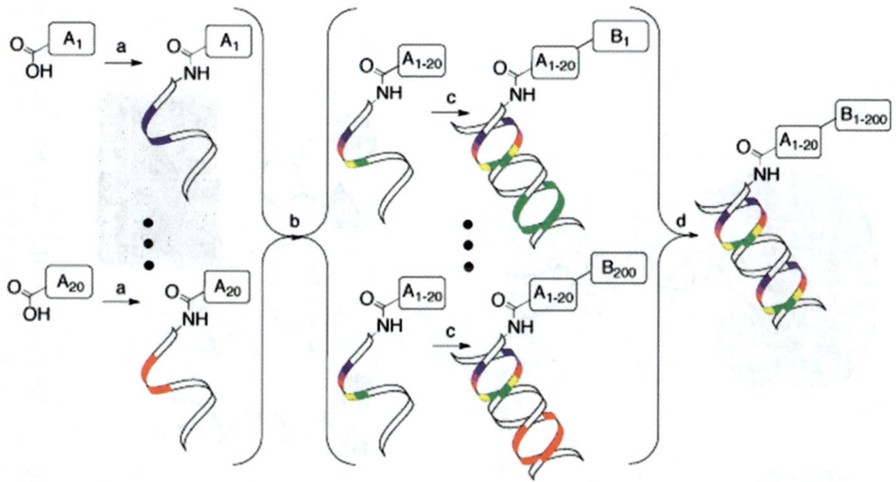

**Figure 8.22** Neri's DNA-recorded library synthesis scheme. The encoding is done by enzymatic extension.[78] Reprinted, with permission, from Buller *et al.*[78] © 2008 Elsevier.

principle, DNA triplex or quadruplex can also be formed to display three or four sets of pharmacophores combinatorially.

Neri's group has prepared a series of ESAC libraries and interrogated a collection of targets including streptavidin,[89] trypsin,[85,87] MMP-3,[87] bovine carbonic anhydrase II,[83] and albumin.[90] Binders discovered have $K_d$ values in a range from single nanomolar to micromolar. It is noteworthy that Neri and co-workers have accomplished "affinity maturation" of know binders as illustrated in Figure 8.23b. For example, the $IC_{50}$ of a weakly trypsin-binding benzamidine has been improved from 90 μM to 98 nM, nearly a $10^3$-fold of improvement.[85]

The ESAC approach brings significant benefit to DNA-encoded libraries since there is no templated reaction, no split–pool-based partitioning, and no enzymatic encoding, it can be considered as a minimalist's version of the DNA-encoded library, because the only step is mixing the sets of DNA-conjugated fragments together. The fragment–DNA conjugates can be prepared by simple bio-conjugation chemistry without sophisticated linker design and usable building blocks are widely available. In addition, this approach combines the power of dynamic combinatorial library by letting the protein target select its own preferred combination of fragments, but without the disadvantage of dynamic library in which only building blocks with very specific functional groups can be used.[91]

However, one of the disadvantages of ESAC libraries is that selected fragments have to be linked to form the full target-binding structure. As shown in one of Neri's ESAC selection to discover MMP-3 inhibitors, fragment linking is difficult since the linker structure, length, flexibility, and position all have

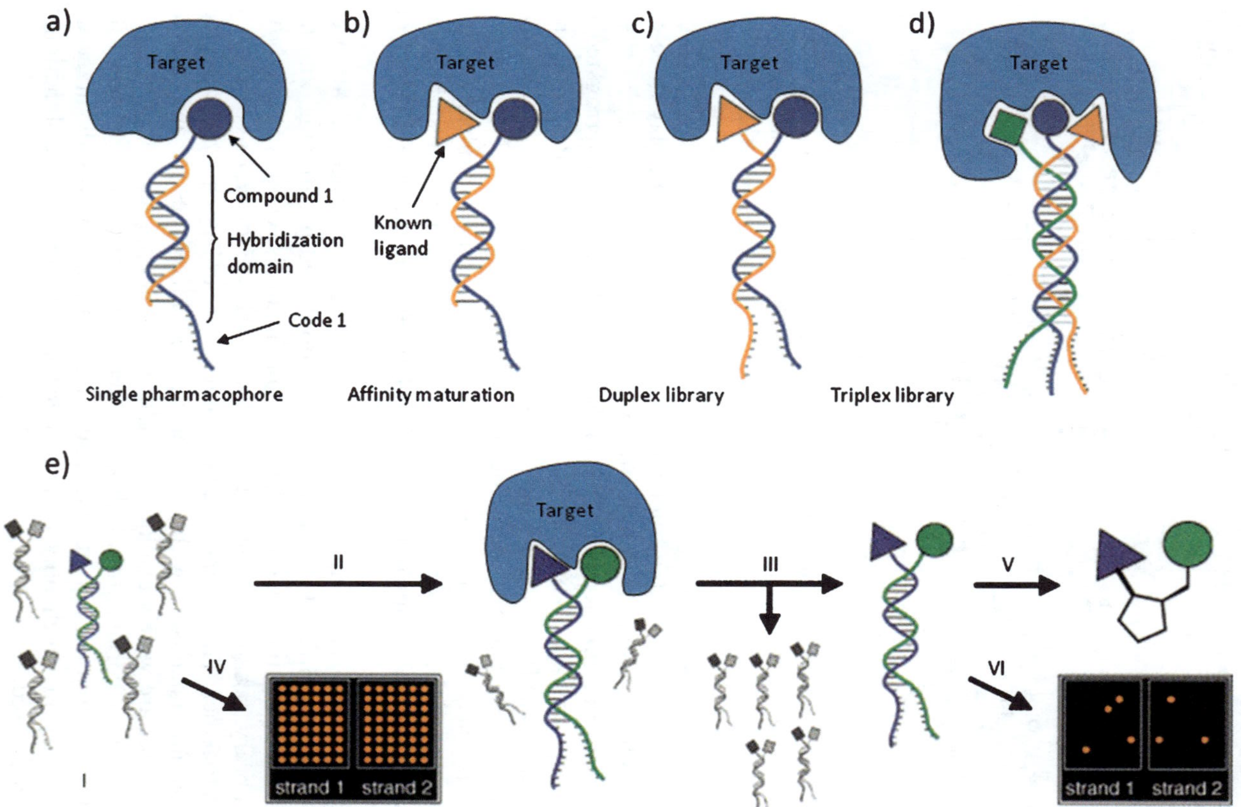


**Figure 8.23** The principles of ESAC libraries. (a–d) Various formats of ESAC; (e) Schematic illustration of the assembly, selection and decoding of an ESAC library against target. Reprinted from http://en.wikipedia.org/wiki/File:Luca5.png with GNU Free Documentation License.

significant impact on the binding properties.[87] Second, the selection results from ESAC libraries currently can only generate enrichment information on individual fragments; the connectivity information between fragments is lost. For example, in the MMP-3 selection, the enriched fragments from each sub-libraries have to be combinatorially linked with each other searching for the optimal binding structure.[87] In addition to the variations of linkers, the post-selection fragment linking from ESAC selection could be a time-consuming and unpredictable task. Finally, at higher diversity dimensions with DNA triplex and quadruplex, decoding from selections and fragment-linking could be even more challenging.

## 8.4 Discussion

The concept of using DNA to encode for combinatorial libraries was suggested by Brenner and Lerner in 1992.[92] However, it was not until the early 2000s when Liu and co-workers pioneered the idea of using DNA templates to direct specific chemical reactions, this field started attracting attentions and seeing substantial developments. In the following years, the work from Liu and co-workers progressively paved a path to evolving small molecules with DNA molecules, analogous to the translation from DNA sequences to protein functions. The construction and selection of Liu's macrocycle libraries can be considered a milestone and the industrialization of DNA-Programmed Chemistry (DPC™) by Ensemble Discovery is another significant progress in achieving this goal. In the same time period, powered by ingenious designs and alternative concepts, many other research groups and companies have developed a variety of DNA-encoded libraries and related technologies. Again, most of these developed technologies have been adapted into industrial platforms for hit identification and lead discovery in pharmaceutical companies.

DNA-encoded libraries come in many different forms, but they share many common features, such as chemical reactions used, selection and amplification method, hit identification, *etc*. These approaches are different in their solutions to address these following questions:

- How are the templates formatted? – Already diversified or to be diversified in the process of chemical reactions.
- How are the building blocks encoded? – By specific DNA hybridization, spatial encoding, or enzymatic encoding.
- How are the chemical reactions of building blocks controlled? – Templated by DNA, spatially controlled, or no templated reaction.

The answers to these questions bring a mix of advantages and disadvantages into every library approach. A recent review nicely summarized and compared these approaches.[76] However, the ultimate judgment should not arise from parametric comparisons but should come from their performance in actual drug discovery programs. By this criteria, as a matter of fact, currently there is no one technology obviously superior to others. Therefore in the following discussion, we will try to look at these technologies globally as a whole, review what has and what has not been achieved, and finally, discuss the prospects of DNA-encoded library in the future of drug discovery.

## 8.4.1   The Advantages of DNA-encoded Library

### 8.4.1.1   Miniaturization of High-throughput Screening

At current stage of development, DNA-encoded library is still mostly a miniaturized version of HTS. Instead of separating each compound spatially as in most HTS libraries, all member of molecules are virtually separated by individual DNA templates. Miniaturization brings the advantages of significantly reduced costs on material, compound synthesis, compound management, and screening, especially with the ever-increasing size of modern small molecule libraries in pharmaceutical companies. The utilization of DNA as nano-devices makes the synthesis and "screening (selection)" of millions of compounds in a normal research laboratory a routine practice.

### 8.4.1.2   Expansion of High-throughput Screening Diversities

Albeit on a miniaturized scale, the DNA-encoded libraries can reach a diversity level far beyond normal HTS libraries. For example, a library by Harbury's DNA routing approach has 100 million compounds; Nuevolution announced on their website that their Chemetics™ library passed the 1 billion compound mark; while GSK's DEL-B library has more than 800 million different structures. These numbers are on a par with biopolymer libraries prepared by genetics techniques such as phage-display and mRNA display, while the largest synthetic combinatorial library is around merely ~2 million.[20] Therefore, DNA-encoded library is truly an expansion of HTS into a much higher level of molecular diversity.

### 8.4.1.3   Molecular Evolution

The reason to construct libraries encoded by DNA, originally, was the intrigue of using Nature's strategy to perform direct evolution on the function of small molecules, which nicely fits the practical request of obtaining optimal protein target binders in drug discovery. Even though most of these technologies are only *selectable*, not yet *amplifiable* and *evolvable*, they can enrich high affinity binders even though these molecules only exist in low abundances in the original library. It is worth noting that Vipergen's rolling translation is one step closer to a true evolution as it reconstructs a second generation library from the selected population.

### 8.4.1.4   Access to Unique Chemical Spaces

Even with limited usable chemical reactions, in some cases, DNA-encoded libraries can be a unique tool to access chemical structures previously difficult to prepare. For example, Ensemble Discovery's DPC® platform enables effective synthesis of macrocyclic peptide libraries (Ensemblins™), a class of molecules known to be difficult to prepare, especially in a library format. These molecules have been considered to have special binding properties against protein–protein interaction targets;[63] and Ensemble Discovery's initial selections of macrocycle libraries against Bcl-X$_L$ and TNFα have generated a series of high-affinity binders.[68]

## 8.4.1.5 Comprehensive Structure–Activity Relationships

All the molecules in a DNA-encoded library are selected simultaneously. As shown in Ensemble Discovery and GSK's data analyses, a comprehensive view of SAR can be rapidly generated; from the lines and planes formed in the SAR plot, high priority structural elements can be identified and applied to the next generation of library design, or to the synthesis of discrete compounds. In addition, a clear correlation between related structural elements with binding potency strongly validates the selection itself.

## 8.4.2 Prospects and Outlook

In less than 10 years, the DNA-encoded library has gained rapid developments on all fronts. Even though currently there is no drug lead in clinical trials discovered by DNA-encoded libraries, the prospect is promising as evidenced by its extensive implementations in pharmaceutical companies as an alternative tool in drug discovery. Time and experiment will give the final judgment but we believe the DNA-encoded library would embrace more power and versatility in discovering drug leads and addressing unmet medical problems, especially with improvements on these important aspects discussed in the sections below.

## 8.4.2.1 Chemistry

One of the most critical limitations of a DNA-encoded library is DNA itself. It requires an aqueous media and mild environmental conditions to ensure DNA's chemical stability and hybridization fidelity. Naturally, many useful chemical reactions cannot be directly applied. Liu and co-workers have adapted a variety of organic reactions into DNA-compatible format; however, a survey of the structures synthesized by DNA encoding indicates the compounds made by DNA-templated chemistry are still being confined within the paradigm of peptidic structures, an undesired structural feature for drug molecules.

Therefore, expanding DNA-compatible chemical reactions is one of the most critical tasks for further development of this technology for DNA-encoded library to generate hit molecules with drug-like or even lead-like structures. Liu and co-workers have attempted DNA-templated reaction in organic solvents assisted by detergents[45] and achieved some success, which may be a viable approach to address this issue.

### 8.4.2.2   Selectable Targets

Currently, all selections with DNA-encoded libraries are affinity-based and purified protein targets are immobilized either before or after selections. However, the structures and functions of proteins vary significantly in different environments. Hits generated from an *in vitro* selection do not always work *in vivo*; as a matter of fact, this discrepancy is one of the major reasons many drug candidates failed the clinical trials in the past decade.[93] Therefore if a DNA-encoded library can be used to interrogate protein targets in their native environment, such as cell lysates, or even within live cells, the impact of this technology will be tremendously improved. DNA-conjugated small molecules have poor cell permeability and stability, therefore, possible DNA surrogates, such as PNA and other analogs, may be an interesting alternative to be developed to address this issue.

### 8.4.2.3   True Evolution

To become a technology platform to truly evolve, the selection of active small molecules is another important aspect for further development. First, it should be able to construct a second-generation library directly from a selected library population without multiple steps of manipulations as in Vipergen's "rolling translation"; second, it should be able to introduce mutations and recombination to the "chemical genes" for library diversification; and finally, it would ideally be a dynamic system so that non-binders are removed and the selection is iterated automatically.

   Naturally, not all the goals above can be feasibly achieved with the current state of the art, but we believe technological advancement and ingenious human input, as they have already done, will eventually accomplish these tasks and make DNA-encoded libraries indispensable tools in drug discovery.

## Acknowledgements

We thank all the research groups and companies discussed above for their work to make the contents of this chapter possible. We thank Professor David R. Liu and Ensemble Discovery for providing figures and permissions. We thank Dr Nick Terrett and Dr. Frank Favaloro for their critical reading of the manuscript.

## References

1. K. H. Altmann, J. Buchner, H. Kessler, F. Diederich, B. Krautler, S. Lippard, R. Liskamp, K. Muller, E. M. Nolan, B. Samori, G. Schneider, S. L. Schreiber, H. Schwalbe, C. Toniolo, C. A. A. van Boeckel, H. Waldmann and C. T. Walsh, *ChemBioChem*, 2009, **10**, 16–29.
2. A. L. Hopkins, C. R. Groom and A. Alex, *Drug Discovery Today*, 2004, **9**, 430–431.

3. G. Schneider and U. Fechner, *Nat. Rev. Drug Discovery*, 2005, **4**, 649–663.
4. B. A. Posner, *Curr. Opin. Drug Discovery Dev.*, 2005, **8**, 487–494.
5. M. H. Flight, *Nat. Rev. Drug Discovery*, 2008, **7**, 388–388.
6. W. J. Dower and L. C. Mattheakis, *Curr. Opin. Chem. Biol.*, 2002, **6**, 390–398.
7. J. A. Bittker, K. J. Phillips and D. R. Liu, *Curr. Opin. Chem. Biol.*, 2002, **6**, 367–374.
8. V. A. Petrenko, *Expert Opin. Drug Delivery*, 2008, **5**, 825–836.
9. L. R. Pepper, Y. K. Cho, E. T. Boder and E. V. Shusta, *Comb. Chem. High Throughput Screening*, 2008, **11**, 127–134.
10. G. P. Smith and V. A. Petrenko, *Chem. Rev.*, 1997, **97**, 391–410.
11. J. Hanes and A. Pluckthun, *Proc. Natl Acad. Sci. U. S. A.*, 1997, **94**, 4937–4942.
12. J. M. Kim, H. J. Shin, K. Kim and M. S. Lee, *Mol. Biotechnol.*, 2007, **36**, 32–37.
13. C. A. Valencia, S. W. Cotten, B. A. Dong and R. H. Liu, *Biotechnol. Prog.*, 2008, **24**, 561–569.
14. D. S. Wilson, A. D. Keefe and J. W. Szostak, *Proc. Natl Acad. Sci. U. S. A.*, 2001, **98**, 3750–3755.
15. A. D. Keefe and J. W. Szostak, *Nature*, 2001, **410**, 715–718.
16. R. W. Roberts and J. W. Szostak, *Proc. Natl Acad. Sci. U. S. A.*, 1997, **94**, 12297–12302.
17. A. Sergeeva, M. G. Kolonin, J. J. Molldrem, R. Pasqualini and W. Arap, *Adv. Drug Deliv. Rev.*, 2006, **58**, 1622–1654.
18. M. Silacci, S. Brack, G. Schirru, J. Marlind, A. Ettorre, A. Merlo, F. Viti and D. Neri, *Proteomics*, 2005, **5**, 2340–2350.
19. D. S. Wilson and J. W. Szostak, *Annu. Rev. Biochem.*, 1999, **68**, 611–647.
20. D. S. Tan, M. A. Foley, B. R. Stockwell, M. D. Shair and S. L. Schreiber, *J. Am. Chem. Soc.*, 1999, **121**, 9073–9087.
21. G. E. Joyce, *Angew. Chem. Int. Ed. Engl.*, 2007, **46**, 6420–6436.
22. A. P. Silverman and E. T. Kool, *Chem. Rev.*, 2006, **106**, 3775–3789.
23. N. C. Seeman, *Nature*, 2003, **421**, 427–431.
24. U. Feldkamp and C. M. Niemeyer, *Angew. Chem. Int. Ed. Engl.*, 2006, **45**, 1856–1876.
25. Anon., *Nature Nanotechnol.*, 2009, **4**, 203.
26. G. F. Joyce, *Nature*, 1989, **338**, 217–224.
27. Y. Z. Xu and E. T. Kool, *J. Am. Chem. Soc.*, 2000, **122**, 9040–9041.
28. L. E. Orgel, *Acc. Chem. Res.*, 1995, **28**, 109–118.
29. X. Li, Z. Y. Zhan, R. Knipe and D. G. Lynn, *J. Am. Chem. Soc.*, 2002, **124**, 746–747.
30. M. K. Herrlein, J. S. Nelson and R. L. Letsinger, *J. Am. Chem. Soc.*, 1995, **117**, 10151–10152.
31. X. Y. Li and D. R. Liu, *Angew. Chem. Int. Ed. Engl.*, 2004, **43**, 4848–4870.
32. R. K. Bruick, P. E. Dawson, S. B. Kent, N. Usman and G. F. Joyce, *Chem. Biol.*, 1996, **3**, 49–56.
33. M. Bolli, R. Micura and A. Eschenmoser, *Chem. Biol.*, 1997, **4**, 309–320.

34. Z. J. Gartner, M. W. Kanan and D. R. Liu, *Angew. Chem. Int. Ed. Engl.*2002, **41**, 1796–1800.
35. R. E. Kleiner, Y. Brudno, M. E. Birnbaum and D. R. Liu, *J. Am. Chem. Soc.*, 2008, **130**, 4646–4659.
36. D. M. Rosenbaum and D. R. Liu, *J. Am. Chem. Soc.*, 2003, **125**, 13924–13925.
37. Y. Brudno and D. R. Liu, *Chem. Biol.*, 2009, **16**, 265–276.
38. K. Sakurai, T. M. Snyder and D. R. Liu, *J. Am. Chem. Soc.*, 2005, **127**, 1660–1661.
39. J. M. Heemstra and D. R. Liu, *J. Am. Chem. Soc.*, 2009, **131**, 11347–11349.
40. T. Li and K. C. Nicolaou, *Nature*, 1994, **369**, 218–221.
41. A. T. Poulin-Kerstien and P. B. Dervan, *J. Am. Chem. Soc.*, 2003, **125**, 15811–15821.
42. C. T. Calderone and D. R. Liu, *Angew. Chem. Int. Ed. Engl.*, 2005, **44**, 7383–7386.
43. C. T. Calderone, J. W. Puckett, Z. J. Gartner and D. R. Liu, *Angew. Chem. Int. Ed. Engl.*, 2002, **41**, 4104–4108.
44. M. M. Rozenman, M. W. Kanan and D. R. Liu, *J. Am. Chem. Soc.*, 2007, **129**, 14933–14938.
45. M. M. Rozenman and D. R. Liu, *ChemBioChem*, 2006, **7**, 253–256.
46. X. Y. Li and D. R. Liu, *J. Am. Chem. Soc.*, 2003, **125**, 10188–10189.
47. M. W. Kanan, M. M. Rozenman, K. Sakurai, T. M. Snyder and D. R. Liu, *Nature*, 2004, **431**, 545–549.
48. D. J. Gorin, A. S. Kamlet and D. R. Liu, *J. Am. Chem. Soc.*, 2009, **131**, 9189–9191.
49. Z. J. Gartner, M. W. Kanan and D. R. Liu, *J. Am. Chem. Soc.*, 2002, **124**, 10304–10306.
50. X. Y. Li, Z. J. Gartner, B. N. Tse and D. R. Liu, *J. Am. Chem. Soc.*, 2004, **126**, 5090–5092.
51. Z. J. Gartner, B. N. Tse, R. Grubina, J. B. Doyon, T. M. Snyder and D. R. Liu, *Science*, 2004, **305**, 1601–1605.
52. Z. J. Gartner, R. Grubina, C. T. Calderone and D. R. Liu, *Angew. Chem. Int. Ed. Engl.*, 2003, **42**, 1370–1375.
53. T. C. Bruice and U. K. Pandit, *J. Am. Chem. Soc.*, 1960, **82**, 5858–5865.
54. Z. J. Gartner and D. R. Liu, *J. Am. Chem. Soc.*, 2001, **123**, 6961–6963.
55. T. M. Snyder, B. N. Tse and D. R. Liu, *J. Am. Chem. Soc.*, 2008, **130**, 1392–1401.
56. L. H. Eckardt, K. Naumann, W. M. Pankau, M. Rein, M. Schweitzer, N. Windhab and G. von Kiedrowski, *Nature*, 2002, **420**, 286–286.
57. M. H. Hansen, P. Blakskjaer, L. K. Petersen, T. H. Hansen, J. W. Hojfeldt, K. V. Gothelf and N. J. V. Hansen, *J. Am. Chem. Soc.*, 2009, **131**, 1322–1327.
58. J. W. Hojfeldt, P. Blakskjaer and K. V. Gothelf, *J. Org. Chem.*, 2006, **71**, 9556–9559.
59. B. N. Tse, T. M. Snyder, Y. H. Shen and D. R. Liu, *J. Am. Chem. Soc.*, 2008, **130**, 15611–15626.

60. J. B. Doyon, T. M. Snyder and D. R. Liu, *J. Am. Chem. Soc.*, 2003, **125**, 12372–12373.
61. R. B. Woodward, E. Logusch, K. P. Nambiar, K. Sakan, D. E. Ward, B. W. Auyeung, P. Balaram, L. J. Browne, P. J. Card, C. H. Chen, R. B. Chenevert, A. Fliri, K. Frobel, H. J. Gais, D. G. Garratt, K. Hayakawa, W. Heggie, D. P. Hesson, D. Hoppe, I. Hoppe, J. A. Hyatt, D. Ikeda, P. A. Jacobi, K. S. Kim, Y. Kobuke, K. Kojima, K. Krowicki, V. J. Lee, T. Leutert, S. Malchenko, J. Martens, R. S. Matthews, B. S. Ong, J. B. Press, T. V. Rajanbabu, G. Rousseau, H. M. Sauter, M. Suzuki, K. Tatsuta, L. M. Tolbert, E. A. Truesdale, I. Uchida, Y. Ueda, T. Uyehara, A. T. Vasella, W. C. Vladuchick, P. A. Wade, R. M. Williams and H. N. C. Wong, *J. Am. Chem. Soc.*, 1981, **103**, 3213–3215.
62. B. B. Liau, V. Gnanadesikan and E. J. Corey, *Org. Lett.*, 2008, **10**, 1055–1057.
63. E. M. Driggers, S. P. Hale, J. Lee and N. K. Terrett, *Nat. Rev. Drug Discovery*, 2008, **7**, 608–624.
64. L. A. Wessjohann and E. Ruijter, *Top. Curr. Chem.*, 2005, **243**.
65. D. R. Schmidt, O. Kwon and S. L. Schreiber, *J. Comb. Chem.*, 2004, **6**, 286–292.
66. E. Marsault, H. R. Hoveyda, M. L. Peterson, C. Saint-Louis, A. Landry, M. Vezina, L. Ouellet, Z. G. Wang, M. Ramaseshan, S. Beaubien, K. Benakli, S. Beauchemin, R. Deziel, T. Peeters and G. L. Fraser, *J. Med. Chem.*, 2006, **49**, 7190–7197.
67. S. Hale, The Rapid Creation and Screening of Conformation-Restricted Synthetic Libraries against Protein–Protein Interaction Targets, in *Drug Discovery Chemistry: Tools, Targets, and Therapies*, San Diego, La Jolla, CA, 2008.
68. M. D. Taylor, Ensemblins: A Novel Class of Therapeutics for Addressing Protein–Protein Interactions, in *BIO International Convention*, Atlanta, GA, 2008.
69. S. J. Wrenn, R. M. Weisinger, D. R. Halpin and P. B. Harbury, *J. Am. Chem. Soc.*, 2007, **129**, 13137–13143.
70. D. R. Halpin, J. A. Lee, S. J. Wrenn and P. B. Harbury, *Plos Biol.*, 2004, **2**, 1031–1038.
71. D. R. Halpin and P. B. Harbury, *PLoS Biol.*, 2004, **2**, 1015–1021.
72. D. R. Halpin and P. B. Harbury, *PLoS Biol.*, 2004, **2**, 1022–1030.
73. H. Pedersen, *et al.* U.S. Patent application 12/179,323 Pat., 2008.
74. H. Pedersen, *et al.* U.S. Patent application 10/175,539 Pat., 2004.
75. T. Franch, *et al.* U.S. Patent application 10/549,619 Pat., 2004.
76. T. R. Heitner and N. J. V. Hansen, *Exp. Opin. Drug Discovery*, 2009, **4**, 1201–1213.
77. M. A. Clark, R. A. Acharya, C. C. Arico-Muendel, S. L. Belyanskaya, D. R. Benjamin, N. R. Carlson, P. A. Centrella, C. H. Chiu, S. P. Creaser, J. W. Cuozzo, C. P. Davie, Y. Ding, G. J. Franklin, K. D. Franzen, M. L. Gefter, S. P. Hale, N. J. V. Hansen, D. I. Israel, J. W. Jiang, M. J. Kavarana, M. S. Kelley, C. S. Kollmann, F. Li, K. Lind, S. Mataruse, P. F. Medeiros, J. A.

Messer, P. Myers, H. O'Keefe, M. C. Oliff, C. E. Rise, A. L. Satz, S. R. Skinner, J. L. Svendsen, L. J. Tang, K. van Vloten, R. W. Wagner, G. Yao, B. G. Zhao and B. A. Morgan, *Nat. Chem. Biol.*, 2009, **5**, 647–654.
78. F. Buller, L. Mannocci, Y. X. Zhang, C. E. Dumelin, J. Scheuermann and D. Neri, *Bioorg. Med. Chem. Lett.*, 2008, **18**, 5926–5931.
79. F. Buller, Y. X. Zhang, J. Scheuermann, J. Schafer, P. Buhlmann and D. Neri, *Chem. Biol.*, 2009, **16**, 1075–1086.
80. L. Mannocci, Y. X. Zhang, J. Scheuermann, M. Leimbacher, G. De Bellis, E. Rizzi, C. Dumelin, S. Melkko and D. Neri, *Proc. Natl Acad. Sci. U. S. A.*, 2008, **105**, 17670–17675.
81. P. J. Hajduk and J. Greer, *Nat. Rev. Drug Discovery*, 2007, **6**, 211–219.
82. D. C. Rees, M. Congreve, C. W. Murray and R. Carr, *Nat. Rev. Drug Discovery*, 2004, **3**, 660–672.
83. S. Melkko, J. Scheuermann, C. E. Dumelin and D. Neri, *Nat. Biotechnol.*, 2004, **22**, 568–574.
84. S. Melkko, C. E. Dumelin, J. Scheuermann and D. Neri, *Drug Discovery Today*, 2007, **12**, 465–471.
85. S. Melkko, Y. Zhang, C. E. Dumelin, J. Scheuermann and D. Neri, *Angew. Chem. Int. Ed. Engl.*, 2007, **46**, 4671–4674.
86. S. Melkko, C. E. Dumelin, J. Scheuermann and D. Neri, *Chem. Biol.*, 2006, **13**, 225–231.
87. J. Scheuermann, C. E. Dumelin, S. Melkko, Y. X. Zhang, L. Mannocci, M. Jaggi, J. Sobek and D. Neri, *Bioconjugate Chem.*, 2008, **19**, 778–785.
88. K. I. Sprinz, D. M. Tagore and A. D. Hamilton, *Bioorg. Med. Chem. Lett.*, 2005, **15**, 3908–3911.
89. C. E. Dumelin, J. Scheuermann, S. Melkko and D. Neri, *Bioconjugate Chem.*, 2006, **17**, 366–370.
90. C. E. Dumelin and D. Neri, U.K. Patent 0621973,7, 2006.
91. O. Ramstrom and J. M. Lehn, *Nat. Rev. Drug Discovery*, 2002, **1**, 26–36.
92. S. Brenner and R. A. Lerner, *Proc. Natl Acad. Sci. U. S. A.*, 1992, **89**, 5381–5383.
93. I. Kola and J. Landis, *Nat. Rev. Drug Discovery*, 2004, **3**, 711–715.

# Subject Index

Note: Page numbers in *italics* refer to Figures or Tables.